In this chapter you will learn how to …

- use non-calculator methods to calculate with positive and negative numbers.
- perform operations in the correct order based on mathematical conventions.
- recognise inverse operations and use them to simplify and check calculations.

For more resources relating to this chapter, visit GCSE Mathematics Online.

Using mathematics: real-life applications

Everyone uses numbers on a daily basis often without really thinking about them. Shopping, cooking, working out bills, paying for transport and measuring all rely on a good understanding of numbers and calculation skills.

 Tip

You probably already know most of the concepts in this chapter. They have been included so that you can revise concepts if you need to and check that you know them well.

"Number puzzles and games are very popular and there are mobile apps and games available for all age groups. I use an app with my GCSE classes where they have to work in the correct order to solve different number puzzles." *(Secondary School Teacher)*

Before you start …

KS3	You should be able to add, subtract, multiply and divide positive and negative numbers.	1	Copy and complete each statement to make it true. Use only $<$, $=$ or $>$. **a** $2 + 3 \square 4 - 7$ **b** $-3 + 6 \square 4 - 7$ **c** $-1 - 4 \square 20 \div -4$ **d** $-6 \times 2 \square -7 - (-5)$
KS3	You should know the rules for working when more than one operation is involved in a calculation (BODMAS).	2	Spot the mistake in each calculation and correct the answers. **a** $3 + 8 + 3 \times 4 = 56$ **b** $3 + 8 \times 3 + 4 = 37$ **c** $3 \times (8 + 3) \times 4 = 130$

KS3	You should understand that addition and subtraction, and multiplication and division are inverse operations.	3	Identify the inverse operation by choosing the correct option.
			a $14 \times 4 = 56$
			   A $56 \times 4 = 14$ **B** $14 \div 4 = 56$ **C** $56 \div 4 = 14$
			b $200 \div 10 = 20$
			   A $200 \div 20 = 10$ **B** $200 = 10 \times 20$ **C** $10 \times 200 = 2000$
			c $27 + 53 = 80$
			   A $80 = 4 \times 20$ **B** $80 - 27 = 53$ **C** $80 + 27 = 107$

Find answers at: cambridge.org/ukschools/gcsemaths-studentbookanswers

Assess your starting point using the Launchpad

STEP 1

1 Calculate, without using a calculator, and show your working.
  a 647 + 786
  b 1406 − 289
  c 45 × 19
  d 414 ÷ 23

GO TO Section 1: Basic calculations

STEP 2

2 Choose the correct answer.
  a 9 ÷ (2 + 1) − 2
    A 9   **B** $3\frac{1}{2}$ **C** 1   **D** 0
  b (3 × 8) ÷ 4 + 8
    A 2   **B** 30   **C** 16   **D** 14
  c 12 − 6 × 2 + 11
    A 78   **B** 23   **C** 1   **D** 11
  d [5 × (9 + 1)] − 3
    A 53   **B** 47   **C** 40   **D** 43
  e (6 + 5) × 2 + (15 − 2 × 3) − 6
    A 40   **B** 20   **C** 32   **D** 25

GO TO Section 2: Order of operations

STEP 3

3 The perimeter of a square is equal to four times the length of a side. If the perimeter is 128 cm, what is the length of a side?

4 What should you add to 342 to get 550?

5 If a number divided by 45 is 30, what is the number?

GO TO Section 3: Inverse operations

GO TO Chapter review

CAMBRIDGE

Foundation

MATHEMATICS
GCSE for Edexcel
Student Book

Karen Morrison, Julia Smith, Pauline McLean, Rachael Horsman and Nick Asker

CAMBRIDGE
UNIVERSITY PRESS

University Printing House, Cambridge CB2 8BS, United Kingdom

Cambridge University Press is part of the University of Cambridge.

It furthers the University's mission by disseminating knowledge in the pursuit of education, learning and research at the highest international levels of excellence.

www.cambridge.org
Information on this title:
www.cambridge.org/9781107448025 (Paperback)
www.cambridge.org/9781107449718 (1 year Online Subscription)
www.cambridge.org/9781107449671 (2 year Online Subscription)
www.cambridge.org/9781107447912 (Paperback + Online Subscription)

© Cambridge University Press 2015

This publication is in copyright. Subject to statutory exception and to the provisions of relevant collective licensing agreements, no reproduction of any part may take place without the written permission of Cambridge University Press.

First published 2015

Printed in Dubai by Oriental Press

A catalogue record for this publication is available from the British Library

ISBN 978-1-107-44802-5 (Paperback)
ISBN 978-1-107-44971-8 (1 year Online Subscription)
ISBN 978-1-107-44967-1 (2 year Online Subscription)
ISBN 978-1-107-44791-2 (Paperback + Online Subscription)

Additional resources for this publication at www.cambridge.org/ukschools

Cambridge University Press has no responsibility for the persistence or accuracy of URLs for external or third-party internet websites referred to in this publication, and does not guarantee that any content on such websites is, or will remain, accurate or appropriate. Information regarding prices, travel timetables, and other factual information given in this work is correct at the time of first printing but Cambridge University Press does not guarantee the accuracy of such information thereafter.

....

NOTICE TO TEACHERS IN THE UK
It is illegal to reproduce any part of this work in material form (including photocopying and electronic storage) except under the following circumstances:
(i) where you are abiding by a licence granted to your school or institution by the Copyright Licensing Agency;
(ii) where no such licence exists, or where you wish to exceed the terms of a licence, and you have gained the written permission of Cambridge University Press;
(iii) where you are allowed to reproduce without permission under the provisions of Chapter 3 of the Copyright, Designs and Patents Act 1988, which covers, for example, the reproduction of short passages within certain types of educational anthology and reproduction for the purposes of setting examination questions.

....

In order to ensure that this resource offers high quality support for the associated Edexcel qualification, it has been through a review process by the awarding body to confirm that it fully covers the teaching and learning content of the specification or part of a specification at which it is aimed, and demonstrates an appropriate balance between the development of subject skills, knowledge and understanding, in addition to preparation for assessment.

While the publishers have made every attempt to ensure that advice on the qualification and its assessment is accurate, the official specification and associated assessment guidance materials are the only authoritative source of information and should always be referred to for definitive guidance.

Edexcel examiners have not contributed to any sections in this resource relevant to examination papers for which they have responsibility.

No material from this endorsed GCSE Mathematics Foundation Student Book will be used verbatim in any assessment by Edexcel.

Endorsement of this GCSE Mathematics Foundation Student Book does not mean that the resource is required to achieve this Edexcel qualification, nor does it mean that it is the only suitable material available to support the qualification, and any resource lists produced by the awarding body shall include this and other appropriate resources.

Contents

INTRODUCTION	vi

1 Calculations — 1
- Section 1: Basic calculations — 3
- Section 2: Order of operations — 8
- Section 3: Inverse operations — 11

2 Shapes and solids — 14
- Section 1: 2D shapes — 16
- Section 2: Symmetry — 20
- Section 3: Triangles — 23
- Section 4: Quadrilaterals — 27
- Section 5: 3D objects — 30

3 2D representations of 3D shapes — 36
- Section 1: 3D objects and their nets — 38
- Section 2: Drawing 3D objects — 43
- Section 3: Plan and elevation views — 48

4 Properties of whole numbers — 53
- Section 1: Reviewing number properties — 55
- Section 2: Prime factors — 58
- Section 3: Multiples and factors — 59

5 Introduction to algebra — 65
- Section 1: Using algebraic notation — 67
- Section 2: Simplifying expressions — 69
- Section 3: Multiplying out brackets — 72
- Section 4: Factorising expressions — 74
- Section 5: Using algebra to solve problems — 74

6 Fractions — 78
- Section 1: Equivalent fractions — 80
- Section 2: Operations with fractions — 82
- Section 3: Fractions of quantities — 87

7 Decimals — 92
- Section 1: Review of decimals and fractions — 93
- Section 2: Calculating with decimals — 97

8 Powers and roots — 103
- Section 1: Index notation — 104
- Section 2: The laws of indices — 108
- Section 3: Working with powers and roots — 109

9 Rounding, estimation and accuracy — 115
- Section 1: Approximate values — 117
- Section 2: Approximation and estimation — 122
- Section 3: Limits of accuracy — 124

10 Mensuration — 128
- Section 1: Standard units of measurement — 130
- Section 2: Compound units of measurement — 137
- Section 3: Maps, scale drawings and bearings — 141

11 Perimeter — 150
- Section 1: Perimeter of simple and composite shapes — 152
- Section 2: Circumference of a circle — 158
- Section 3: Problems involving perimeter and circumference — 164

12 Area — 169
- Section 1: Area of polygons — 171
- Section 2: Area of circles and sectors — 177
- Section 3: Area of composite shapes — 179

13 Further algebra — 187
- Section 1: Multiplying two binomials — 188
- Section 2: Factorising quadratic expressions — 192
- Section 3: Apply your skills — 197

14 Equations — 201
- Section 1: Linear equations — 203
- Section 2: Quadratic equations — 210
- Section 3: Simultaneous equations — 215
- Section 4: Using graphs to solve equations — 222

15 Functions and sequences — 230
- Section 1: Sequences and patterns — 232
- Section 2: Finding the nth term — 234
- Section 3: Functions — 237
- Section 4: Special sequences — 239

Find answers at: cambridge.org/ukschools/gcsemaths-studentbookanswers

16 Formulae — 244
- Section 1: Writing formulae — 246
- Section 2: Substituting values into formulae — 249
- Section 3: Changing the subject of a formula — 251
- Section 4: Working with formulae — 254

17 Volume and surface area — 259
- Section 1: Prisms and cylinders — 261
- Section 2: Cones and spheres — 266
- Section 3: Pyramids — 271

18 Percentages — 274
- Section 1: Review of percentages — 276
- Section 2: Percentage calculations — 279
- Section 3: Percentage change — 283

19 Ratio — 288
- Section 1: Introducing ratios — 289
- Section 2: Sharing in a given ratio — 292
- Section 3: Comparing ratios — 293

20 Probability basics — 299
- Section 1: The probability scale — 301
- Section 2: Calculating probability — 303
- Section 3: Experimental probability — 306
- Section 4: Mixed probability problems — 311

21 Construction and loci — 318
- Section 1: Using geometrical instruments — 320
- Section 2: Ruler and compass constructions — 324
- Section 3: Loci — 328
- Section 4: Applying your skills — 330

22 Vectors — 334
- Section 1: Vector notation and representation — 336
- Section 2: Vector arithmetic — 337
- Section 3: Mixed practice — 340

23 Straight-line graphs — 343
- Section 1: Plotting graphs — 345
- Section 2: Gradient and intercepts of straight-line graphs — 346
- Section 3: Parallel lines — 353
- Section 4: Working with straight-line graphs — 355

24 Graphs of functions and equations — 360
- Section 1: Review of linear graphs — 362
- Section 2: Quadratic functions — 364
- Section 3: Other polynomials and reciprocals — 369
- Section 4: Plotting, sketching and recognising graphs — 373

25 Angles — 377
- Section 1: Angle facts — 379
- Section 2: Parallel lines and angles — 381
- Section 3: Angles in triangles — 383
- Section 4: Angles in polygons — 384

26 Probability – combined events — 391
- Section 1: Representing combined events — 393
- Section 2: Theoretical probability of combined events — 400

27 Standard form — 409
- Section 1: Expressing numbers in standard form — 411
- Section 2: Calculators and standard form — 414
- Section 3: Working in standard form — 416

28 Similarity — 421
- Section 1: Similar triangles — 423
- Section 2: Enlargements — 427
- Section 3: Similar shapes — 434

29 Congruence — 439
- Section 1: Congruent triangles — 441
- Section 2: Applying congruency — 445

30 Pythagoras' theorem — 450
- Section 1: Finding the length of the hypotenuse — 452
- Section 2: Finding the length of any side — 454
- Section 3: Proving whether a triangle is right-angled — 457
- Section 4: Using Pythagoras' theorem to solve problems — 459

31 Trigonometry — 465
- Section 1: Trigonometry in right-angled triangles — 467
- Section 2: Exact values of trigonometric ratios — 474
- Section 3: Solving problems using trigonometry — 476

32 Growth and decay	482
Section 1: Simple and compound growth	483
Section 2: Depreciation and decay	487

33 Proportion	492
Section 1: Direct proportion	494
Section 2: Algebraic and graphical representations	498
Section 3: Inverse proportion	501

34 Algebraic inequalities	506
Section 1: Expressing inequalities	508
Section 2: Number lines	509
Section 3: Solving inequalities	510
Section 4: Working with inequalities	512

35 Sampling and representing data	515
Section 1: Populations and samples	517
Section 2: Tables and graphs	519
Section 3: Pie charts	526
Section 4: Line graphs for time-series data	529

36 Data analysis	536
Section 1: Summary statistics	538
Section 2: Misleading graphs	545
Section 3: Scatter diagrams	548

37 Interpretation of graphs	555
Section 1: Graphs of real-world contexts	557
Section 2: Gradients	559

38 Transformations	564
Section 1: Reflections	566
Section 2: Translations	571
Section 3: Rotations	576

GLOSSARY	584
INDEX	587
ACKNOWLEDGEMENTS	592

Note
The colour of each chapter corresponds to the area of maths that it covers:

- Number
- Algebra
- Ratio, proportion and rates of change
- Geometry and measures
- Probability
- Statistics

Find answers at: cambridge.org/ukschools/gcsemaths-studentbookanswers

Introduction

This book has been written by experienced teachers to help build your understanding and enjoyment of the maths you will meet at GCSE.

Each chapter opens with a list of skills that are covered in the chapter. The **real-life applications** section describes an example of how the maths is used in real life.

All chapters build on knowledge that you will have learned in previous years. You might need to revise some topics before starting a chapter. To check your knowledge, answer the questions in the **Before you start…** table. You can check your answers using the free answer booklet available at **www.cambridge.org/ukschools/gcsemaths-studentbookanswers**. If you answer any questions incorrectly, you might need to revise the topic from your work in earlier years.

The chapters are divided into sections, each covering a single topic. Some chapters may cover topics that you already know and understand. You can use the **Launchpad** to identify the best section for you to start with. Answer the questions in each step. If you find a question difficult to answer correctly, the step suggests the section that you should look at.

Throughout the book, there are features to help you build knowledge and improve your skills:

 This means you might need a calculator to work through a question.

 This means you should work through a question without using a calculator.

 This shows the question is from a past exam paper.

Tip
Tip boxes provide helpful hints.

Calculator tip
Calculator tips help you to use your calculator.

Learn this formula
Learn this formula boxes contain formulae that you need to know.

Key vocabulary
Important maths terms are written in **green**. You can find what they mean in **Key vocabulary** boxes and also in the **Glossary** at the back of the book.

Did you know?
Did you know? boxes contain interesting maths facts.

WORK IT OUT
Work it out boxes contain a question with several worked solutions. Some of the solutions contain common mistakes. Try to spot the correct solution and check the free answer booklet available at **www.cambridge.org/ukschools/gcsemaths-studentbookanswers** to see if you're right.

WORKED EXAMPLE
Worked examples guide you through model answers to help you understand methods of answering questions.

Some chapters contain a **Problem-solving framework**, which sets a problem and then shows how you can go about answering it.

Checklist of learning and understanding
At the end of a chapter, use the **Checklist of learning and understanding** to check whether you have covered everything you need to know.

Chapter review
You can check whether you have understood the topics using the **Chapter review,** which contains questions from the whole chapter.

A booklet containing answers to all exercises is free to download at **www.cambridge.org/ukschools/gcsemaths-studentbookanswers**.

You can find more resources, including interactive widgets, games and quizzes on **GCSE Mathematics Online**.

Section 1: Basic calculations

You will not always have a calculator so it is useful to know how to do calculations using mental and written strategies.

It is best to use a method that you are confident with and always **show your working**.

Remember that when a question asks you to find the:

- **sum** you need to add
- **difference** you need to subtract the smaller number from the larger number
- **product** you need to multiply
- **quotient** you need to divide.

> **Tip**
>
> Some examination papers will not allow you to use your calculator.

WORK IT OUT 1.1

Look at these calculations carefully.

Discuss with a partner what methods these students have used to find the answer.

Which method would you use to do each of these calculations? Why?

① $489 + 274$
$400 + 200 \rightarrow 600$
$80 + 70 \rightarrow 150$
$9 + 4 \rightarrow 13$
763

② $284 - 176$
$2\overset{7}{\cancel{8}}\overset{1}{\cancel{4}}$
-176
108

③ 29×17
$\rightarrow 30 \times 17 - 17$
$\rightarrow 3 \times 170 - 17$
$\rightarrow 510 - 17$
$\rightarrow 493$

④ 15×62
$= 30 \times 31 \qquad 310$
$= 930 \qquad 310$
$ \underline{310}$
$ 930$

⑤ 207×47

×	200	0	7
40	8000	0	280
7	1400	0	49

$9400 + 0 + 329$
$= 9729$

⑥ $2394 \div 42$

2394
$-1680 \quad (40) \qquad 42 \times 10 = 420$
$714 \qquad 42 \times 20 = 840$
$-420 \quad (10) \qquad 42 \times 40 = 1680$
$294 \qquad 42 \times 5 = 210$
$-210 \quad (5) \qquad 42 \times 2 = 84$
84
$-84 \quad (2)$
$= 57$

Problem-solving strategies

There are some useful strategies and techniques that you can use to break down complex problems to help you solve them more easily.

If you follow these steps each time you are faced with a problem, you will become more confident at problem solving and more able to check that your answers are sensible.

These are important skills both for your GCSE courses and for everyday life.

Problem-solving framework

Sally buys, repairs and sells used furniture at a market.

Last week she bought a table for £32 and a bench for £18.

She spent £12 on wood, nails, varnish and glue to fix them up.

She then sold the two items on her stall for £69.

How much profit did she make on the two items?

Steps for solving problems	What you would do for this example
Step 1: Work out what you have to do. Start by reading the question carefully.	Find the profit on the two items.
Step 2: What information do you need? Have you got it all?	Cost of items = £32 + £18 Cost of repairs = £12 Selling price = £69
Step 3: Is there any information that you don't need?	In this problem you don't need to know what she spent the money on, you just need to know how much she spent. Many problems contain extra information that you don't need so as to test your understanding.
Step 4: Decide what maths you can do.	Profit = selling price − cost You can add the costs and subtract them from the selling price.
Step 5: Set out your solution clearly. Check your working and make sure your answer is reasonable.	Cost = £32 + £18 + £12 = £62 Profit = £69 − £62 = £7 Sally made £7 profit.
Step 6: Check that you have answered the question.	Yes. You needed to find the profit and you have found it.

EXERCISE 1A

Solve these problems using written methods.

Set out your solutions clearly to show the methods you chose.

1 Nola checked the prices of pens at three different supermarkets. She found that the cheapest pack of pens was £3.90 for three. She bought fifteen pens.

How much did she pay in total and how much did she pay for each pen?

  a What two things are you asked to find here?

  b How many packs of pens did she buy? Why do you need to know this?

  c What operation would you do to find the total cost? Why?

  d How would you work out the cost of each pen?

  e Does a price of £1.30 for a pen seem reasonable to you?

Tip

You don't always need to write something for the first few steps in the problem-solving framework, but you should still consider these steps mentally when approaching a problem in order to help you decide what to do. You should **always** show your working fully.

2 Sandra bought a pair of jeans for £34, a scarf for £9.50 and a top for £20.

If she had saved £100 to buy these items, how much money would she have left?

3 How many 16-page brochures can you make from 1030 pages?

4 Jason can type 48 words per minute.

  a How many words can he type in an hour and a half?

  b Approximately how long would it take him to type an article of 2000 words?

5 At the start of a year the population of Greenside Village was 56 309.

During the year 617 people died, 1835 babies were born, 4087 people left the village and 3099 people moved into the village.

What was the population at the end of the year?

6 The Amazon River is 6448 km long, the Nile River is 6670 km and the Severn is 354 km long.

  a How much longer is the Nile than the Amazon?

  b How much shorter is the Severn than the Amazon?

Did you know?

The Severn is the longest river in the UK.

7 What is the combined sum of 132 and 99 plus the product of 36 and 127?

8 What is the result when the difference between 8765 and 3087 is added to the result of 1206 divided by 18?

Working with negative and positive integers

Key vocabulary

integers: whole numbers belonging to the set {… −3, −2, −1, 0, 1, 2, 3, …}; they are sometimes called directed numbers because they have a negative or positive sign.

Tip

You will be expected to work with negative and positive values in algebra, so it is important to make sure you can do this early on in your GCSE course.

When doing calculations involving positive and negative **integers**, you need to remember the following:

- Adding a negative number is the same as subtracting the number:
  $4 + -3 = 1$
- Subtracting a negative number is the same as adding a positive number:
  $5 - -3 = 8$
- Multiplying or dividing the same signs gives a positive answer:
  $-4 \times -2 = 8$ and $\dfrac{-4}{-2} = 2$
- Multiplying or dividing different signs gives a negative answer:
  $4 \times -2 = -8$ and $\dfrac{-4}{2} = -2$

EXERCISE 1B

1 Calculate.
  a $12 - 5 + 8$
  b $-3 - 4 - 8$
  c $3 + 5 - 6$
  d $-2 - 8 + 5$
  e $14 - 3 - 9$
  f $9 - 3 - 4$
  g $-34 + 18 - 12$
  h $25 - 19 - 42$

2 Calculate.
  a $-9 - (-7)$
  b $-3 - (-10)$
  c $-4 - (-12)$
  d $8 - (-9)$
  e $9 - (-8)$
  f $-14 - (-14)$
  g $-3 - 8 - (-9)$
  h $-12 + 4 - (-8)$

3 Calculate.
  a $-2 \times -4 \times -4$
  b $-4 \times 3 \times -6$
  c $-3 \times -4 \times 3$
  d $-4 \times -8 \times 3$
  e $3 \times 6 \times -4$
  f $12 \times 2 \times -3$
  g $1 \times -1 \times 10$
  h $-3 \times -8 \times 9$

4 Calculate.
  a $24 \div 3$
  b $-24 \div 3$
  c $-48 \div -6$
  d $400 \div -40$
  e $-22 \div -22$
  f $-33 \div 11$
  g $45 \div -9$
  h $-64 \div -8$

5 Calculate.
  a $\dfrac{-40}{5}$
  b $\dfrac{-28}{-4}$
  c $\dfrac{30}{-5}$
  d $\dfrac{12}{-2}$
  e $\dfrac{-48}{-6}$
  f $\dfrac{-63}{7}$
  g $\dfrac{-60}{-20}$
  h $\dfrac{60}{-6}$

6 Apply the operations in the first row to the given number to complete each table.

a

	− 10	× −2	+ 4	÷ −2	− 8	+ 1
−5						

b

	× −4	÷ −5	+ 8	− 3	× 2	− 9
10						

c

	− 10	× −2	+ 4	÷ −2	− 8	+ 1
0						

7 Here are some bank transactions.

Calculate the new balance in each case.

a Balance of £230, withdraw £100.

b Balance of £250.50, withdraw £300.

c Balance of −£450, deposit £900 then withdraw £300.

d Balance of −£100, deposit £2000 then withdraw £550.

8 The damaged opening of an oil well, 5000 feet below sea level, caused a massive oil spill in the Gulf of Mexico in 2010.

The oil well itself extended to a depth of 13 000 feet.

Express the answers to these questions as directed numbers.

a How deep was the deepest part of the oil well below the sea bed?

b How far did oil travel from the bottom of the well to reach the surface of the water?

c The oil company involved estimated that they were losing money at the rate of $15 000 000 per day. Use an integer to express the change in the company's balance after:

  i one week ii thirteen weeks.

Tip

The foot (plural feet) is a standard unit of imperial measurement for length; the metric measurement for length is metres. You will learn about **metric** measurements in Chapter 10.

9 Here is a set of integers.

{−8, −6, −3, 1, 3, 7}

a Find two numbers with a difference of 9.

b Find three numbers with a sum of 1.

c Find two numbers whose product is −3.

d Find two numbers which, when divided, will give an answer of −6.

10 One more than −6 is added to the product of 7 and 6 less than 3.
What is the result?

11 The temperature in Inverness is 4 °C at 7 pm at night.

By 1 am the same night, it has dropped by 12 degrees.

a What is the temperature at 1 am?

b What is the average hourly change in the temperature?

c By noon the next day, the temperature is 7 °C. How many degrees warmer is this than it was at 1 am?

Section 2: Order of operations

Jose posted this calculation on his wall on social media.

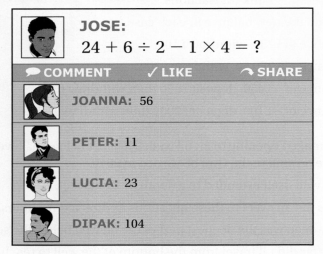

Within minutes, his friends had posted four different answers.

Which one (if any) do you think is correct? Why?

There is a set of rules that tell you the order in which you need to work when there is more than one operation.

The order of operations is:

1 Do any operations in brackets first.

2 If there are any '**powers of**' or '**fractions of**' in the calculation, do them next.

3 Do division and multiplication next, working from left to right.

4 Do addition and subtraction last, working from left to right.

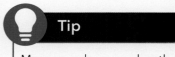

Tip

Many people remember these rules using the letters **BODMAS**. (or sometimes BIDMAS).
Brackets
Of ('powers **of**' or 'fractions **of**'; in BIDMAS, I stands for Indices)
Divide and/or **M**ultiply
Add and/or **S**ubtract.

Brackets and other grouping symbols

Brackets are used to group operations. For example:

$(3 + 7) \times (30 \div 2)$

When there is more than one set of brackets, you work from the **innermost set** to the **outermost set**.

WORKED EXAMPLE 1

Calculate $2((4 + 2) \times 2 - 3(1 - 3) - 10)$

$2\,((4 + 2) \times 2 - 3(1 - 3) - 10)$ Highlight the different pairs of brackets to help if you need to.

$2((4 + 2) \times 2 - 3(1 - 3) - 10)$
$= 2(6 \times 2 - 3(-2) - 10)$
$= 2(6 \times 2 - 3 \times -2 - 10)$

The red brackets are the innermost, so do the calculations inside these ones first. There are two lots of red brackets, so work from left to right. **Note** that you can leave -2 inside brackets if you prefer because $3(-2)$ is the same as 3×-2.

$2(6 \times 2 - 3 \times -2 - 10)$
$= 2(12 - -6 - 10)$
$= 2(8)$
$= 2 \times 8$
$= 16$

Blue brackets are next. Do the multiplications first from left to right, then the subtractions from left to right.

Often a different style of bracket will be used to make it easier to identify each pair.

For example, the following different types of brackets have been used below: $(\,), [\,], \{\,\}$.

$\{2 - [4(2 - 7) - 4(3 + 8)] - 2\} \times 8$

Other symbols can also be used to group operations.

For example:

Fraction bars: $\dfrac{5 - 12}{3 - 8}$

Roots: $\sqrt{16 + 9}$

These symbols are treated like brackets when you do a calculation.

> **Tip**
>
> $\dfrac{5 - 12}{3 - 8}$ is the same calculation as $(5 - 12) \div (3 - 8)$.

Calculator tip

Most modern calculators are programmed to use the correct order of operations. Check your calculator by entering $2 + 3 \times 4$. You should get 14.

If the calculation has brackets, you need to enter the brackets into the calculator to make sure it does these first.

WORK IT OUT 1.2

Which of the solutions is correct in each case?
Find the mistakes in the incorrect option.

	Option A	Option B
1	$7 \times 3 + 4$ $= 21 + 4$ $= 25$	$7 \times 3 + 4$ $= 7 \times 7$ $= 49$
2	$(10 - 4) \times (4 + 9)^2$ $= 6 \times 16 + 81$ $= 96 + 81$ $= 177$	$(10 - 4) \times (4 + 9)^2$ $= 6 \times (13)^2$ $= 6 \times 169$ $= 1014$
3	$45 - [20 \times (4 - 3)]$ $= 45 - [20 \times 1]$ $= 45 - 21$ $= 24$	$45 - [20 \times (4 - 3)]$ $= 45 - 20 \times 1$ $= 45 - 20$ $= 25$
4	$30 - 4 \div 2 + 2$ $= 26 \div 2 + 2$ $= 13 + 2$ $= 15$	$30 - 4 \div 2 + 2$ $= 30 - 2 + 2$ $= 30$
5	$\dfrac{18 - 4}{4 - 2}$ $= \dfrac{18}{2}$ $= 9$	$\dfrac{18 - 4}{4 - 2}$ $= \dfrac{14}{2}$ $= 7$
6	$\sqrt{36 \div 4} + 40 \div 4 + 1$ $= \sqrt{9} + 10 + 1$ $= 3 + 11$ $= 14$	$\sqrt{36 \div 4} + 40 \div 4 + 1$ $= \sqrt{9} + 40 \div 5$ $= 3 + 8$ $= 11$

EXERCISE 1C

1. Calculate, showing the steps in your working.

   a $5 \times 10 + 3$
   b $5 \times (10 + 3)$
   c $2 + 10 \times 3$
   d $(2 + 10) \times 3$
   e $23 + 7 \times 2$
   f $6 \times 2 \div (3 + 3)$
   g $10 - 4 \times 5$
   h $12 + 6 \div 2 - 4$
   i $3 + 4 \times 5 - 10$
   j $18 \div 3 \times 5 - 3 + 2$
   k $5 - 3 \times 8 - 6 \div 2$
   l $7 + 8 \div 4 - 1$
   m $\dfrac{15 - 5}{2 \times 5}$
   n $(17 + 1) \div 9 + 2$
   o $\dfrac{16 - 4}{4 - 1}$

p $17 + 3 \times 21$ q $48 - (2 + 3) \times 2$ r $12 \times 4 - 4 \times 8$
s $15 + 30 \div 3 + 6$ t $20 - 6 \div 3 + 3$ u $10 - 4 \times 2 \div 2$

2 Check whether these answers are correct.

If the answer is wrong, work out the correct answer.

a $12 \times 4 + 76 = 124$
b $8 + 75 \times 8 = 698$
c $12 \times 18 - 4 \times 23 = 124$
d $(16 \div 4) \times (7 + 3 \times 4) = 76$
e $(82 - 36) \times (2 + 6) = 16$
f $(3 \times 7 - 4) - (4 + 6 \div 2) = 12$

3 Use the numbers listed to make each number sentence true.

a □ − □ ÷ □ = □   0, 2, 5, 10
b □ − □ ÷ □ = □   9, 11, 13, 18
c □ ÷ (□ − □) − □ = □ 1, 3, 8, 14, 16
d (□ + □) − (□ − □) = □ 4, 5, 6, 9, 12

Section 3: Inverse operations

The four operations (add, subtract, multiply and divide) are related to each other.

Operations are inverses of each other if one undoes (cancels out) the effect of the other.

- Adding is the inverse of subtracting, e.g. add 5 is undone by subtract 5.
- Multiplying is the inverse of dividing, e.g. multiply by 2 is undone by divide by 2.
- Taking a square root is the inverse of squaring a number, e.g. 4^2 is undone by $\sqrt{16}$.
- Taking the cube root is the inverse of cubing a number, e.g. 2^3 is undone by $\sqrt[3]{8}$.

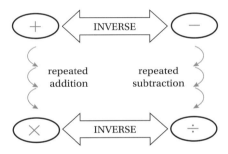

Additive inverse

The additive inverse of 1 is −1 (add 1 is undone by subtract 1).

The additive inverse of −5 is 5 (subtract 5 is undone by add 5).

When you add a number to its inverse the answer is always 1.

Multiplicative inverse

The multiplicative inverse of 2 is $\frac{1}{2}$ (multiply by 2 is undone by divide by 2).

The multiplicative inverse of $\frac{1}{4}$ is 4 (divide by 4 is undone by multiply by 4).

When you multiply a number by its inverse, the answer is always 1.

Inverse operations are useful for checking the results of your calculations because carrying out the inverse operation gets you back to the number you started with.

For example, is $4320 - 500 = 3820$ correct?

Check by adding 500 back to the result (i.e. doing the inverse operation) to see if you get 4320.

$3820 + 500 = 4320$ so the original calculation is correct.

Tip

The multiplicative inverse of a number is also called its **reciprocal**. For example, $\frac{1}{3}$ is the reciprocal of 3.

Tip

You will use inverse operations to solve equations and when you deal with functions, so it is important that you understand how they work.

EXERCISE 1D

1 Find the additive inverse of each of these numbers.
  a 5 **b** 2 **c** 100 **d** −3 **e** −16 **f** −12

2 By what number would you multiply each of these to get an answer of 1?
  a 5 **b** 10 **c** −3 **d** $\frac{1}{3}$ **e** 9 **f** $\frac{1}{9}$

3 Use inverse operations to check each calculation. Correct those that are wrong.
  a 6172 − 3415 = 2757 **b** 488 − 156 = 322 **c** 219 − 361 = −142
  d 264 + 469 = 723 **e** 4019 + 217 = 4235 **f** 617 + 728 = 1345

4 Use inverse operations to check each calculation. Correct those that are wrong.
  a 512 ÷ 4 = 43 **b** 672 ÷ 12 = 56 **c** 1275 ÷ 15 = 75
  d 3840 ÷ 30 = 128 **e** 30 × 125 = 3770 **f** 214 × 8 = 1732
  g $\sqrt{900} = 30$ **h** $\sqrt{15\,625} = 120$ **i** $400^2 = 16\,000$

5 Use inverse operations to find the missing values in each of these calculations.
  a ☐ + 217 = 529 **b** ☐ + 388 = 490 **c** ☐ − 218 = 182
  d 121 × ☐ = −605 **e** −6 × ☐ = 870 **f** ☐ ÷ 40 = 5400

Checklist of learning and understanding

Basic calculations
- Written methods are important for when you do not have a calculator.
- You can use any method as long as you show your working.
- Negative and positive numbers can be added, subtracted, multiplied and divided as long as you apply the rules to get the correct sign in the answer.

Order of operations
- In maths there is a conventional order for working when there is more than one operation.
- Always work out brackets (or other grouping symbols) first, then powers. Multiply and/or divide next, then add and/or subtract.
- A useful memory aid for the order of operations is BODMAS.

Inverse operations
- An inverse operation undoes the previous operation.
- Addition is the inverse of subtraction.
- Multiplication is the inverse of division.
- Squaring is the inverse of taking the square root.

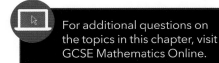

Chapter review

1 These are the solutions to a cross-number puzzle.

The clues are all calculations that involve using the correct order of operations.

Write a set of clues that would give these results.

2 Use integers and operations to write ten different calculations that give an answer of −17.

3 a Work out 2 × (8 − 3) *(1 mark)*

b Work out 32 + 4 × 5 *(2 marks)*

©Pearson Education Ltd 2013

4 On a page of a magazine there are three columns of text.

Each column contains 42 rows.

If there is an average of 32 letters per column row, approximately how many letters are there on a page?

5 A stadium has seats for 32 000 people.

How many rows of 125 is this?

6 Two numbers have a sum of −15 and a product of −100.

What are the numbers?

7 The sum of two numbers is 1, but their product is −20.

What are the numbers?

8 Josie's bank account was overdrawn.

She deposited £1000 and this brought her balance to £432.

By how much was her account overdrawn to start with?

Find answers at: cambridge.org/ukschools/gcsemaths-studentbookanswers

2 Shapes and solids

In this chapter you will learn how to …
- use the correct geometrical terms to talk about lines, angles and shapes.
- recognise and name common 2D shapes and 3D objects.
- describe the symmetrical properties of various polygons.
- classify triangles and quadrilaterals and use their properties to identify them.
- use properties of shapes to find missing angles and sides.

For more resources relating to this chapter, visit GCSE Mathematics Online.

Using mathematics: real-life applications

Artists, craftspeople, builders, designers, architects and engineers use shape and space in their jobs, but almost everyone uses lines, angles, patterns and shapes in different ways every day.

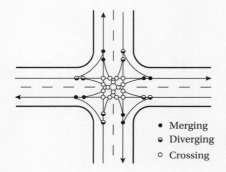

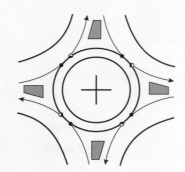

- Merging
- Diverging
- Crossing

"I use a CAD package to plot lines and angles and show the direction of traffic flow when I design new road junctions." *(Civil engineer)*

Before you start …

KS3	You need to use geometrical terms correctly.	① Choose the correct label for each letter on the diagram. base — vertex — acute angle — point edge — right angle — height — face
KS3	You should be able to recognise and name different types of shapes.	② Identify three different shapes in this diagram and use letters to name them correctly. ③ *ABCE* is one face of a solid with eight faces. What type of solid could it be?

Assess your starting point using the Launchpad

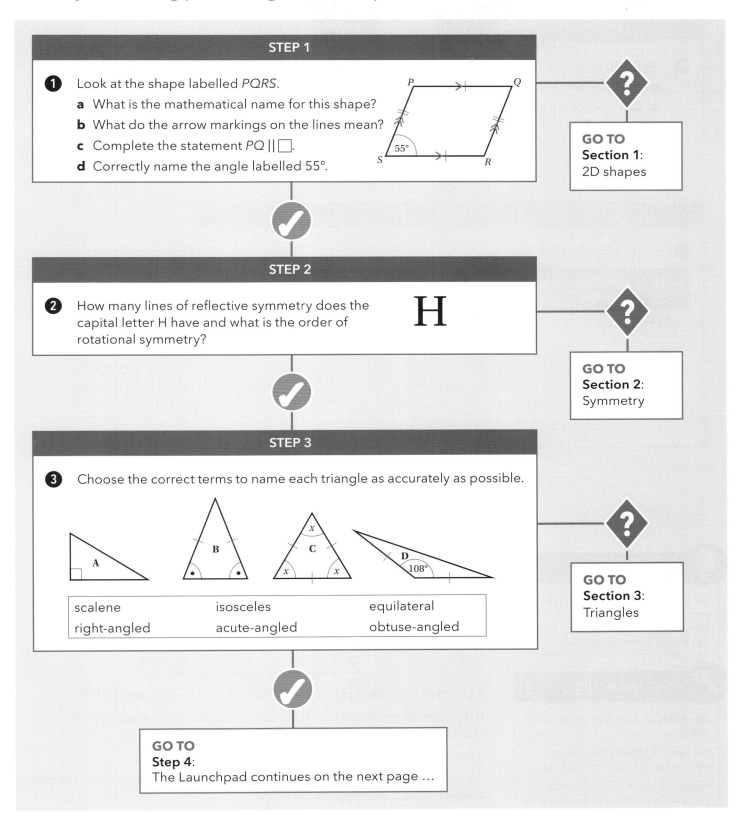

STEP 1

1 Look at the shape labelled *PQRS*.
   a What is the mathematical name for this shape?
   b What do the arrow markings on the lines mean?
   c Complete the statement *PQ* ‖ ☐.
   d Correctly name the angle labelled 55°.

GO TO Section 1: 2D shapes

STEP 2

2 How many lines of reflective symmetry does the capital letter H have and what is the order of rotational symmetry?

GO TO Section 2: Symmetry

STEP 3

3 Choose the correct terms to name each triangle as accurately as possible.

scalene   isosceles   equilateral
right-angled   acute-angled   obtuse-angled

GO TO Section 3: Triangles

GO TO Step 4: The Launchpad continues on the next page …

Find answers at: cambridge.org/ukschools/gcsemaths-studentbookanswers

Launchpad continued …

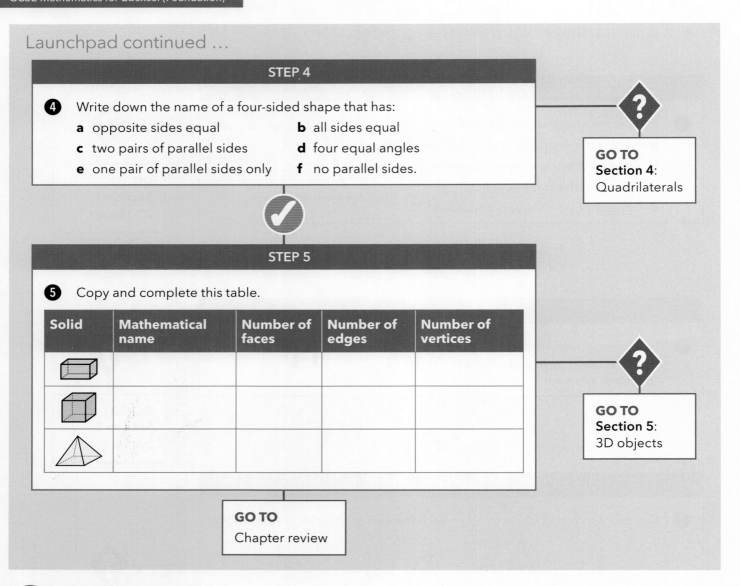

STEP 4

4 Write down the name of a four-sided shape that has:
- **a** opposite sides equal
- **b** all sides equal
- **c** two pairs of parallel sides
- **d** four equal angles
- **e** one pair of parallel sides only
- **f** no parallel sides.

GO TO Section 4: Quadrilaterals

STEP 5

5 Copy and complete this table.

Solid	Mathematical name	Number of faces	Number of edges	Number of vertices

GO TO Section 5: 3D objects

GO TO Chapter review

Section 1: 2D shapes

> **Tip**
> Circles and ellipses (ovals) are plane shapes, but they do not have straight sides, so they are not classified as polygons.

Flat shapes are called **plane shapes** or two-dimensional (2D) shapes.

A **polygon** is a closed plane shape with three or more straight sides.

If the sides of a polygon are all the same length and the angles between the sides (interior angles) are equal, then it is a **regular polygon**.

An equilateral triangle is an example of a regular polygon.

A polygon that does not have all sides equal or does not have equal interior angles is called an **irregular polygon**.

A rectangle is an irregular polygon because its sides are not all equal in length.

The fact that the angles of a rectangle are all equal to 90° does not make it a regular polygon.

> **Key vocabulary**
>
> **plane shape:** a flat, two-dimensional shape.
>
> **polygon:** a closed plane shape with three or more straight sides.
>
> **regular polygon:** a polygon with equal straight sides and equal angles.
>
> **irregular polygon:** a polygon that does not have equal sides or equal angles.

Naming polygons

Polygons can be named according to the number of sides they have.

Name of polygon	Number of sides	Regular polygon	Irregular polygon
triangle	3	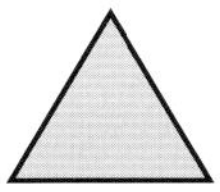	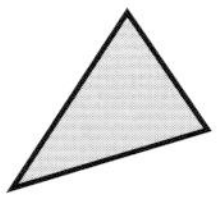
quadrilateral	4		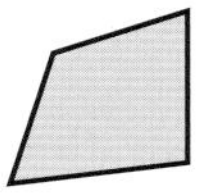
pentagon	5	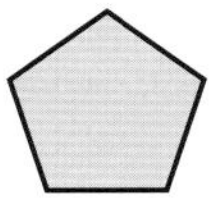	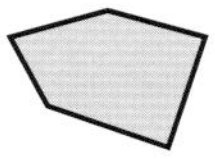
hexagon	6	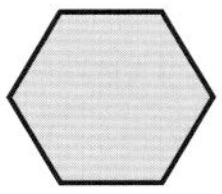	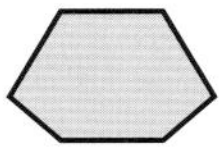
heptagon	7	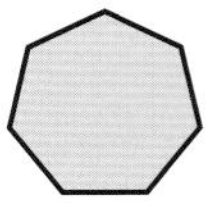	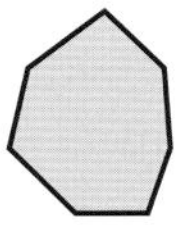
octagon	8	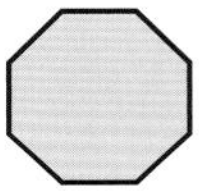	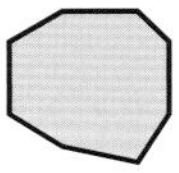
nonagon	9	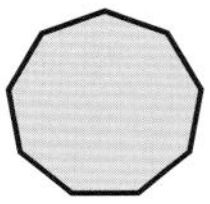	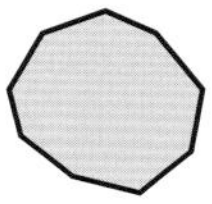
decagon	10	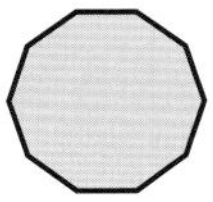	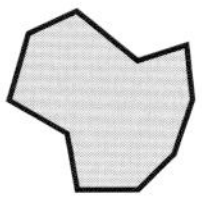

EXERCISE 2A

1 What is the correct mathematical name for each of the following shapes?

  a A plane shape with three equal sides.

  b A polygon with five equal sides.

  c A polygon with six vertices and six equal angles.

  d A plane shape with eight equal sides and eight equal internal angles.

2 Which of the following is not a polygon?

  A square   **B** sector   **C** parallelogram   **D** triangle

3 Where might you find the following in real life?

  a a regular octagon

  b a circle

  c a regular quadrilateral

  d an irregular pentagon

Find answers at: cambridge.org/ukschools/gcsemaths-studentbookanswers

Perpendicular and parallel lines

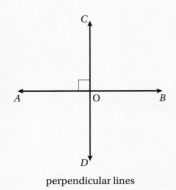

perpendicular lines

Perpendicular lines meet at right angles (90°).

The symbol ⊥ means 'perpendicular to'.

In the diagram AB ⊥ CD.

The shortest distance from a point to a line is the perpendicular distance between them.

The sides of shapes are perpendicular if they form a 90° angle.

Lines are parallel if they are the same perpendicular distance apart at any point along their length.

You can say that parallel lines are equidistant along their length, i.e. always an equal distance apart and never meeting.

The symbol || means 'parallel to'.
In the diagram $AB \parallel CD$ and $MN \parallel PQ$.

Small arrow symbols are drawn on lines to indicate that they are parallel to each other

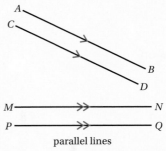

parallel lines

When there is more than one pair of parallel lines in a diagram, each pair is usually given a different set of arrow markings.

In this diagram $AB \parallel DC$, $AD \parallel EG$ and $EF \parallel HC$.

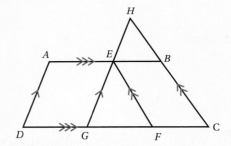

Drawing and labelling diagrams

Mathematical diagrams are drawn and labelled in standard ways so that their meaning is clear to anyone who uses them.

Sides and angles

Shapes are labelled using capital letters on each vertex, which are usually written in alphabetical order as you move round the shape.

The shape would be called $\triangle ABC$.

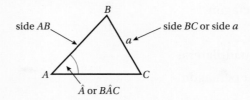

Each side of this triangle can be named using the capital letters on the vertices: *AB*, *BC* and *CA*.

Sometimes single letters are used to name the sides.

In this example, side *BC* can also be called side *a* because it is opposite angle *A*.

This convention is often used when you work with Pythagoras' theorem and in trigonometry, which explore angles and lengths of sides in triangles.

The angles can be named in different ways.

The angle at vertex *A* can be named *A*, *BAC* or *CAB*.

Symbols can be used to label angles. For example, ∠BAC or BÂC.

Marking equal sides and angles

Small lines can be drawn on the sides of a shape to show whether the sides are equal or not.

Curved lines and symbols such as dots or letters can be used to show whether angles are equal or not.

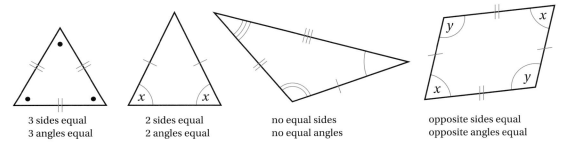

3 sides equal 2 sides equal no equal sides opposite sides equal
3 angles equal 2 angles equal no equal angles opposite angles equal

EXERCISE 2B

1 Read each clue in Column A. Match it to a term in Column B.

	Column A	Column B
a	a shape that has two fewer sides than an octagon	decagon
b	a shape that has two sides more than a triangle	hexagon
c	a shape with four sides	equilateral triangle
d	a traffic stop sign is an example of this shape	two-dimensional
e	a figure that has length and height	pentagon
f	a closed plane shape with all sides *x* cm long and all angles the same size	quadrilateral
g	a ten-sided figure	square
h	another name for a regular four-sided polygon	regular polygon
i	the more common name for a regular three-sided polygon	octagon

2 Look at this diagram.

Say whether each of the following statements is true or false.

a $AF \parallel EC$.

b $\triangle BFD$ is isosceles.

c $CE \perp BC$.

d $AE \parallel BD$.

e $ABCE$ is a regular polygon.

f $GB \parallel BC$.

g In $\triangle DHJ$, angle H = angle J = angle D.

h $\triangle GHJ$ is a regular polygon.

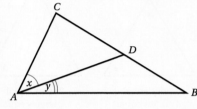

3 Draw and correctly label a sketch of each of the following shapes.

a A triangle, ABC with angle B = angle C and side $AB \perp AC$.

b A regular four-sided polygon $DEFG$.

c Quadrilateral $PQRS$ such that $PQ \parallel SR$ but $PQ \neq SR$ and $\angle PSR = \angle QRS$.

4 a In the diagram, why can you not use just the single letter A to represent the angle labelled x?

b How should it be labelled?

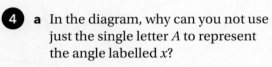

Section 2: Symmetry

Some shapes and objects are symmetrical.

Symmetry is a property that can be used to help identify shapes, find missing lengths and angles, and solve problems.

You need to recognise two types of symmetry in plane shapes: line symmetry and rotational symmetry.

Line symmetry

If you can fold a shape in half to create a mirror image (**reflection**) on either side of the fold the shape has line symmetry.

The fold is known as the **line of symmetry**.

Each half of the shape is a reflection of the other half so this type of symmetry is also called reflection symmetry.

Triangle A has line symmetry. The dotted line is the line of symmetry.

If you fold the shape along this line the two parts will fit onto each other exactly.

Triangle B is not symmetrical. You cannot draw a line to divide it into two identical parts.

triangle A

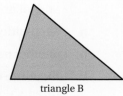
triangle B

> **Tip**
>
> The line of symmetry is sometimes called the mirror line. If you place a small mirror on the line of symmetry you will see the whole shape reflected in the mirror.

> **Key vocabulary**
>
> **reflection**: an exact image of a shape about a line of symmetry.
>
> **line of symmetry**: a line that divides a plane shape into two identical halves, each the reflection of the other.

A shape can have more than one line of symmetry.

A regular pentagon has five lines of symmetry.

Lines of symmetry can be horizontal, vertical or diagonal.

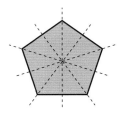

Rotational symmetry

A shape has **rotational symmetry** if you rotate it around a fixed point and it looks identical in different positions.

The **order** of rotational symmetry tells you how many times the shape will look identical before it returns to the starting point.

If you can only rotate the shape a full 360° before it appears identical again then it does **not** have rotational symmetry.

 Key vocabulary

rotational symmetry: symmetry by turning a shape around a fixed point so that it looks the same from different positions.

 Tip

You will deal with reflections in mirror lines and rotations about a fixed point again in Chapter 38 when you deal with transformations using coordinates.

The order of rotational symmetry of a regular polygon depends on the number of sides it has.

A square has an order of rotational symmetry of 4 around its centre.

The star is just to show the position of one vertex of the square as it rotates.

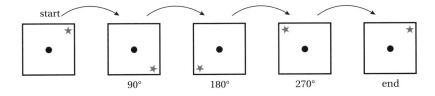

This is the national symbol for the Isle of Man.

It has an order of rotational symmetry of 3 about its centre.

Find answers at: cambridge.org/ukschools/gcsemaths-studentbookanswers

EXERCISE 2C

1 Which of the dotted lines in each figure are lines of symmetry?

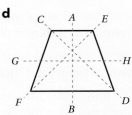

2 By sketching, investigate and work out the number of lines of symmetry and the order of rotational symmetry of each shape.

Copy and complete the table to summarise your results.

Shape	Number of lines of symmetry	Order of rotational symmetry
square		
rectangle		
isosceles triangle		
equilateral triangle		
parallelogram		
regular hexagon		
regular octagon		

3 Give an example of a shape which has rotational symmetry of order 3 but which is not a triangle.

4 Which of the following letters have rotational symmetry?

C   H   A   R

5 Describe the symmetrical features of this design in as much detail as possible.

Use sketches if you need to.

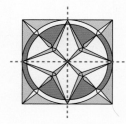

6 Metal alloy rims for tyres are very popular on modern cars.

Find and draw five alloy rim designs that you like.

For each one, state its order of rotational symmetry.

7 Sketch five different symmetrical designs or logos that you can find in your environment.

Label your sketches to indicate how the design is symmetrical.

Section 3: Triangles

Triangles are three-sided polygons which are given special names according to their properties.

Type of triangle	Properties
scalene	No equal sides.
	No equal angles.
	No line of symmetry.
	No rotational symmetry.
isosceles	Two equal sides.
	Angles at the base of the equal sides are equal.
	One line of symmetry.
	Line of symmetry is the perpendicular height.
	No rotational symmetry.
equilateral	All sides equal.
	Three equal angles, each is 60°.
	Three lines of symmetry.
	Rotational symmetry of order 3.
acute-angled	All angles are less than 90° (acute).
right-angled	One angle is a right angle (90°).
obtuse-angled	One angle is greater than 90° (obtuse).

Find answers at: cambridge.org/ukschools/gcsemaths-studentbookanswers

Triangles can be a combination of types.

For example, an isosceles triangle could be a right-angled isosceles triangle, an acute-angled isosceles triangle or an obtuse-angled isosceles triangle depending on the size of the angles.

Triangle MNO is a right-angled isosceles triangle.

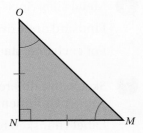

Angle properties

The angles inside a triangle are called interior angles.

The three interior angles of any triangle always add up to 180°.

If you extend the length of one side of a triangle you form another angle outside the triangle, called an exterior angle.

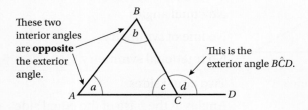

If you tear off the angles of any triangle and place them against a straight edge (180°) you can see that the interior angles add up to 180°.

You can also see that the exterior angle is equal to the sum of the two interior angles that are opposite it.

> **Tip**
>
> Rearranging pieces of a triangle demonstrates the angle properties. You will learn how to prove these properties using mathematical principles in Chapter 25.

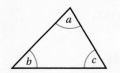  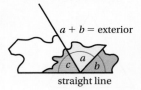

Using the properties of triangles to solve problems

You can use the properties of triangles to solve problems involving unknown angles and lengths of sides.

Problem-solving framework

Triangle ABC is isosceles with perimeter 85 mm.

$AB = BC$ and $AC = 25$ mm. Angle $ABC = 48°$.

Calculate:

a the length of each equal side

b the size of each equal angle.

Continues on next page ...

Steps for solving problems	What you would do for this example
Step 1: Work out what you have to do. Start by reading the question carefully.	You need to use the properties of isosceles triangles and the given information to find the length of two sides and the size of two angles.
Step 2: What information do you need? Have you got it all?	You should draw a labelled sketch to see whether you have the information you need. *(Triangle ABC with angle B = 48°, base AC = 25 mm, angles at A and C labelled x, sides AB and BC marked equal)*
Step 3: Decide what maths you can do.	You can use the values you already have to make equations to find the missing values.
Step 4: Set out your solution clearly. Check your working and that your answer is reasonable.	**a** Perimeter $= AC + AB + BC = 85$ mm So, $85 = 25 + AB + BC$ $85 - 25 = AB + BC$ $60 = AB + BC$ But $AB = BC$, so $AB = BC = 60 \div 2 = 30$ mm Check: $30 + 30 + 25 = 85$. **b** Let each equal angle be $x°$. $48 + 2x = 180$ (angle sum of triangle) $2x = 180 - 48$ $2x = 132$ $x = 66$ Check: $66 + 66 + 48 = 180$
Step 5: Check that you've answered the question.	Each equal side is 30 mm long. Each equal angle is 66°.

Tip

You will use these properties often when you deal with trigonometry in Chapter 31.

In many problems you will have to find the size of unknown angles before you can move on and solve the problem.

EXERCISE 2D

1 What type of triangle is this? Explain how you decided without measuring.

2 What type of triangle is this?

Choose the correct answer.

A obtuse-angled scalene

B right-angled isosceles

C acute-angled isosceles

D obtuse-angled isosceles

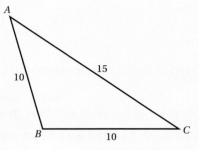

3 Which of the following triangles are not possible?

Explain why.

a An isosceles triangle with a right angle.

b A scalene triangle with two angles > 90°.

c A scalene triangle with three angles 34°, 64° and 92°.

d An obtuse equilateral triangle.

e An isosceles triangle with side lengths 6.5 cm, 7 cm and 7.5 cm.

4 Two angles in a triangle are 38° and 104°.

a What is the size of the third angle?

b What type of triangle is this?

5 Find the size of angles *a* to *e*.

Show your working and give mathematical reasons for any deductions you make.

  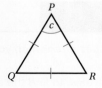 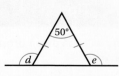

6 Isosceles triangle *DEF* with *DE* = *EF* has a perimeter of 50 mm. Find the length of *EF* if:

a $DF = 15$ mm

b $DE = \sqrt{130}$ mm.

Section 4: Quadrilaterals

A quadrilateral is a four-sided plane figure.

In the photographs to the right you can find rectangles, squares and more than one trapezium.

Quadrilaterals are classified and named according to their properties.

- A trapezium has one pair of opposite sides parallel.
- A kite has two pairs of adjacent sides equal.
- A parallelogram has both pairs of opposite sides parallel.
- A rhombus is a parallelogram with all sides equal.
- A rectangle is a parallelogram with angles of 90°.
- A square is a rectangle with all sides equal.

Some properties overlap, for example, a rectangle is actually a special type of parallelogram.

The rectangle meets the definition of a parallelogram (it has both pairs of opposite sides parallel and equal in length), so all rectangles are parallelograms.

In this case the reverse of the statement is not true. All parallelograms are not rectangles.

Quadrilateral	Properties	
trapezium		• One pair of opposite sides are parallel.
kite		• Two pairs of **adjacent** sides are equal. • Diagonals are perpendicular. • One diagonal **bisects** the other at right angles (and this is a line of symmetry). • One diagonal bisects the angles.
parallelogram		• Both pairs of opposite sides are parallel. • Both pairs of opposite sides are equal. • Both pairs of opposite angles are equal. • Diagonals bisect each other.
rhombus		As for parallelogram, plus: • All sides are equal. • Diagonals bisect at right angles. • Diagonals bisect the angles.

Continues on next page …

Key vocabulary

adjacent: next to each other; in shapes, sides that meet at a common vertex.

bisect: to divide exactly into two halves.

Quadrilateral		Properties
rectangle		As for parallelogram, plus: • All angles are 90°. • Diagonals are equal in length.
square		As for a rectangle, Plus: • All sides are equal. • Diagonals bisect at right angles. • Diagonals bisect the angles.

Using the properties of quadrilaterals to solve problems

You can use the given or marked properties of a quadrilateral to identify and name it.

You should always state what properties you are using to justify your answer.

WORKED EXAMPLE 1

A plane shape has two diagonals.

The diagonals are perpendicular.

a What shape(s) could this be?

b The diagonals are not the same length. Which shape(s) could it not be?

a Two diagonals means that the shape is a quadrilateral.

Only the square, rhombus and kite have diagonals that intersect at 90°.

The shape could be a square, rhombus or kite.

b Of the three shapes, only the square has diagonals that are equal in length.

Therefore, it could not be a square.

The angle sum of quadrilaterals

All quadrilaterals have two (and only two) diagonals. If you draw in one diagonal you divide the quadrilateral into two triangles.

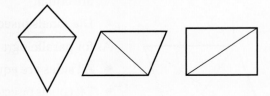

You already know that the interior angles of a triangle add up to 180°.

Therefore, the interior angles of a quadrilateral are equal to
$2 \times 180° = 360°$.

WORKED EXAMPLE 2

a Find the size of unknown angle x.   b Find the size of unknown angle y.

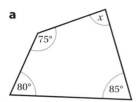
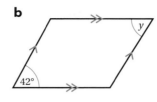

a $75° + 80° + 85° + x = 360°$
    $240° + x = 360°$
    $x = 360° - 240°$
    $x = 120°$

> Shape is a quadrilateral, angles of a quadrilateral add up to 360°.

b $y = 42°$

> Shape is a parallelogram, so the opposite angles are equal.

EXERCISE 2E

1 Identify the quadrilateral from the description.

There may be more than one correct answer.

  a All angles are equal.
  b Diagonals are equal in length.
  c Two pairs of sides are equal and parallel.
  d No sides are parallel.
  e The only regular quadrilateral.
  f Diagonals bisect each other.

2 You can identify a quadrilateral by considering its diagonals.

Copy and complete this table.

Shape	Diagonals are equal in length	Diagonals bisect each other	Diagonals are perpendicular
rhombus			
parallelogram			
square			
kite			
rectangle			

3 What is the most obvious difference between a square and a rhombus?

Find answers at: cambridge.org/ukschools/gcsemaths-studentbookanswers

4. Millie says that a quadrilateral has all four sides the same length.

   Elizabeth says it must be a square.

   Is Elizabeth correct? Give an explanation for your answer.

5. A kite has one angle of 47° and one of 133°. What sizes are the other two angles?

6. State whether each statement is always true, sometimes true or never true.

   Give a reason for your answer.

   a A square is a rectangle.
   b A rectangle is a square.
   c A rectangle is a rhombus.
   d A rhombus is a parallelogram.
   e A parallelogram is a rhombus.

Section 5: 3D objects

Solids

Solids are three-dimensional (3D) objects.

The parts of a solid are given specific names.

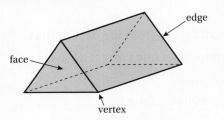

> **Tip**
>
> 3D means an object has three dimensions or measurements: length, width and height.

A solid, such as a triangular prism, that has only flat (plane) surfaces can also be called a **polyhedron**.

Cylinders, spheres and cones are not polyhedra. They are solids with a curved surface.

> **Key vocabulary**
>
> **polyhedron** (plural: polyhedra): a solid shape with flat faces that are polygons.

Two faces of a solid meet at an edge.

Three or more faces meet at a point called a vertex. (The plural of vertex is vertices.)

Each type of polyhedron has some properties that are not shared by the others.

Cubes and cuboids

> **Tip**
>
> You need to know the properties of the basic polyhedra and other 3D solids.
>
> You will use these properties to draw plans, elevation and nets of solids in Chapter 3 and you will apply them when you solve problems relating to volume and surface area in Chapter 17.

Cubes and cuboids are box-shaped polyhedra.

They have six faces, twelve edges and eight vertices.

A cuboid has six rectangular faces.

A cube also has six faces, but to be classified as a cube, the faces must all be **congruent** squares.

All cubes are cuboids, but not all cuboids are cubes.

Prisms

A prism is a 3D object with two congruent, parallel faces.

If the prism is sliced parallel to one of these faces the cross-section will always be the same shape.

The diagram (right) shows a prism with two triangular end faces.

You can see that slicing it anywhere along its length gives a triangular cross-section.

The parallel faces of a prism can be any shape.

If the prism is a polyhedron all of the other faces are rectangular.

Prisms are named according to the shape of their parallel faces.

pentagonal prism	**hexagonal prism**	**octagonal prism**
2 pentagonal faces	2 hexagonal faces	2 octagonal faces
5 rectangular faces	6 rectangular faces	8 rectangular faces

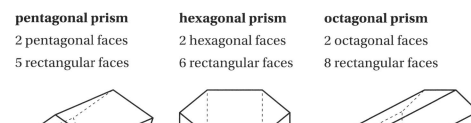

Although you do not normally refer to them as prisms, a cube is a square prism and a cuboid is a rectangular prism.

Pyramids

A pyramid is a polyhedron with a base and triangular faces which meet at a vertex (sometimes called the apex of the pyramid).

Pyramids are named according to the shape of their base.

triangular pyramid   **square pyramid**   **pentagonal pyramid**   **hexagonal pyramid**

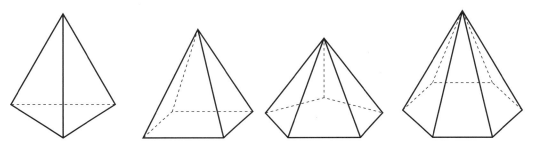

The number of sides of the base can be used to work out how many triangular faces the pyramid has.

Key vocabulary

congruent: identical in shape and size.

A square has four sides, so a square pyramid has four triangular faces.

The Louvre Museum in Paris is famous for the massive glass square pyramid at its entrance.

There is another smaller, inverted square pyramid in the underground shopping mall behind the museum.

A regular polyhedron is a solid whose faces are all congruent regular polygons.

There are only five regular polyhedra.

Other solids

Cylinders, cones and spheres are also 3D objects.

They do not have straight edges or flat faces that are polygons so they are not polyhedral.

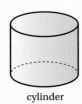  

cylinder   cone   sphere

A cylinder has two circular end faces and a curved surface along its length.

A cone has a circular base and a curved surface that forms a point.

A sphere is shaped like a ball. It has only one continuous curved surface.

EXERCISE 2F

1 Sketch an example of each of the following solids.
   a a tall thin cylinder   **b** a cube
   c a rectangular prism   **d** an octagonal pyramid
   e a prism with a parallelogram-shaped cross-section

2 How many faces, edges and vertices does a triangular prism have?

3 What are the polygons that form a triangular prism?

4 What is the difference between a sphere and a circle?

5 Compare a cone and a cylinder. How are they similar? How are they different?

6 Name two solids that each have six flat faces.

7 Copy and complete this table.

3D shape	Faces	Vertices	Edges
cube			
cuboid			
triangular pyramid			
square pyramid			
triangular prism			
hexagonal prism			

8 In total how many faces, edges and vertices does the tower have?

 Checklist of learning and understanding

Types of shapes
- Polygons are closed plane shapes with straight sides.
- Triangles, quadrilaterals, pentagons and hexagons are all polygons.
- Circles and ovals are plane shapes, but they are not polygons.

Symmetry
- Shapes have line symmetry if they can be folded along a line of symmetry to produce two identical mirror images.
- Shapes have rotational symmetry if they fit onto themselves more than once during a 360° rotation.

Triangles
- Triangles are three-sided polygons.
- Triangles can be classified and named using their side and angle properties.
- The sum of the interior angles of a triangle is 180°.

Find answers at: cambridge.org/ukschools/gcsemaths-studentbookanswers

Quadrilaterals

- Quadrilaterals are four-sided polygons.
- Quadrilaterals can be classified and named using their side, angle and diagonal properties.
- The sum of the interior angles of a quadrilateral is 360°.

Properties of 3D objects

- 3D objects are solids with length, breadth and height.
- Polyhedra are solids with flat faces and straight edges.
- Prisms and pyramids are polyhedral.
- Cylinders, cones and spheres are 3D objects but they are not polyhedral.

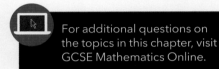

Chapter review

1. True or false?
   a. A slice of pizza can be accurately described as a triangle.
   b. A triangular pyramid has 4 vertices, 4 faces and 6 edges.
   c. A pair of lines that are equidistant and never meet are described as being perpendicular.

2. Describe the symmetrical features of a regular hexagon as fully as possible.

3. Find the values of the missing angles in this trapezium.

   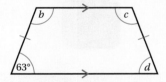

4. Is the missing angle a right angle? Explain.

   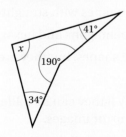

5 Find the value of the angle marked *x*.

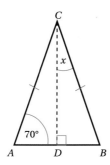

6 A quadrilateral has two pairs of equal sides. Opposite sides are not equal.

What shape is it?

7 What is the difference between a polygon and a polyhedron?

8 A solid has two surfaces and no straight edges.

What is the name of the solid?

9 Here are some solid 3D shapes.

    A   B   C   D   E

  a Write down the letter of the shape that is a sphere. *(1 mark)*

  b Write down the mathematical name of shape **A**. *(1 mark)*

  c How many faces does shape **B** have? *(1 mark)*

  d How many edges does shape **D** have? *(1 mark)*

©Pearson Education Ltd 2012

3 2D representations of 3D shapes

In this chapter you will learn how to …
- apply what you already know about the properties of 3D objects.
- work with 2D representations of 3D objects.
- construct and interpret plans and elevations of 3D objects.

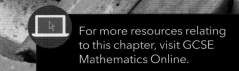
For more resources relating to this chapter, visit GCSE Mathematics Online.

Using mathematics: real-life applications

Buildings, engine parts, vehicles and packaging are all carefully planned and designed before they are built or made. Most design work starts on paper or screen using two-dimensional images to represent the final three-dimensional objects.

"No one will buy an apartment that isn't built yet if they don't know what it is going to look like. When we sell a development we show people floor plans as well as elevations from all four sides. Sometimes we also have a 3D scale model of the development."

(Estate agent)

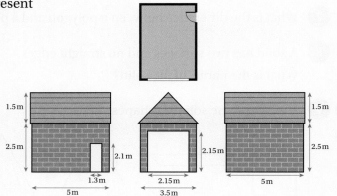

Before you start …

Ch 2	You must be able to identify and name some common 3D objects.	**1**	Name each of these 3D objects. **a** **b** **c** **d**
Ch 2	You should know the basic properties of polygons and other 3D objects.	**2**	True or false? Correct the false statements. **a** A cube has 4 faces. **b** A cube has 12 edges. **c** A cuboid has 8 vertices.
KS3	You must be able to accurately use a ruler, protractor and compasses to draw shapes.	**3**	Draw a triangle with a base of 8 cm and angles at the ends of 45° and 60°. What are the lengths of the other two sides?

Assess your starting point using the Launchpad

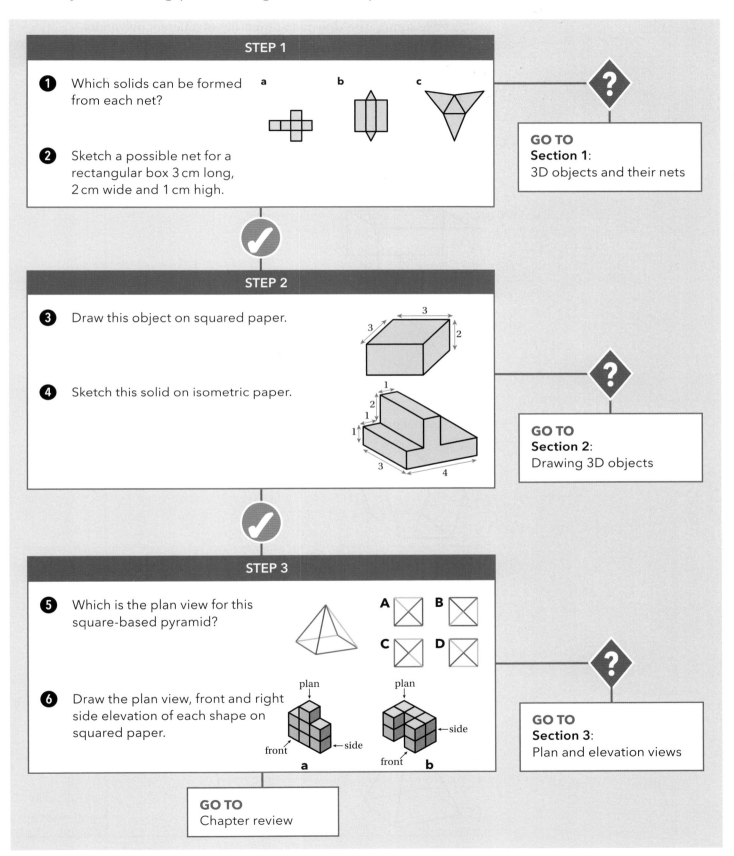

Section 1: 3D objects and their nets

This table summarises the main properties of different polyhedra.

Polyhedron	Faces	Vertices	Edges
cube (square prism)	6 square faces	8	12
cuboid (rectangular prism)	3 pairs of congruent rectangular faces	8	12
triangular prism	2 congruent triangular end faces 3 rectangular faces	6	9
pentagonal prism	2 congruent pentagonal end faces 5 rectangular faces	10	15
triangular pyramid	1 triangular base 3 triangular faces that meet at an apex	4	6
square-based pyramid	1 square base 4 triangular faces that meet at an apex	5	8

3 2D representations of 3D shapes

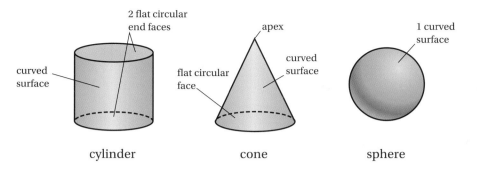

Tip

Cylinders, cones and spheres are also 3D shapes but they are not classified as polyhedra because they are not formed of flat faces that are polygons.

Nets of 3D objects

A **net** is a 2D representation of a 3D shape. You can fold up a net to make the 3D shape.

For printed packaging the design is printed onto the net, and then the net is folded up to make the box itself.

A cube has six square faces. There are 11 possible ways of arranging the faces to make the net of a cube. These are the two nets most commonly used.

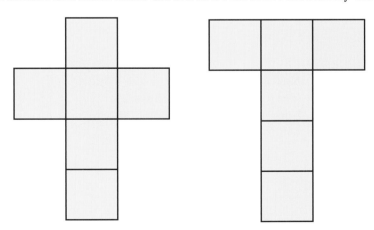

Did you know?

The parts of a car chassis are cut from sheets of steel or aluminium. The way in which multiple nets can be placed on the sheet metal prevents wastage and helps a company to be efficient.

You can see from the net on the right that the object has one square face and four triangular faces. When this net is folded up, the triangular faces will meet at a common point.

You know that a square-based pyramid has a square base and four triangular faces that meet at an apex. So this must be the net of a square-based pyramid.

You can use the properties of a 3D shape to help you recognise nets and identify the shapes they will make. You can also use the properties to sketch or construct the net of a shape.

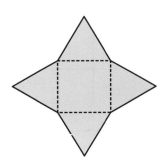

Find answers at: cambridge.org/ukschools/gcsemaths-studentbookanswers

WORKED EXAMPLE 1

Construct an accurate net of this rectangular prism.

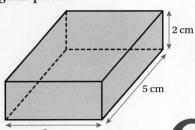

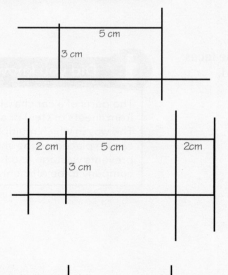

You know that a cuboid has six rectangular faces, so your net will have six faces.

Draw a rough sketch.

> **Tip**
>
> When you draw a net, always start in the middle of the page to give yourself space to construct all the faces.

To construct an accurate net you need to use the measurements given on the diagram.

Use a ruler and pencil to draw the net.

Draw the bottom face first. This is a rectangle 3 cm wide and 5 cm long.

Next construct two of the sides that join onto the bottom. These are both rectangles 3 cm long and 2 cm wide.

Construct the other two sides that join onto the bottom. These are both rectangles 5 cm long and 2 cm wide.

Continues on next page …

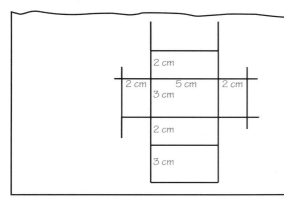

Lastly, draw the top face. This can be joined onto either the back or the front face or either of the sides. It is a rectangle 3 cm wide and 5 cm long.

EXERCISE 3A

1 You have seen the two most commonly used nets of a cube. Draw the other nine.

2 Which of the 3D shapes shown below can be created from this net?

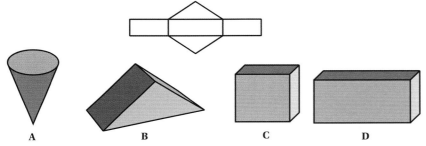

3 Which of these nets could be used to make a cylinder?

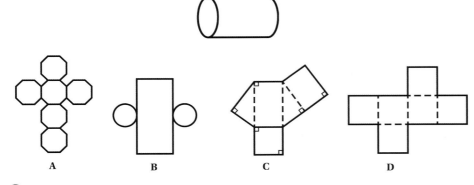

4 Which dice is represented by this net?

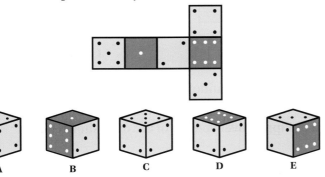

Tip

The numbers on the opposite faces of a dice add up to 7.

Find answers at: cambridge.org/ukschools/gcsemaths-studentbookanswers

5. Which 3D shapes can be formed from these nets?

a   b

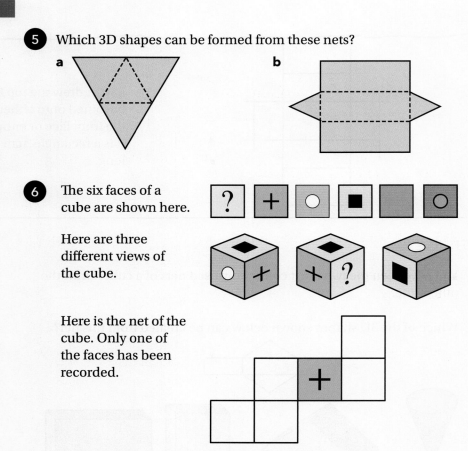

6. The six faces of a cube are shown here.

Here are three different views of the cube.

Here is the net of the cube. Only one of the faces has been recorded.

Copy and complete the net.

7. Here are some possible arrangements of five square faces.

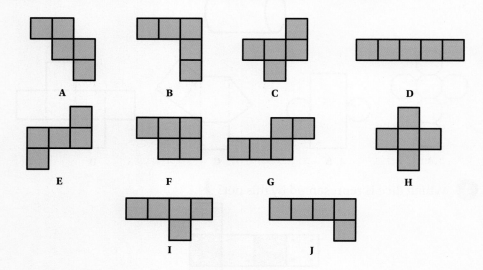

a Which of these nets cannot be folded up to form an open box?

b Draw one more net for an open box. Make sure your net is not just a turned or flipped over version of the ones shown here.

Section 2: Drawing 3D objects

You need to be able to make drawings of 3D objects and to interpret and make sense of drawings of 3D objects from different perspectives.

It can be challenging to draw a 3D object because you are trying to show three dimensions on a two-dimensional plane (your paper).

There are a number of ways of drawing 3D objects to show their features in 2D. As you read through each method, try it out on rough paper.

Prisms and cylinders using end faces

For prisms and cylinders, visualise the position of their end faces and draw these first.

A cuboid (rectangular prism):

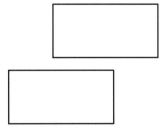 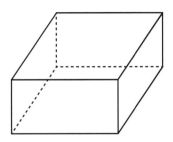

First draw two rectangular faces.   Then draw lines to join the vertices.

> **Tip**
> Once you've drawn the end faces, you can join them by drawing lines to represent the edges. For prisms make sure you match up the corresponding vertices.

A cylinder:

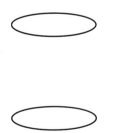 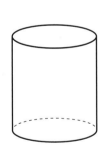 

First draw the two circular end faces by drawing ovals.   Then draw in two lines to join the end faces.   Shading can make the cylinder look more realistic.

> **Tip**
> When you draw an upright cylinder you use ovals for the faces to get a more realistic drawing.

A triangular prism:

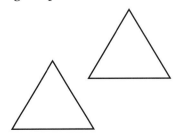 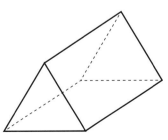

First draw the two triangular end faces.   Then draw in lines to join the vertices.

Tip

You can shade your completed figures to make them look more realistic.

Any shaped prism:

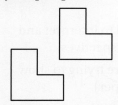 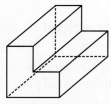 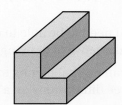

First draw the end faces.   Then draw lines to join the vertices.

Prisms and pyramids from parallel lines

You can draw square and rectangular prisms and square-based pyramids using two pairs of parallel lines as a starting point.

To draw a prism:

 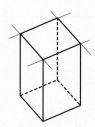

Begin by drawing two pairs of parallel lines that intersect.

Then draw four lines of equal length down (or up) from the intersections. Complete the shape by joining the ends to make a prism.

To draw a square-based pyramid:

 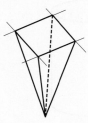

Start by drawing two pairs of parallel lines.

Mark a point and draw lines from three intersections of the parallel lines to the point.

Drawing shapes on squared or isometric grids

3D objects can be drawn on squared or isometric grids. The grid may be made from dots or from faintly printed lines.

This diagram shows a cube and a cuboid drawn on a square grid.

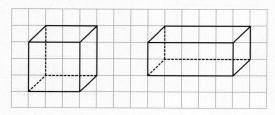

This diagram shows the same objects drawn on an **isometric grid**.

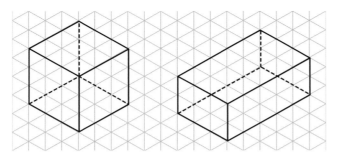

> **Key vocabulary**
>
> **isometric grid**: special drawing paper based on an arrangement of equilateral triangles.

The vertical lines on the grid are used to represent the vertical edges of the 3D object. You draw along the lines at an angle on the paper to represent the horizontal edges of the 3D object.

When you draw shapes on both square and isometric grids you use broken lines to show the edges that would not be seen if you viewed the shape from that angle.

Isometric drawings

Isometric drawings are used to visually represent three-dimensional objects in two dimensions in technical and engineering drawings.

The diagram on the right shows the design of an engineering component on isometric paper.

Isometric paper is very useful for drawing solids built from cubes.

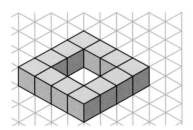 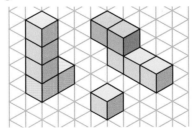

WORKED EXAMPLE 2

Draw this shape on the grid provided.

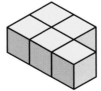

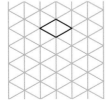

Start by drawing the horizontal face of one of the cubes. Here, one of the top yellow faces has been drawn.

Continues on next page …

45

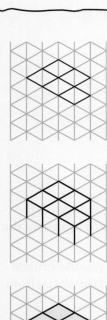

Use that face to draw in the other horizontal faces.

Use the vertical lines on the grid to draw in the vertical edges.

Use the angled lines to draw in the bottom horizontal edges.

WORK IT OUT 3.1

Students were asked to draw this view of a shape on an isometric grid.

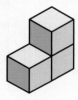

This is how they started their sketches.

Student A **Student B** **Student C**

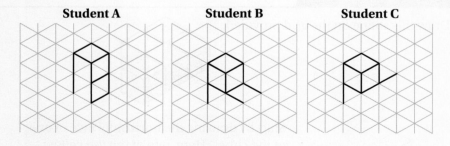

Which student is likely to end up with the correct view of the shape?

What are the others doing incorrectly?

EXERCISE 3B

1 Draw the following objects without using a grid.

a 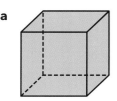 b 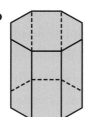 c 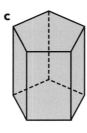 d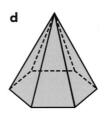

2 One of the parallel end faces of each of three different prisms is shown here.

i   ii   iii

a Sketch each prism on squared paper.

b Use isometric paper to draw each prism.

c Compare the two drawings of each prism. How does the grid affect what your drawing looks like?

3 Draw the following shapes on an isometric grid.

a 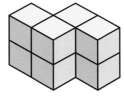 b 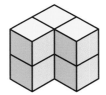 c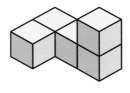

4 The diagrams below show different shapes made from cubes. If there are no cubes missing from the layers you cannot see, how many cubes would you need to build each shape?

a 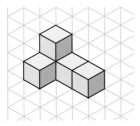 b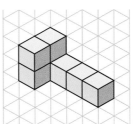

c 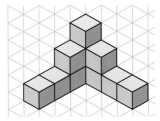 d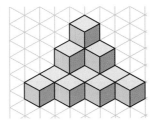

Find answers at: cambridge.org/ukschools/gcsemaths-studentbookanswers

Section 3: Plan and elevation views

Key vocabulary

plan view: the view of an object from directly above.

elevation view: a view of an object from the front, side or back (front elevation, side elevation or back elevation).

A **plan view** shows a 3D object from above. That is the top view of the object that you would see if you looked at it from directly above.

You can also view objects from the front, sides or back.

The **front elevation** is the view from the front of the object.

The **side elevation** is the view from the side of the object.

This diagram shows the plan view, front elevation and the left side elevation of a shape built out of cubes.

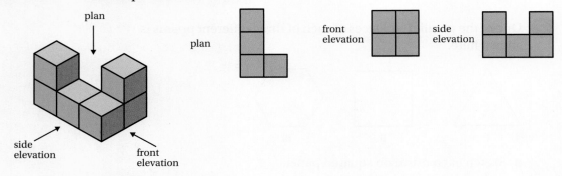

The shape below is a prism with trapezium-shaped ends. The front is higher than the back.

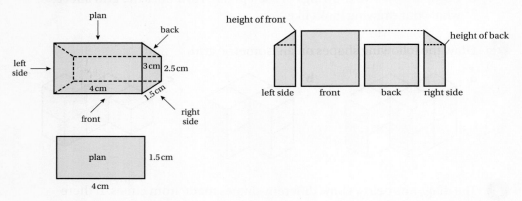

The plan view is a rectangle. Even though the top of the object slopes down to the back, if you look at it from above, it will look like a rectangle. The plan view is normally drawn above the front view because the two views will be the same width.

The left and right elevations are reflections of each other. They are drawn on the left and right side of the front elevation.

The front elevation is a rectangle. So is the back, but it is not as tall as the front, so it looks different when you draw it accurately.

You get different information from different views because each one shows two of the three dimensions of the solid.

- The plan view shows the length and width of the solid.
- The front view shows the length and height of the solid.
- The side views show the width and height of the solid.

When you draw plans or elevations you show any immediate changes in height as solid lines. Use dotted lines to indicate any hidden edges.

3 2D representations of 3D shapes

WORKED EXAMPLE 3

Draw the plan, front and side elevations of this solid.

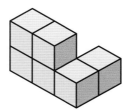

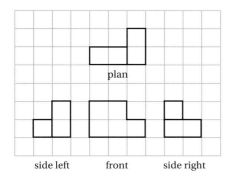

Start by drawing the plan view.

Next, draw the front view below it. It will be the same width as the plan view.

Work out what the right side will look like if you view it face on.

Draw it next to the front view. It will be the same height.

You can't see the left view, so you have to visualise it.

Draw it in the correct place.

When you draw views of a shape, you have to think quite carefully about what the parts you cannot see clearly will look like.

For example, the shape on the right is built from **four** cubes.

You can only see three cubes. You have to work out that the fourth one is supporting the 'top' cube.

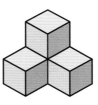

EXERCISE 3C

1. Select the correct plan view of each object.

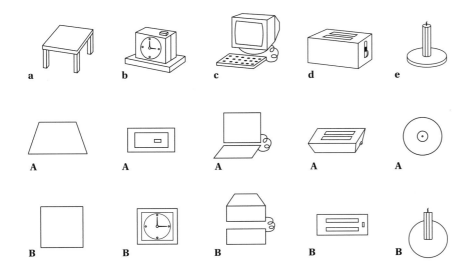

2 **a** Match each shape to its plan and elevation image.

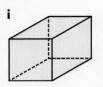 i   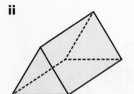 ii   iii   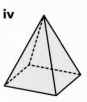 iv

A
plan   elevation

B
plan   elevation

C
plan   elevation

D
plan   elevation

b Sketch and label the elevations that are not shown for each shape.

3 For each set of cubes, draw plan, front and side elevations:

a 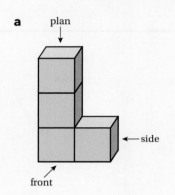   b 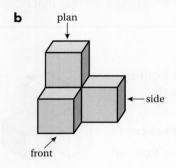   c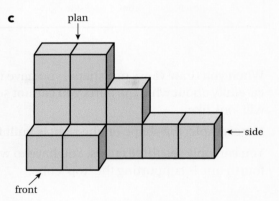

4 Draw the plan, the front elevation and the side elevation of the shape below.

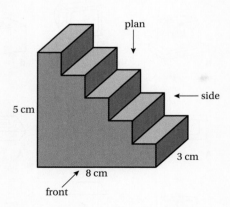

5 The plan view and elevations of different solids are shown below. Use these to work out what each solid looks like and draw it on an isometric grid.

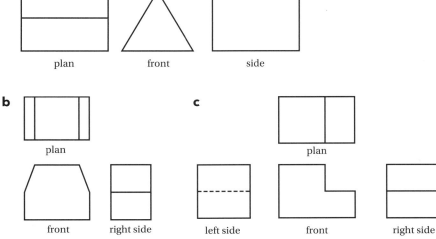

Checklist of learning and understanding

Properties of 3D shapes

- Prisms are shapes with congruent polygonal faces and a regular cross-section.
- Pyramids have a base and triangular sides that meet at an apex.
- Cylinders, cones and spheres are 3D shapes, but they are not polyhedra.
- The number and shape of the faces and the number of edges and vertices can be used to identify and name shapes.

2D representations of 3D shapes

- 3D shapes can be drawn on squared or isometric grids.
- Hidden edges are shown as dotted lines.

Plans and elevations

- A plan is a view from above a shape.
- An elevation is a view from the front, sides or back of a shape.

Chapter review

1 Which 3D objects can be formed from these nets?

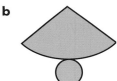

For additional questions on the topics in this chapter, visit GCSE Mathematics Online.

Find answers at: cambridge.org/ukschools/gcsemaths-studentbookanswers

2 **a** Match each block of cubes to the correct plan and elevation.
   b Identify any missing elevations and draw them for each shape.

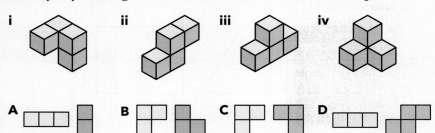

3 The plan and elevation of a solid built from cubes are shown below. Work out what the solid looks like and sketch it accurately on an isometric grid.

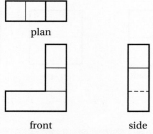

4 The front elevation and the side elevation of a cuboid are drawn on the grid. On a copy of the grid, draw the plan of the cuboid. *(1 mark)*

©Pearson Education Ltd 2012

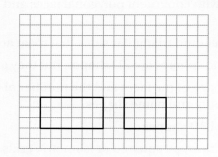

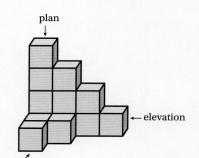

5 **a** Draw the plan, front and side elevation for the solid on the left.
   b How many cubes are in the solid?

6 On the right is a solid built from cubes.
   a Draw one possible plan view of this shape.
   b Draw a plan view that is not possible for this shape.
   c What is the least and greatest number of cubes that the shape could be built from to have these elevations?

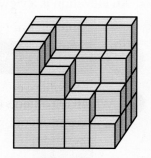

4 Properties of whole numbers

In this chapter you will learn how to …
- identify the properties of different types of number and use the correct words to talk about them.
- identify prime numbers and write any whole number as a product of its prime factors.
- find the HCF and LCM of two numbers by listing and by prime factorisation.

 For more resources relating to this chapter, visit GCSE Mathematics Online.

Using mathematics: real-life applications

People use numbers and basic calculations on a daily basis. A market stall holder has to quickly calculate the cost of a customer's order; a logistics manager has to order stock and divide the supplies so that they are never over or under stocked. There are many applications of basic calculation.

 Tip

You probably already know most of the concepts in this chapter. They have been included so you can revise concepts if you need to and check that you know them well.

"Counting in multiples saves quite a bit of time. If I know that each shelf has 15 boxes and each box contains 5 reams of paper, then I know straight away that I have 75 reams on each shelf without having to count each ream." *(Logistics manager)*

Before you start …

KS3	You should be able to recognise and find the factors of a number and to list multiples of a number.	**1** If these are the factors, what is the number? **a** 1, 5, 25 **b** 1, 2, 3, 6 **c** 1, 11 **2** If these are all multiples of a number (that isn't 1), what is the number? **a** 8, 10, 12, 14 **b** 18, 21, 27, 33 **c** 5, 20, 35, 60
KS3	You should know the first few prime numbers, square numbers and cube numbers.	**3** Which of the numbers below are: **a** prime numbers **b** square numbers **c** cube numbers? 0, 1, 2, 3, 4, 5, 6, 7, 8, 9, 10, 11, 12, 13, 14, 15, 16, 17, 18, 19, 20
KS3	You will need to be able to write a number as a product of its prime factors.	**4** Match each number to the product of its prime factors. **a** 450 **b** 180 **c** 120 **d** 72 **A** $2 \times 2 \times 2 \times 3 \times 3$ **B** $2 \times 2 \times 3 \times 3 \times 5$ **C** $2 \times 2 \times 2 \times 3 \times 5$ **D** $2 \times 3 \times 3 \times 5 \times 5$

Find answers at: cambridge.org/ukschools/gcsemaths-studentbookanswers

Assess your starting point using the Launchpad

STEP 1

1 Say whether each statement is true or false.
   a 1 is the smallest prime number.
   b If you square 7 you get 14.
   c 8 is the cube of 2.
   d Any whole number that ends in 1 is an odd number.
   e 33, 43 and 53 are prime numbers.
   f 7, 14 and 21 are factors of 7.

2 There is one incorrect number in each of the following sets.
Work out what the set is and find the incorrect number.
   a 20, 22, 24, 26, 28, 29, 30
   b 11, 22, 33, 44, 56, 66
   c 1, 2, 3, 4, 8, 12
   d 27, 30, 33, 36, 39, 41
   e 1, 2, 3, 4, 6, 9, 12, 18, 24, 36
   f 12, 24, 48, 60, 72, 86
   g 2, 3, 5, 7, 9, 11, 13, 17, 19

GO TO
Section 1: Reviewing number properties

STEP 2

3 Choose the correct way of writing each number using prime factors.
   a 48
      A $2 \times 2 \times 2 \times 3 \times 3$ **B** $2 \times 2 \times 2 \times 2 \times 3$
   b 100
      A $2 \times 2 \times 5 \times 5$ **B** $2 \times 5 \times 5$

GO TO
Section 2: Prime factors

STEP 3

4 Given that $72 = 2 \times 2 \times 2 \times 3 \times 3$ and $120 = 2 \times 2 \times 2 \times 3 \times 5$, choose the correct answers.
   a The HCF of 72 and 120 is:
      A 360   **B** 12   **C** 24   **D** 5
   b The LCM of 72 and 120 is:
      A 30   **B** 2   **C** 120   **D** 360

GO TO
Section 3: Multiples and factors

GO TO
Chapter review

Section 1: Reviewing number properties

Mathematical terms and their meanings

You need to remember the correct mathematical terms for the different types of numbers shown in the table.

Mathematical term	Definition	Example
Odd number	A whole number that cannot be divided exactly by 2; it has a remainder of 1.	1, 3, 5, 7, …
Even number	A whole number that can be divided exactly by 2 (no remainder).	0, 2, 4, 6, 8, …
Prime number	A whole number greater than 1 that can only be divided exactly by itself and by 1. (It only has two factors.)	2, 3, 5, 7, 11, 13, 17, 19, …
Square number	The product when an integer is multiplied by itself. For example $2 \times 2 = 4$, so 4 is a square number.	1, 4, 9, 16, … 3×3 can be written using powers as 3^2.
Cube number	The product when an integer is multiplied by itself twice. For example $2 \times 2 \times 2 = 8$, so 8 is a cube number.	1, 8, 27, 64, … $5 \times 5 \times 5$ can be written using powers as 5^3.
Root ($\sqrt{\ }$)	The number that produces a square number when it is multiplied by itself is a square root. The number that produces a cubed number when it is multiplied by itself and then by itself again is a cube root.	The square root of 25 is 5. $\sqrt{25} = 5 \quad (5 \times 5 = 25)$ The cube root of 8 is 2. $\sqrt[3]{8} = 2 \quad (2 \times 2 \times 2 = 8)$
Factor (also called divisor)	A number that divides exactly into another number, without a remainder.	Factors of 6 are 1, 2, 3 and 6. Factors of 7 are 1 and 7. Factors of 25 are 1, 5 and 25.
Multiple	A multiple of a number is found when you multiply that number by a whole number. Your times tables are really just lists of multiples.	Multiples of 3 are 3, 6, 9, 12, … Multiples of 7 are 7, 14, 21, …
Common factor	A common factor is a factor shared by two or more numbers. 1 is a common factor of all numbers.	Factors of 6 are 1, 2, 3, and 6. Factors of 12 are 1, 2, 3, 4, 6 and 12. 1, 2, 3 and 6 are common factors of 6 and 12.
Common multiple	A common multiple is a multiple shared by two or more numbers. It is in both of their times tables.	Multiples of 2 are 2, 4, 6, 8, 10, 12, … Multiples of 3 are 3, 6, 9, 12, … 6 and 12 are common multiples of 2 and 3.

Did you know?

Mathematicians use the following arguments to define zero as an even number:
- When divided by two, it results in zero; and there is no remainder.
- In a list of consecutive numbers, an even number has an odd number before and after it; −1 and 1 are either side of zero.
- When an even number is added to another even number it will give an even result, but when added to an odd number it will give an odd result; when zero is added to any number, the result is the number you started with.
- Numbers that end in 0 are even.
- It is the next number in a pattern of even numbers: 8, 6, 4, 2, …

Key vocabulary

consecutive: following each other in order. For example 1, 2, 3 or 35, 36, 37.

EXERCISE 4A

1 Here is a set of numbers.

1	2	3	4	5	6	7	8	9	10
11	12	13	14	15	16	17	18	19	20
21	22	23	24	25	26	27	28	29	30

Choose and list the numbers from the box that are:

- **a** odd
- **b** even
- **c** prime
- **d** square
- **e** cube
- **f** factors of 24
- **g** multiples of 3
- **h** common factors of 8 and 12
- **i** common multiples of 3 and 4.

2 Write down:
- **a** the next four odd numbers after 207
- **b** four **consecutive** even numbers between 500 and 540
- **c** the square numbers between 20 and 70
- **d** the factors of 23
- **e** the next four prime numbers greater than 13
- **f** the first ten cube numbers
- **g** the first five multiples of 8
- **h** the factors of 36.

3 Say whether the results of the following calculations will be odd or even.
- **a** The sum of two odd numbers.
- **b** The sum of two even numbers.
- **c** The difference between two even numbers.
- **d** The square of an odd number.
- **e** The product of an odd and an even number.
- **f** The cube of an odd number.

4
- **a** Write down all the factors of: **i** 4 **ii** 9 **iii** 16.
- **b** Write down all the factors of: **i** 6 **ii** 12 **iii** 20.
- **c** What do you notice about the number of factors and the properties of the numbers from your answers to **a** and **b**?

5 From your answer to question **2h** above (write down all the factors of 36), which factors are:
- **a** even
- **b** prime
- **c** square?

Place value

Consider the number 22*2*222.

Each of the 2s in the number has a different place value.

The place value tells you the value of the digit; the underlined 2 in the number has a value of 2 thousands or 2000.

Hundred thousands 100 000	Ten thousands 10 000	Thousands 1000	Hundreds 100	Tens 10	Ones/units 1
2	2	2	2	2	2

Each column in the place-value table is 10 times the value of the place to the right of it.

EXERCISE 4B

1. Write these sets of numbers in order from smallest to biggest.
   - **a** 432   456   348   843   654
   - **b** 606   660   607   670   706
   - **c** 123   1231   312   1321   231
   - **d** 12 700   71 200   21 700   21 007

2. The prices of some second hand cars are given. Write the list in order, starting with the most expensive.

   £5490   £3645   £5250   £3700   £4190

3. The table shows the population of five small towns. Write the list in order starting with the most populous.

Town	Population
Besbrough	467 542
Attleton	793 963
Witten	340 415
Thetham	351 000
Pullinge	627 250

4. What is the value of the 5 in each of these numbers?
   - **a** 35
   - **b** 534
   - **c** 256
   - **d** 25 876
   - **e** 50 346 987
   - **f** 1 532 980
   - **g** 5 678 432
   - **h** 356 432

5. What is the biggest and smallest number you can make with each set of digits?

   Use each digit only once in each number.
   - **a** 4, 0 and 6
   - **b** 5, 7, 3 and 1
   - **c** 1, 0, 3, 4, 6 and 2

Tip

It will help you work faster if you can learn to recognise all the prime numbers up to 100.

Key vocabulary

prime factor: a factor which is also a prime number.

Tip

Remember that 1 is **not** a prime number because it only has one factor, and 2 is the only even prime number.

Section 2: Prime factors

The number 1 is not a prime number because it only has one factor.

2 is the only even prime number.

If a factor of a number is a prime number it is called a **prime factor**.

Every whole number greater than 1 can be written as a product of its prime factors.

Finding the prime numbers that multiply together to make a given number is known as **prime factorisation**.

You can find the prime factors of a number by **repeatedly dividing by prime numbers**, or by using **factor trees**.

WORKED EXAMPLE 1

Express 48 as a product of its prime factors by:

a division **b** using a factor tree.

a

```
2 | 48
2 | 24
2 | 12
2 |  6
3 |  3
```

$2 \times 2 \times 2 \times 2 \times 3$

Divide by prime numbers.

Start with the lowest divisor that is prime (always try 2 first).

Continue dividing, moving to higher prime numbers as necessary.

Write the prime numbers you have used for dividing as a product.

b

$2 \times 2 \times 2 \times 2 \times 3$

Write the number as a product of any two of its factors (but **not** 1 and the number itself).

Keep doing this for the factors until you cannot divide a factor any more (i.e. until you get to a prime factor).

Write the prime numbers you have used for dividing as a product.

Tip

The **unique factorisation theorem** in mathematics states that each number can be written as a product of prime factors in one way only.

It means that different numbers cannot have the same product of prime factors.

Both methods give the same result because a whole number can only be expressed in terms of its prime factors in one way. This is called the **unique factorisation theorem**.

Even if you do the division in a different order and split the factors differently in the factor tree, you will always get the same result for a given number.

Here are three ways of finding the prime factors of 280.

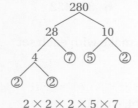

$2 \times 2 \times 2 \times 5 \times 7$

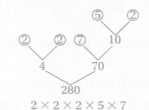

$2 \times 2 \times 2 \times 5 \times 7$

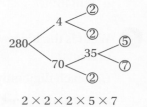

$2 \times 2 \times 2 \times 5 \times 7$

The examples show the product of factors in expanded form, but you can write them in a more efficient way using powers.

$2 \times 2 \times 2 \times 5 \times 7 = 2^3 \times 5 \times 7$

expanded form   index form using powers

EXERCISE 4C

1 Identify the prime numbers in each set.

   a 1, 2, 3, 4, 5, 6, 7, 8, 9, 10

   b 50, 51, 52, 53, 54, 55, 56, 57, 58, 59, 60

   c 95, 96, 97, 98, 99, 100, 101, 102, 103, 104, 105

2 Write each of the following numbers as a product of their prime factors. Use the method you prefer. Write your final answers using powers.

   a 36   **b** 65   **c** 64   **d** 84

   e 80   **f** 1000   **g** 1270   **h** 1963

3 A number is expressed as $2^3 \times 3^3 \times 5$.

   a What is the number?

   b Could it be any other number? Explain why.

Section 3: Multiples and factors

The lowest common multiple (LCM)

The lowest common multiple (LCM) of two or more numbers is the smallest number that is a multiple of all the given numbers.

To find the LCM, list the multiples of the given numbers until you find the first multiple that appears in all the lists.

WORKED EXAMPLE 2

Find the LCM of 4 and 7.

M_4 = 4, 8, 12, 16, 20, 24, 28, 32, …   Begin to list the multiples of 4.

M_7 = 7, 14, 21, 28, …   Begin to list the multiples of 7.

LCM of 4 and 7 = 28   Stop listing at 28 as it appears in both lists.

The highest common factor (HCF)

The highest common factor (HCF) of two or more numbers is the largest number that is a factor of all the given numbers.

To find the HCF, list all the factors of each number then pick out the highest number that appears in all the lists.

WORKED EXAMPLE 3

Find the HCF of 8 and 24.

$F_8 = \underline{1}, \underline{2}, \underline{4}, \underline{8}$ — Write out all the factors of 8 and 24.

$F_{24} = \underline{1}, \underline{2}, 3, \underline{4}, 6, \underline{8}, 12, 24$ — Underline the common factors.

HCF of 8 and 24 = 8 — The highest underlined number is the HCF.

Tip

The LCM is used to find the lowest common denominator when you add or subtract fractions.

The HCF is useful for cancelling fractions. You will also use them in Chapter 5 to factorise algebraic expressions.

With word problems you need to work out whether to use the LCM or HCF to find the answers.

- Problems involving the LCM usually include repeating events. You might be asked how many items you need to 'have enough' or when something will happen again at the same time.
- Problems involving the HCF usually involve splitting things into smaller pieces or arranging things in equal groups or rows.

Using prime factors to find the HCF and LCM

When you work with larger numbers you can find the HCF and LCM by writing the numbers as products of prime factors (i.e. by prime factorisation).

Once you've done that you can use the factors to quickly find the HCF and LCM.

WORKED EXAMPLE 4

Find **a** the HCF and **b** the LCM, of 72 and 120.

a $72 = \underline{2} \times \underline{2} \times \underline{2} \times \underline{3} \times 3$
$120 = \underline{2} \times \underline{2} \times \underline{2} \times \underline{3} \times 5$

First express each number as a product of prime factors.

Underline the common factors.

$2 \times 2 \times 2 \times 3 = 24$
HCF of 72 and 120 = 24

Write down the common factors and multiply them out.

Continues on next page …

b $72 = \underline{2} \times \underline{2} \times \underline{2} \times \underline{3} \times \underline{3}$
$120 = 2 \times 2 \times 2 \times 3 \times \underline{5}$

First express each number as a product of prime factors.

Underline the largest set of multiples of each factor across **both** lists.

Here, 2 appears three times in each list, so underline the 2s in one of the lists. 3 appears twice in the first list, but only once in the second, so underline the top two 3s. 5 only appears in the bottom list, so underline that one.

$2 \times 2 \times 2 \times 3 \times 3 \times 5 = 360$
LCM of 72 and 120 is 360

Write down each set of multiples and multiply them out.

Tip

Use the letters to help you remember what to do. LCM requires the **L**argest set of **M**ultiples.

EXERCISE 4D

1 Find the LCM of the given numbers.
   a 9 and 18   **b** 12 and 18   **c** 15 and 18   **d** 24 and 12
   e 36 and 9   **f** 4, 12, and 8   **g** 3, 9 and 24   **h** 12, 16 and 32

2 Find the HCF of the given numbers.
   a 12 and 18   **b** 18 and 36   **c** 27 and 90   **d** 12 and 15
   e 20 and 30   **f** 19 and 45   **g** 60 and 72   **h** 250 and 900

3 Find the LCM and the HCF of the following numbers by using prime factors.
   a 27 and 14   **b** 85 and 15   **c** 96 and 27   **d** 24 and 60
   e 450 and 105   **f** 234 and 66   **g** 550 and 128   **h** 315 and 275

4 Sian has two rolls of cotton fabric.

One roll has 72 metres on it, the other has 90 metres on it.

She wants to cut the rolls to make pieces of equal length without wasting any of it.

What is the longest possible length the pieces can be?

5 In a shopping centre promotion every 30th shopper gets a £10 voucher and every 120th shopper gets a free meal.

How many shoppers must enter the mall before one receives both a voucher and a free meal?

6 Amanda has 40 pieces of fruit and 100 sweets to share among the students in her class.

She is able to give each student an equal number of pieces of fruit and an equal number of sweets.

What is the largest possible number of students in her class?

7 Samir and Li start to walk in opposite directions around a walking track.

They start at the same point at the same time.

It takes Samir 5 minutes to walk round the track and it takes Li 4 minutes.

If they walk at this pace, how long will it be before they meet again at the starting point?

8 In a group of four cyclists, Lana cycles every 2nd day, Pete cycles every 3rd day, Karen cycles every 4th day and Anna cycles every 5th day.

They all cycle on 1 January this year.

a How many days will pass before they all cycle on the same day again?

b How many times a year will they all cycle on the same day?

9 Mr Abbot has three pieces of ribbon with lengths 2.4 m, 3.18 m and 4.26 m.

He wants to cut the ribbons into pieces that are all the same length.

He doesn't want any of the ribbon left over.

What is the greatest possible length for the pieces?

10 Two warning lights in a tunnel flash every 20 seconds and 30 seconds, respectively.

They flashed together at 4.30 pm.

When will they next flash at the same time?

Checklist of learning and understanding

Properties of whole numbers

- Even numbers are multiples of 2, odd numbers are not.
- Factors are numbers that divide exactly into a number.
- Prime numbers have only two factors, 1 and the number itself.
- Square numbers are the product of a number and itself ($n \times n = n^2$).
- Cube numbers are the product of a number multiplied by itself twice ($n \times n \times n = n^3$).
- The value of a digit depends on its place in the number.

Prime factors

- If a factor is a prime number it is called a prime factor.
- Whole numbers can be written as the product of their prime factors.

You find the prime factors by:

- repeated division by prime numbers (starting from 2 and working upwards)
- using a factor tree and breaking down factors until they are prime numbers.

Factors and multiples

- The lowest common multiple (LCM) of two numbers can be found by:
  - listing the multiples of both numbers and selecting the lowest multiple that appears in both lists
  - finding the largest set of multiples of each of the prime factors and multiplying them together.
- The highest common factor (HCF) of two numbers can be found by:
  - listing the factors of both numbers and selecting the highest factor that appears in both lists
  - finding the common prime factors and multiplying them together.

Chapter review

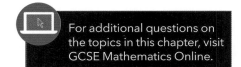

For additional questions on the topics in this chapter, visit GCSE Mathematics Online.

1 Complete the crossword puzzle provided by your teacher.

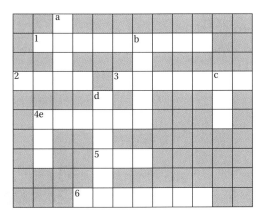

Across

1. The times tables are examples of these.
2. Whole numbers divisible by 2.
3. Another word used for factor.
4. Numbers in the sequence 1, 4, 9, 16, …
5. An even prime number.
6. The result of a multiplication.

Down

a. $n \times n \times n$ is the __ of n.
b. Numbers with only two factors.
c. Whole numbers that are not exactly divisible by 2.
d. Number that divides into another with no remainder.
e. HCF of 12 and 18.

Find answers at: cambridge.org/ukschools/gcsemaths-studentbookanswers

2 Here is a list of numbers.

5   15   30   50   60   90   100   125

From the numbers in the list, write down

   i two different numbers that add up to an even number

   ii a multiple of 20

   iii a factor of 45

   iv a cube number

(4 marks)

©*Pearson Education Ltd 2013*

3 Is 149 a prime number? Explain how you decided.

4 Find the HCF and the LCM of 20 and 35 by listing the factors and multiples.

5 Express 800 as a product of prime factors, giving your final answer using powers.

6 Determine the HCF and LCM of the following by prime factorisation.

   a 72 and 108   **b** 84 and 60

7 Jo, Mo and Jenny were jumping up a flight of stairs.

Jo jumped 2 steps at a time, Mo jumped 3 steps at time while Jenny managed 4 steps at a time.

They started together on the first step. What is the next step they will all jump on together?

5 Introduction to algebra

In this chapter you will learn how to ...
- use algebraic notation and write algebraic expressions.
- simplify and manipulate algebraic expressions.
- multiply out (expand) algebraic expressions with brackets.
- use common factors to factorise expressions.
- use algebra to solve problems in different contexts.

For more resources relating to this chapter, visit GCSE Mathematics Online.

Using mathematics: real-life applications

Algebra lets you describe and represent patterns using concise mathematical language.

This is useful in many different careers including accounting, navigation, building, plumbing, health, medicine, science and computing.

"You are unlikely to think about algebra when you watch cartoons or play video games, but animators use complex algebra to program the characters and make objects move." *(Games designer)*

Before you start ...

KS3	You need to understand the basic conventions of algebra.	**1**	Choose the correct way to write each of these.
			a $n \times n$
			  A $2n$ **B** n^2 **C** 2^n **D** $2(n)$
			b c multiplied by 3 and then added to 5
			  A $3c + 5$ **B** $3(c + 5)$ **C** $c + 15$
			c n squared and then multiplied by 2
			  A $2n^2$ **B** $(2n)^2$ **C** $4n^2$

KS3	You should be able to substitute numbers for letters and evaluate expressions.	**2**	**a** Evaluate the following expressions for $n = 5$ and $n = -5$. **i** $3n + 4$ **ii** $3(n + 4)$ **b** What is the value of $\dfrac{(2 + 4)^2}{6}$?
Ch 4	You should be able to find the highest common factor in a group of terms.	**3**	Write down the HCF of: **a** $12xy$ and $18y^2$ **b** $45x$ and $50xy$.

Find answers at: cambridge.org/ukschools/gcsemaths-studentbookanswers

Assess your starting point using the Launchpad

STEP 1

1 Use the correct notation and conventions to write each statement as an algebraic expression.
  a Multiply n by 3 and add 4 to the result.
  b Subtract 4 from n and multiply the result by 3.
  c Multiply n squared by 4, add 3 and divide the result by 2.

GO TO Section 1: Using algebraic notation

STEP 2

2 Simplify these expressions by collecting like terms.
  a $3a + 2b + 2a - b$
  b $4x + 7 + 3x - 3 - x$
  c $4a^2 + 8ab - 10a^2 - 5ab$

GO TO Section 2: Simplifying expressions

STEP 3

3 Multiply out the brackets and simplify.
  a $m(n - p)$
  b $3(x + 5) + 4(x + 2)$
  c $2z(z + 4) - z(z + 5)$

GO TO Section 3: Multiplying out brackets

STEP 4

4 Complete the following.
  a $3x + 12 = \square(x + 4)$
  b $5x + 10y = \square(x + 2y)$
  c $x^2 - 3x = \square(x - 3)$
  d $ab - ac = a(\square - \square)$
  e $-x + 7x^2 = -x(\square \square \square)$

5 Factorise each expression and write it as a product of its factors.
  a $2x + 4y$
  b $-3x - 9$
  c $5x + 5y$

GO TO Section 4: Factorising expressions

Section 5: Using algebra to solve problems

GO TO Chapter review

Section 1: Using algebraic notation

In algebra letters are used to represent unknown numbers.

For example, $x + y = 20$.

The letters can represent many different values so they are called **variables**.

Letters and numbers can be combined with operation signs to form an **expression**, such as $5a^3 - 2xy + 3$.

This expression has three **terms**.

Terms are separated by + or − signs; the sign belongs to the term that follows it.

Terms should always be written in the shortest, simplest way:

$2 \times h$ is written as $2h$ and $x \times x \times y$ is written as x^2y.

$4x \div 3$ is written as $\dfrac{4x}{3}$ and $(x + 4) \div 2$ is written as $\dfrac{x+4}{2}$.

To find a **product**, you multiply the factors.

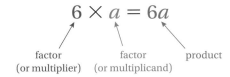

factor (or multiplier)   factor (or multiplicand)   product

In algebraic notation, multiplication is shown by writing the factors next to each other.

You write $a \times b$ as ab and $5 \times z$ as $5z$.

$a \times b \equiv ab$ and $5 \times z \equiv 5z$ are called **identities**.

The symbol $\equiv$ means exactly the same as, or identical to.

For division, you write $a \div b$ as $\dfrac{a}{b}$ and $5 \div z$ as $\dfrac{5}{z}$.

Key vocabulary

variable: a letter representing an unknown number.

expression: a group of numbers and letters linked by operation signs.

term: a combination of letters and/or numbers.

product: the result of multiplying numbers and/or terms together.

identity: an equation that is true no matter what values are chosen for the variables.

Tip

When you have numbers and letters in a term, the number is written first and letters are usually written in alphabetical order. So, you write $5x$ not $x5$ and $3xy$ not $3yx$.

EXERCISE 5A

1 **a** Are these true or false?

   i $2n \equiv n + 2$ **ii** $2n \equiv n + n$ **iii** $2n \equiv n^2$ **iv** $2n^2 \equiv (2n)^2$

   b Are there any values for n so that $2n \equiv n^2$?

2 Write the algebraic expression for:

   a x multiplied by 3 and added to y multiplied by 7

   b 4 subtracted from x squared and the result multiplied by 5

   c x cubed added to y squared and the result divided by 4

   d 6 added to x, the result multiplied by 4 and then y subtracted

   e x multiplied by itself and then divided by 2.

3 Match each statement to the correct algebraic expression.

a	Take a number and multiply it by 3 then add 2 to it.	i	$\dfrac{6+x}{2}$
b	Take a number and add 3 to it, then double it.	ii	$3x + 2$
c	Take a number, multiply it by itself then add 3 to it.	iii	$5(x-4)$
d	Add 6 to a number then divide it by 2.	iv	$9x^2$
e	Subtract 4 from a number and multiply the result by 5.	v	$x^2 + 3$
f	Square a number then multiply it by 9.	vi	$2x^2 - 3x^3$
g	Square a number and multiply it by 2, then subtract the same number cubed and multiplied by 3.	vii	$2(x+3)$

4 Use algebra to write these in as short a form as possible.

- **a** $2 \times 3a$
- **b** $4b \times 5$
- **c** $d \times (-9)$
- **d** $4a \times 3b$
- **e** $5c \times 2d$
- **f** $-3m \times 4n$
- **g** $-2p \times (-3q)$
- **h** $a \times a$
- **i** $m \times m$
- **j** $2a \times 4a$
- **k** $-3a \times 5a$
- **l** $-2m \times (-4m)$
- **m** $7a \times 8ab$
- **n** $-6cd \times (-2de)$
- **o** $2a \times 2a \times 2a$

5 Rewrite each division using algebraic conventions. Simplify them if possible.

- **a** $15x \div 5$
- **b** $27y \div 3$
- **c** $24a^2 \div 8$
- **d** $7 \times 15p \div 21$
- **e** $24x \div (8 \times 3)$
- **f** $18y \div (6 \times 2)$
- **g** $-18x^2 \div 9$
- **h** $-16a^2 \div (-4)$
- **i** $15 \div (3 \times n \times n)$

6 Write each of the following without multiplication or division signs.

- **a** $2 \times 5n$
- **b** $p \times q$
- **c** $a \times (b+c)$
- **d** $(x+y) \div z$
- **e** $n^2 \times n^3$
- **f** $p^3 \times p$

7 Write expressions to represent the perimeter and area of each shape:

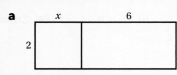
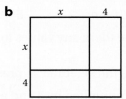
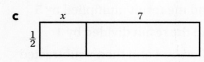
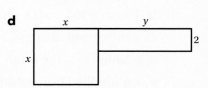

8 A man is x years old.

- **a** How old will he be ten years from now?
- **b** How old was he ten years ago?
- **c** His daughter is a third of his age. How old is his daughter?

Substitution

You can evaluate expressions if you are told what values the letters represent.
If $x = -2$, then:
$2x + 1 = 2 \times -2 + 1 = -4 + 1 = -3$.

WORKED EXAMPLE 1

Given that $a = -2$ and $b = 8$, evaluate:

a ab **b** $3b - 2a$ **c** $2a^3$ **d** $2(a + b)$

a $ab = a \times b$
$= -2 \times 8$
$= -16$

b $3b - 2a = 3 \times b - 2 \times a$
$= 3 \times 8 - 2 \times -2$
$= 24 - (-4) = 28$

c $2a^3 = 2 \times a^3 = 2 \times (-2)^3$
$= 2 \times -8 = -16$

d $2(a + b) = 2 \times (a + b)$
$= 2 \times (-2 + 8)$
$= 2 \times 6$
$= 12$

Remember to do the calculation in brackets first.

Tip

When you substitute values into a term such as $2y$ you need to remember that $2y$ means $2 \times y$. So, if $y = 6$, you need to write $2y$ as 2×6 and not as 26.

Tip

Substitution is an important skill. You will need to substitute values for letters when you work with formulae for perimeter, area and volume of shapes and when you solve problems involving Pythagoras' theorem.

EXERCISE 5B

1 Given that $x = 3$ and $y = 6$, evaluate these expressions.
  a $2x + 3y$ **b** $3x + 2y$ **c** $10y - 2x$ **d** $x + 2y$

2 Given that $x = 2$ and $y = 7$, evaluate these expressions.
  a $6x + y$ **b** $5x - 5y$ **c** $2xy$ **d** $\frac{1}{2}xy$

3 Find the value of each expression when $a = -2$ and $b = 5$.
  a $-5ab + 10$ **b** $-3ab - 6$ **c** $\frac{10}{b}$ **d** $\frac{400}{a}$

4 Find the value of each expression when $a = 2$ and $b = -3$.
  a $\frac{6}{a} - \frac{15}{b}$ **b** $\frac{15}{b} - \frac{24}{2a}$ **c** $8 - 2a + 2b$ **d** $7a - 4 + 2b$

5 Evaluate each expression when $a = 3$ and $b = 7$.
  a $2(a + b)$ **b** $3a(15 - 2b)$ **c** $\frac{2}{3}(2b - a + 1)$ **d** $2(b - a) + b(a - 1)$

Section 2: Simplifying expressions

Adding and subtracting like terms

Like terms have exactly the same letters or combination of letters and powers.
You can simplify expressions by adding or subtracting like terms.

$3a$ and $4a$ are **like** terms: $3a + 4a = 7a$

$7xy$ and $2xy$ are **like** terms: $7xy - 2xy = 5xy$

$5x^2$ and $3x^2$ are **like** terms: $5x^2 - 3x^2 = 2x^2$

$5ab^2$ and $2a^2b$ are **not like** terms so $5ab^2 - 2a^2b$ **cannot** be simplified further.

WORK IT OUT 5.1

Here are two terms: $3x^2y$ and $2xy^2$.

Student A said that these two terms are like terms and can be added together to be written as:

$5x^2y^2$

Student B said that these two terms are not like terms and can only be written added together as:

$3x^2y + 2xy^2$

Which student is correct? Why?

When an expression contains many different terms you might be able to simplify it by collecting and then combining like terms.

WORKED EXAMPLE 2

Simplify $2x - 4y + 3x + y$

$2x - 4y + 3x + y$

$= 2x + 3x - 4y + y$ Rearrange the terms so like terms are together. Keep the signs with the terms they belong to.

$= 5x - 3y$ Combine the like terms. Remember $y \equiv 1y$.

Multiplication and division

WORKED EXAMPLE 3

Simplify:

a $5 \times 4a$ **b** $2x \times 6y$ **c** $2a^2 \times 7ab$ **d** $12a \div -4$

e $\dfrac{6x^2}{2}$ **f** $\dfrac{-8xy}{-16}$ **g** $\dfrac{12ab^2}{36ab}$

a $5 \times 4a = 20a$

b $2x \times 6y = 12xy$ Multiply numbers by numbers and write letters in alphabetical order.

c $2a^2 \times 7ab = 14a^3b$ $a^2 = a \times a$, so $a^2 \times a = a \times a \times a = a^3$

Continues on next page …

d $12a \div -4$
$= \dfrac{12a}{-4} = -3a$

> Write the division as a fraction.
> Cancel the fraction to its lowest terms.

e $\dfrac{6x^2}{2} = 3x^2$

> Cancel by 2 to lowest terms.

f $\dfrac{-8xy}{-16} = \dfrac{xy}{2}$

> Cancel by −8 to lowest terms.

g $\dfrac{12ab^2}{36ab} = \dfrac{b}{3}$

> Write the numerator as b not $1b$ by convention.

EXERCISE 5C

1 Say whether each of these pairs are like or unlike terms.
  a $4a$ and $3b$
  b $5b$ and $-3b$
  c $3b$ and $9b$
  d $4p$ and $6p$
  e $8p$ and $-4q$
  f $5a$ and $6b$
  g $7mn$ and $3mn$
  h $4ab$ and $-2ab$
  i $-6xy$ and $-7x$
  j $9ab$ and $3a$
  k $9x^2$ and $6x^2$
  l $6a^2$ and $-7a^2$

2 Simplify.
  a $9x + 4y - 4y - 3x + 5y$
  b $3c + 6d - 6c - 4d$
  c $2xy + 3y^2 - 5xy - 4y^2$
  d $2a^2 - ab^2 + 3ab^2 + 2ab$
  e $5f - 7g - 6f + 9g$
  f $7a^2b + 3a^2b - 4a^2b$
  g $6mn^3 - 2mn^3 + 8mn^3$
  h $3st^2 - 4s^2t + 5s^2t + 6st^2$

3 Find the value of the missing term to complete the following.
  a $2a + \square = 7a$
  b $5b - \square = 2b$
  c $8mn + \square = 12mn$
  d $11pq - \square = 6pq$
  e $4x^2 + \square = 7x^2$
  f $6m^2 - \square = m^2$
  g $8ab - \square = -2ab$
  h $-3st + \square = 5st$

4 Find the value of the missing term to complete the following.
  a $8a \times \square = 16a$
  b $9b \times \square = 18b$
  c $8a \times \square = 16ab$
  d $5m \times \square = 15mn$
  e $3a \times \square = 12a^2$
  f $6p \times \square = 30p^2$
  g $-5b \times \square = 10b^2$
  h $4m \times \square = 12m^2n$

5 Rewrite each expression in the simplest possible form.
  a $7 \times 2x \times -2$
  b $4x \times 2y \times 2z$
  c $2a \times 5 \times a$
  d $ab \times bc \times cd$
  e $-4x \times 2x \times -3y$
  f $\dfrac{1}{4x} \times 4y \times -y$
  g $-9x \div 3$
  h $-24y \div 2x$
  i $18x^2 \div 6$

6 Simplify.

a $\dfrac{4x}{6}$ b $\dfrac{3a}{9}$ c $\dfrac{-12m}{18}$ d $\dfrac{14p}{21}$

e $\dfrac{22x^2}{33}$ f $\dfrac{15xy}{20}$ g $\dfrac{12ab}{a}$ h $\dfrac{2xy}{6xy}$

Section 3: Multiplying out brackets

Removing (multiplying out) brackets is called **expanding** the expression.

To expand an expression such as $2(a + b)$ you multiply each term inside the bracket by the value outside the bracket.

$2(a + b) = 2 \times a + 2 \times b$
$\qquad\quad = 2a + 2b$

$-2(a + b) = -2 \times a + (-2 \times b)$
$\qquad\quad\;\; = -2a - 2b$

Key vocabulary

expanding: multiplying out an expression to get rid of the brackets.

Tip

Remember that $2(a + b)$ means $2 \times (a + b)$. In algebraic notation you don't write the multiplication sign.

Tip

Pay careful attention to the rules for multiplying negative and positive numbers when you multiply out.

If the signs are different the result is negative. If the signs are the same the answer is positive. For example, $-2x \times 3y = -6xy$
$-5 \times -2b = 10b$.

WORKED EXAMPLE 4

Expand:

a $3x(y + 2z)$ b $-2x(4 + y)$ c $-(3x - 2)$

a $3x(y + 2z) = 3x \times y + 3x \times 2z$
$\qquad\qquad\;\; = 3xy + 6xz$

b $-2x(4 + y) = -2x \times 4 + (-2x \times y)$
$\qquad\qquad\;\; = -8x - 2xy$

c $-(3x - 2) = -1 \times 3x - (-1 \times 2)$
$\qquad\qquad = -3x - -2$
$\qquad\qquad = -3x + 2$

Expanding and simplifying

After expanding a bracket, an expression might contain like terms, so it can be simplified.

WORKED EXAMPLE 5

Expand and simplify:

a $6x - 3(2x + 1)$ b $2x(x + y) - x(3x - 4y)$

a $6x - 3(2x + 1)$
$\;\; = 6x - 6x - 3$
$\;\; = -3$

Combine the x terms.

b $2x(x + y) - x(3x - 4y)$
$\;\; = 2x^2 + 2xy - 3x^2 + 4xy$
$\;\; = -x^2 + 6xy$

Combine the x^2 terms and the xy terms.

EXERCISE 5D

1 Some of these expansions are incorrect.
Check each one and correct those that are wrong.
  a $4(a + b) = 4a + b$
  b $5(a + 1) = 5a + 6$
  c $8(p - 7) = 8p - 56$
  d $-3(p - 5) = -3p - 15$
  e $a(a + b) = 2a + ab$
  f $2m(3m + 5) = 6m^2 + 10m$
  g $-6(x - 5) = 6x + 30$
  h $3a(4a - 7) = 12a^2 - 7$
  i $4a(3a + 5) = 12a^2 + 20a$
  j $3x(2x - 7y) = 6x^2 - 21y$

2 Expand and simplify.
  a $2(c + 7) - 9$
  b $(a + 2) + 7$
  c $5(b + 3) + 10$
  d $2(e - 5) + 15$
  e $3(f - 4) - 6$
  f $2a(4a + 3) + 7a$
  g $5b(2b - 3) + 6b$
  h $2a(4a + 3) + 7a^2$
  i $3b(3b - 5) - 7b^2$

3 Expand and simplify.
  a $2(y + 1) + 3(y + 4)$
  b $2(3b - 2) + 5(2b - 1)$
  c $3(a + 5) - 2(a + 7)$
  d $5(b - 2) - 4(b + 3)$
  e $x(x - 2) + 3(x - 2)$
  f $2p(p + 1) - 5(p + 1)$
  g $3z(z + 4) - z(3z + 2)$
  h $3y(y - 4) + y(y - 4)$

4 The expression in each box is obtained by adding the expressions in the two boxes directly below it.

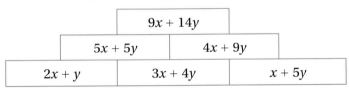

Copy and complete these two pyramids.

a
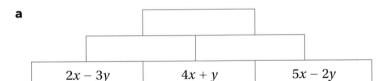

b
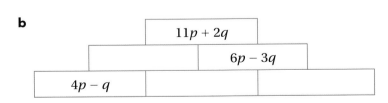

5 Use substitution to show that the following expressions are **not** identities.
  a $a + a$ and a^2
  b $3x + 4 - x + 2$ and $2x + 2$
  c $(m + 2)^2$ and $m^2 + 4$
  d $\dfrac{x+3}{3}$ and $x + 1$

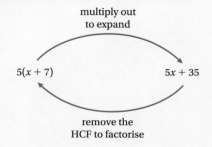

multiply out to expand

$5(x + 7)$ → $5x + 35$

remove the HCF to factorise

Section 4: Factorising expressions

Factorising is the reverse of expanding.

When you factorise an expression you use brackets to write it as a product of its factors.

If you expand $5(x + 7)$ you get $5x + 35$.

To factorise $5x + 35$ you find the highest common factor of the terms.

5 is the HCF of $5x$ and 35, so 5 is written outside the bracket and the remaining factors are written in brackets.

> **Tip**
>
> The highest common factor can be a number or a variable. It can also be a negative quantity.

WORKED EXAMPLE 6

Factorise each expression.

a $10a + 15b$ b $-2x - 8$ c $3x^2 - 6xy$ d $3(m + 2) - n(m + 2)$

a $10a + 15b$
   $10a + 15b = 5(2a + 3b)$

   HCF of 10 and 15 is 5. There are no common variables.

b $-2x - 8$
   $-2x - 8 = -2(x + 4)$

   HCF of -2 and -8 is -2.

c $3x^2 - 6xy$
   $3x^2 - 6xy = 3x(x - 2y)$

   HCF is $3x$.

d $3(m + 2) - n(m + 2)$
   $3(m + 2) - n(m + 2) = (m + 2)(3 - n)$

   *This looks like an expansion, but you are asked to factorise! $(m + 2)$ is common to both terms, so **it** is the HCF.*

EXERCISE 5E

1. Factorise each expression and write it as the product of its factors.

   a $2x + 4$ b $12m - 18n$ c $3a - 3b - 6$
   d $xy - xz$ e $5xy - 15xyz$ f $14ab - 21bc$
   g $pq - pr$ h $x^2 - x$ i $18abc - 12ac$
   j $2x^2 - 4xy$ k $2x^2y - 4xy^2$ l $-6a - 12$
   m $-3a - 9$ n $-xy - 5x$ o $-x^2 + 6x$

> **Tip**
>
> You will learn other methods of factorising expressions in Chapter 13.

2. Factorise.

   a $7x - xy + x^2$ b $2xy + 4xz + 10x$ c $10x - 5y + 15z$
   d $x(x - 2) + 5(x - 2)$ e $a(a - 7) - (a - 7)$ f $(x - 3) - 3(x - 3)$

Section 5: Using algebra to solve problems

You can use algebra to solve problems and test the accuracy of statements.

Read through this example to see how algebra can be used to generalise situations and solve problems.

WORKED EXAMPLE 7

Write down an expression for the sum of any three consecutive numbers.

n
$n + 1$
$n + 2$

> Let the first number be n. If the numbers are consecutive, you know that each number is 1 more than the previous number.
> The next number must be 1 more than n, so let it be $n + 1$.
> The third number is 1 more than $n + 1$, so let it be $n + 1 + 1 = n + 2$.

$n + n + 1 + n + 2 = 3n + 3$.
The sum of the three numbers when the first number, n, is known is $3n + 3$.

> Sum the numbers to write a general rule.

$n = 205$
$3 \times 205 + 3 = 615 + 3 = 618$
Check: $205 + 206 + 207 = 618$

> Test your rule using a set of consecutive numbers, for example 205, 206, 207. Substitute 205 for n in $3n + 3$.

General expressions like that in Worked example 7 are very useful for programmed operations and repeated calculations involving different starting numbers.

EXERCISE 5F

1 Are the following statements true or false?

   a The expression $3z^2 + 5yx - z^2 - 6yx$ simplified is $2z^2 - 11xy$.

   b If you expand the brackets $2p(3p + q)$ you get the expression $6p^2 + 2pq$.

   c This is a correct use of the identity symbol: $4(a + 1) \equiv 4a + 4$.

   d $\dfrac{4}{x}$ always has the same value as $\dfrac{x}{4}$.

   e x squared and added to 7 with the result divided by 3 is $\dfrac{x^2 + 7}{3}$.

2 In a magic square the sum of each row, column and diagonal is the same.
Is the square on the right a magic square?

$m - p$	$m + p - q$	$m + q$
$m + p + q$	m	$m - p - q$
$m - q$	$m - p + q$	$m + p$

3 a Write an expression for each missing length in this rectangle.

   b Write an expression for P, the perimeter of the rectangle.

   c Given that $a = 2.1$ and $b = 4.5$, calculate the area of the rectangle.
   (Area = length × width)

4 a The area of a rectangle is $2x^2 + 4x$. Suggest possible lengths for its sides.

   b If the perimeter of a rectangle is $2x^2 + 4$, what could the lengths of the sides be?

Find answers at: cambridge.org/ukschools/gcsemaths-studentbookanswers

5. Draw two diagrams representing areas to prove that $(3x)^2$ and $3x^2$ are different.

6. Copy and fill in the missing cells. The entry for each cell is formed from the two cells beneath it by addition or subtraction and collecting like terms from the two cells beneath it. Write each expression as simply as possible. The first missing entry for diagram a has been completed in red for you.

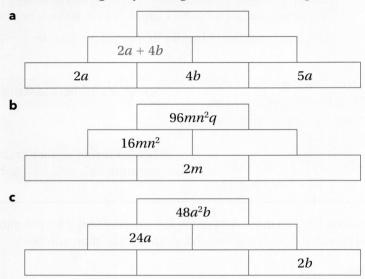

7. **a** Paul plays a 'think of a number game' with his friends and predicts what their answer will be.

   He tells his friends to follow the steps in the box (left).

   Paul then guesses that the answer is 3.

   Use algebra to show why Paul will guess correctly, no matter what number his friends choose as a starting number.

   b Make up a 'think of a number' problem of your own.

   Use algebra to check that it will work and to see which number you end up with.

   Try it out with another student to check that it works.

> Think of a number.
> Double it.
> Add 6.
> Halve it.
> Take away the number you first thought of.

 Checklist of learning and understanding

Algebraic notation
- You can use letters (called variables) in place of unknown quantities in algebra.
- An expression is a collection of numbers, operation signs and at least one variable.
- Each part of an expression is called a term.
- To evaluate an expression you substitute numbers in place of the variables.
- If two expressions are identical, this is called an identity.

Simplifying expressions
- Like terms have exactly the same variables.
- Expressions can be simplified by adding or subtracting like terms.
- You can multiply and divide unlike terms.

Multiplying out brackets

- If an expression contains brackets you multiply them out and then add or subtract like terms to simplify it further.

Factorising

- Factorising involves putting brackets back into an expression.
- If terms have a common factor, write it in front of the bracket and write the remaining terms in the bracket as a factor. (You can check by multiplying out.)

Solving problems

- Algebra allows you to make general rules that apply to any number. This is useful in problem solving.

Chapter review

For additional questions on the topics in this chapter, visit GCSE Mathematics Online.

1 Simplify if possible.

  a $12x - 7x$ **b** $4a - 12b - 3a + 4b$ **c** $5xy \times z \times 2 + 8xyz$

2 The expression $7(x + 4) - 3(x - 2)$ simplifies to $a(2x + b)$. Work out the values of a and b.

3 Check whether each expression has been fully simplified. If not, simplify it further.

  a $5(g + 2) + 8g \qquad = 5g + 10 + 8g$
  b $4z(4z - 2) - z(z + 2) = 15z^2 - 10z$
  c $-5ab \times (-3bc) \qquad = 15ab^2c$
  d $\dfrac{18x^3}{3x} \qquad\qquad\quad = \dfrac{6x^3}{x}$

4 The nth even number is $2n$. The next even number after $2n$ is $2n + 2$.

  a Explain why.

  b Write an expression, in terms of n, for the next even number after $2n + 2$.

  c Show algebraically that the sum of any three consecutive even numbers is always a multiple of 6.

5 Which of the following pairs of expressions are identities? Show how you know.

  a $5(x + 3)$ and $5x + 3$
  b $-3(m - 2)$ and $-3m - 6$
  c $4(y - 3) + 2(y + 4)$ and $6y - 4$

6 **a** Expand and simplify $5(x + 7) + 3(x - 2)$ *(2 marks)*

   b Factorise completely $3a^2b + 6ab^2$ *(2 marks)*

©Pearson Education Ltd 2012

6 Fractions

In this chapter you will learn how to ...
- recognise equivalence between fractions and mixed numbers.
- carry out the four basic operations on fractions and mixed numbers.
- work out fractions of an amount.

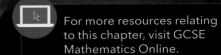

For more resources relating to this chapter, visit GCSE Mathematics Online.

Using mathematics: real-life applications

Nurses and other medical support staff work with fractions, decimals, percentages, rates and ratios every day. They calculate medicine doses, convert between different systems of measurement and set the patients' drips to supply the correct amount of fluid per hour.

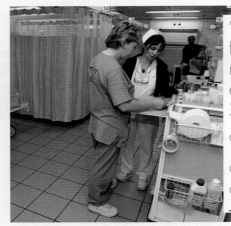

"We have to record how much fluid patients drink when they are recovering from surgery. So, for example, if we give the patient a 300 ml glass of orange juice and they only drink $\frac{2}{3}$ of it, we have to work out that they have taken in 200 ml of fluid." *(Nurse)*

Before you start ...

Ch 4	Check that you can find common factors of sets of numbers.	1	18 24 27 28 30 32 36 Choose numbers from this set that have: **a** a common factor of 9 **b** common factors 2 and 3 **c** common factors 3, 4 and 12 **d** common factors 3 and 6.

Ch 4	Find the lowest common multiple of sets of numbers.	2	Choose the lowest common multiple of each set of numbers.
			 a 5 and 10
			    A 15 **B** 50 **C** 10 **D** 20
			 b 8 and 12
			    A 12 **B** 96 **C** 36 **D** 24
			 c 2, 3 and 5
			    A 1 **B** 30 **C** 10 **D** 6

Ch 1	Know the correct order for performing operations (BODMAS).	3	Which calculation is correct in each pair? Why?

	Student A	Student B
a	$-3 - 2 \times -6 - 4 = 5$	$-3 - 2 \times -6 - 4 = 50$
b	$-60 \div 5 + 3 \times -4 - 8 = 28$	$-60 \div 5 + 3 \times -4 - 8 = -32$
c	$13 - 2 \times -6 - 5 \times 4 = 80$	$13 - 2 \times -6 - 5 \times 4 = 5$

Assess your starting point using the Launchpad

STEP 1

1 Which fraction does not belong in each set?

a $\dfrac{3}{15}, \dfrac{1}{5}, \dfrac{6}{30}, \dfrac{5}{35}, \dfrac{4}{20}$

b $\dfrac{4}{7}, \dfrac{8}{14}, \dfrac{12}{21}, \dfrac{9}{16}, \dfrac{52}{91}$

c $\dfrac{22}{10}, \dfrac{11}{4}, 2\dfrac{3}{4}, \dfrac{33}{12}, 2\dfrac{18}{24}$

GO TO
Section 1: Equivalent fractions

STEP 2

2 Each calculation contains a mistake. Find the mistake and write the correct answer.

a $\dfrac{2}{3} + \dfrac{3}{4} = \dfrac{5}{7}$

b $\dfrac{4}{5} - \dfrac{9}{10} = \dfrac{1}{10}$

c $\dfrac{2}{7} \times \dfrac{4}{5} = \dfrac{6}{35}$

d $30 \div \dfrac{1}{2} = 15$

GO TO
Section 2: Operations with fractions

STEP 3

3 Which is greater in each pair?

a $\dfrac{5}{8}$ of 40 or $\dfrac{3}{5}$ of 60

b $\dfrac{3}{4}$ of 240 or $\dfrac{7}{10}$ of 300

c $\dfrac{1}{4}$ of $\dfrac{1}{2}$ or $\dfrac{1}{2}$ of $\dfrac{3}{4}$

GO TO
Section 3: Fractions of quantities

GO TO Chapter review

Find answers at: cambridge.org/ukschools/gcsemaths-studentbookanswers

Section 1: Equivalent fractions

Fractions tell you what share you have of a quantity.

Fractions that look different can be describing the same share.

For example, you could have $\frac{1}{4}$ of a pizza and a friend could have $\frac{2}{8}$ of a pizza.

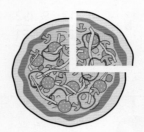

The fractions $\frac{1}{4}$ and $\frac{2}{8}$ are equivalent.

You can find equivalent fractions by **multiplying** the **numerator** and **denominator** by the **same** number.

For example:

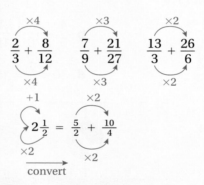

Key vocabulary

numerator: the number at the top of a fraction.

denominator: the number at the bottom of a fraction.

Tip

It doesn't matter what number you use as long as the numerator and denominator are multiplied by the **same** number. What you are really doing is multiplying by 1 because any number divided by itself is equal to 1.

$\frac{4}{4} = 1$, $\frac{5}{5} = 1$ and $\frac{x}{x} = 1$

You can also find equivalent fractions by **dividing** the numerator and denominator by the **same** number.

This is known as simplifying, cancelling or reducing the fraction to its lowest (simplest) terms.

When you give an answer in the form of a fraction, you usually give it in simplest form.

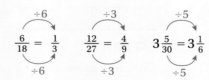

Key vocabulary

common denominator: a number into which all the denominators of a set of fractions divide exactly.

When you are asked to compare fractions that look different, you might need to re-write them both with the same denominator (a **common denominator**) so that you can tell whether they are equivalent or not.

6 Fractions

WORKED EXAMPLE 1

a Is $\frac{5}{6}$ equivalent to $\frac{7}{8}$?   **b** Is $3\frac{3}{4}$ equivalent to $\frac{45}{12}$?

a $\frac{5}{6} = \frac{20}{24}$ and $\frac{7}{8} = \frac{21}{24}$ *The lowest common multiple of 6 and 8 is 24. Write both fractions with the same denominator.*

$\frac{5}{6} \neq \frac{7}{8}$ *It is also easy to tell which one is bigger or smaller. $\frac{7}{8} > \frac{5}{6}$*

b $3\frac{3}{4} = \frac{15}{4}$ *Write the mixed number as an improper fraction.*

$\frac{15}{4} = \frac{45}{12}$ *Write $\frac{15}{4}$ with a denominator of 12, or write $\frac{45}{12}$ with a denominator of 4.*

The fractions are equivalent.

Key vocabulary

mixed number: a number comprising an integer and a fraction.

improper fraction: a fraction where the numerator is greater than (or equal to) the denominator.

Tip

You can use the LCM of the denominators to find a common denominator, but any common denominator works (not just the lowest).

EXERCISE 6A

1 Complete each statement to make a pair of equivalent fractions.

    a $\frac{3}{4} = \frac{\boxed{}}{44}$ **b** $\frac{1}{3} = \frac{1000}{\boxed{}}$ **c** $\frac{1}{2} = \frac{150}{300}$ **d** $\frac{-2}{5} = \frac{-18}{\boxed{}}$

    e $\frac{-6}{-10} = \frac{42}{\boxed{}}$ **f** $\frac{\boxed{}}{5} = \frac{36}{20}$ **g** $\frac{4}{3} = \frac{28}{\boxed{}}$ **h** $\frac{10}{14} = \frac{50}{\boxed{}}$

2 List five fractions that are equivalent to each of these fractions.

    a $\frac{5}{7}$ **b** $\frac{4}{5}$ **c** $\frac{12}{8}$ **d** $\frac{-5}{-3}$

3 Write each mixed number as an improper fraction in its simplest form.

    a $2\frac{1}{3}$ **b** $3\frac{1}{3}$ **c** $5\frac{2}{8}$ **d** $4\frac{6}{12}$

    e $2\frac{14}{21}$ **f** $1\frac{12}{36}$ **g** $2\frac{30}{50}$ **h** $3\frac{35}{45}$

Find answers at: cambridge.org/ukschools/gcsemaths-studentbookanswers

4 Rewrite each fraction as an equivalent mixed number.

a $\dfrac{12}{5}$ b $\dfrac{7}{3}$ c $\dfrac{8}{5}$ d $\dfrac{-11}{5}$

e $\dfrac{12}{11}$ f $\dfrac{13}{9}$ g $\dfrac{-9}{-4}$ h $\dfrac{21}{9}$

5 Determine whether the following pairs of fractions are equivalent (=) or not (≠).

a $\dfrac{2}{5}$ and $\dfrac{3}{4}$ b $\dfrac{2}{3}$ and $\dfrac{3}{4}$ c $\dfrac{3}{8}$ and $\dfrac{5}{12}$ d $\dfrac{2}{11}$ and $\dfrac{1}{10}$

e $\dfrac{3}{5}$ and $\dfrac{9}{15}$ f $\dfrac{10}{25}$ and $\dfrac{4}{10}$ g $\dfrac{6}{24}$ and $\dfrac{5}{20}$ h $\dfrac{11}{9}$ and $\dfrac{121}{99}$

6 Reduce the following fractions to their simplest form.

a $\dfrac{3}{15}$ b $\dfrac{4}{6}$ c $\dfrac{25}{100}$ d $\dfrac{-5}{-10}$

e $\dfrac{4}{12}$ f $\dfrac{7}{21}$ g $\dfrac{36}{24}$ h $\dfrac{60}{100}$

i $\dfrac{-14}{21}$ j $\dfrac{18}{27}$ k $\dfrac{-15}{-21}$ l $\dfrac{18}{42}$

7 Write each set of fractions in ascending order.

a $\dfrac{3}{5}, \dfrac{1}{4}, \dfrac{9}{4}, 1\dfrac{3}{4}, \dfrac{4}{7}$ b $\dfrac{5}{6}, \dfrac{3}{4}, \dfrac{11}{3}, \dfrac{19}{24}, 2\dfrac{2}{3}$ c $2\dfrac{3}{7}, \dfrac{1}{7}, \dfrac{7}{7}, \dfrac{8}{14}, \dfrac{10}{21}, \dfrac{13}{7}$

Section 2: Operations with fractions

Multiplying fractions

This rectangle has been divided into 12 smaller squares, or twelfths.

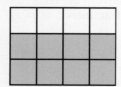 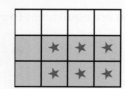

$\dfrac{2}{3}$ of this rectangle is blue. $\dfrac{3}{4}$ of the $\dfrac{2}{3}$ have been marked with a star ★

From the diagram you can see that $\dfrac{3}{4}$ of $\dfrac{2}{3}$ is 6 of the original 12 parts.

$\dfrac{3}{4}$ of $\dfrac{2}{3} = \dfrac{3}{4} \times \dfrac{2}{3} = \dfrac{6}{12}$ which simplifies to $\dfrac{1}{2}$.

To multiply fractions multiply the numerators and then multiply the denominators.

WORKED EXAMPLE 2

a $\dfrac{3}{4} \times \dfrac{2}{7} = \dfrac{3 \times 2}{4 \times 7}$ — Multiply numerators by numerators and denominators by denominators.

$= \dfrac{6}{28}$

$= \dfrac{3}{14}$ — Give the answer in its simplest form.

b $\dfrac{5}{7} \times 3 = \dfrac{5 \times 3}{7 \times 1}$ — Think of a whole number as a fraction with a denominator of 1.

$= \dfrac{15}{7}$ — $\dfrac{15}{7}$ cannot be simplified further but it can be written as a mixed number.

$= 2\dfrac{1}{7}$

c $\dfrac{3}{8} \times 4\dfrac{1}{2} = \dfrac{3}{8} \times \dfrac{9}{2}$ — Rewrite the mixed number as an improper fraction.

$= \dfrac{27}{16}$ — $\dfrac{27}{16}$ cannot be simplified but it can be written as a mixed number.

$= 1\dfrac{11}{16}$

Tip

You can cancel before you multiply to make it easier to simplify the answers.

Adding and subtracting fractions

To add or subtract fractions they must have the same denominators.

Find a common denominator and then find the equivalent fractions before you add or subtract the numerators.

WORKED EXAMPLE 3

a $\dfrac{1}{2} + \dfrac{1}{4} = \dfrac{2}{4} + \dfrac{1}{4}$ — Use 4 as a common denominator. Write $\dfrac{1}{2}$ as its equivalent, $\dfrac{2}{4}$.

$= \dfrac{3}{4}$ — Add the numerators.

b $2\dfrac{1}{2} + \dfrac{5}{6} = \dfrac{5}{2} + \dfrac{5}{6}$ — Rewrite mixed numbers as improper fractions.

$= \dfrac{15}{6} + \dfrac{5}{6}$ — Find a common denominator.

$= \dfrac{20}{6}$ — Add the numerators.

$= \dfrac{10}{3}$ or $3\dfrac{1}{3}$ — Simplify the answer.

Continues on next page …

Tip

Think of $\frac{3}{7}$ and $\frac{2}{7}$ as 3 lots of 7ths and 2 lots of 7ths. If you combine them, you have 5 lots of 7ths, or $\frac{5}{7}$.

You **never add the denominators**.

c $\quad 2\frac{3}{4} - 1\frac{5}{7} = \frac{11}{4} - \frac{12}{7}$

Rewrite mixed numbers as improper fractions.

$\qquad = \frac{77}{28} - \frac{48}{28}$

Find a common denominator.

$\qquad = \frac{29}{28}$ or $1\frac{1}{28}$

Subtract the numverators.
Simplify the answer.

Dividing fractions

Dividing a fraction, such as $\frac{9}{10}$, by another fraction, such as $\frac{1}{2}$, means finding out how many $\frac{1}{2}$s there are in $\frac{9}{10}$.

You are trying to find $\frac{9}{10} \div \frac{1}{2}$ which is the same as the fraction $\frac{\frac{9}{10}}{\frac{1}{2}}$.

If you multiply the numerator and the denominator by 2, you can get rid of the fractional denominator.

$$\frac{\frac{9}{10}}{\frac{1}{2}} = \frac{\frac{9}{10} \times \frac{2}{1}}{\frac{1}{2} \times \frac{2}{1}} = \frac{9}{10} \times \frac{2}{1} = \frac{18}{10} = \frac{9}{5}$$

You can do this calculation faster by just inverting the fraction you are dividing by.

So, $\quad \frac{9}{10} \div \frac{1}{2} = \frac{9}{10} \times \frac{2}{1} = \frac{18}{10} = \frac{9}{5}$

This gives the general method for dividing by a fraction.

To divide one fraction by another fraction you multiply the first fraction by the **reciprocal** of the second fraction.

Key vocabulary

reciprocal: the value obtained by inverting a fraction. Any number multiplied by its reciprocal is 1.

unit fraction: a fraction with numerator 1 and denominator a positive integer.

Tip

The reciprocal of $\frac{3}{4}$ is $\frac{4}{3}$. The reciprocal of a whole number is a **unit fraction**. For example, the reciprocal of 3 is $\frac{1}{3}$.

WORKED EXAMPLE 4

a $\quad \frac{3}{4} \div \frac{1}{2} = \frac{3}{4} \times \frac{2}{1}$

Multiply by the reciprocal of $\frac{1}{2}$

$\qquad = \frac{6}{4}$

$\qquad = \frac{3}{2}$ or $1\frac{1}{2}$

Continues on next page …

b $1\frac{3}{4} \div 2\frac{1}{3} = \frac{7}{4} \div \frac{7}{3}$ ◂ Convert mixed numbers to improper fractions.

$= \frac{\cancel{7}}{4} \times \frac{3}{\cancel{7}}$ ◂ Multiply by the reciprocal of $\frac{7}{3}$. Cancel the 7s.

$= \frac{3}{4}$

c $\frac{6}{7} \div 3 = \frac{6}{7} \times \frac{1}{3}$ ◂ Multiply by the reciprocal of 3.

$= \frac{6}{21}$

$= \frac{2}{7}$

The rules for order of operations and negative and positive signs also apply to calculations with fractions.

EXERCISE 6B

1 Calculate.

- **a** $\frac{3}{4} \times \frac{2}{5}$
- **b** $\frac{1}{5} \times \frac{1}{9}$
- **c** $\frac{5}{7} \times \frac{1}{5}$
- **d** $\frac{7}{10} \times \frac{2}{3}$
- **e** $\frac{4}{7} \times \frac{3}{8}$
- **f** $\frac{6}{11} \times \frac{-5}{6}$
- **g** $\frac{3}{4} \times 24$
- **h** $\frac{4}{9} \times \frac{8}{10}$
- **i** $\frac{3}{8} \times \frac{4}{9}$
- **j** $\frac{7}{25} \times \frac{3}{4}$

2 Find the value of:

- **a** $1\frac{1}{2} \times -10$
- **b** $1\frac{4}{5} \times -6$
- **c** $1\frac{2}{7} \times 3\frac{1}{2}$
- **d** $4\frac{1}{11} \times -3\frac{1}{8}$
- **e** $\frac{7}{25} \times \frac{7}{9}$
- **f** $3\frac{1}{3} \times 9\frac{2}{5}$

3 Simplify.

- **a** $\frac{1}{5} \times \frac{3}{8} \times \frac{-5}{9}$
- **b** $\frac{2}{3} \times \frac{3}{4} \times \frac{4}{5}$
- **c** $\frac{1}{2} \times \frac{2}{3} \times \frac{4}{11}$
- **d** $\frac{5}{8} \times \frac{3}{7} \times \frac{2}{3}$
- **e** $\frac{4}{25} \times \frac{-3}{5} \times \frac{-7}{8}$
- **f** $\frac{9}{20} \times \frac{10}{11} \times \frac{1}{12}$

4 Simplify.

- **a** $\frac{2}{7} + \frac{1}{2}$
- **b** $\frac{1}{2} + \frac{1}{4}$
- **c** $\frac{1}{4} + \frac{3}{8}$
- **d** $\frac{5}{6} + \frac{6}{10}$
- **e** $\frac{5}{8} - \frac{1}{4}$
- **f** $\frac{7}{9} - \frac{1}{3}$
- **g** $\frac{3}{4} + \frac{2}{5}$
- **h** $\frac{3}{4} - \frac{1}{3}$
- **i** $\frac{4}{5} - \frac{1}{3}$
- **j** $\frac{4}{5} - \frac{3}{10}$
- **k** $\frac{3}{4} + \frac{1}{6}$
- **l** $\frac{4}{9} - \frac{1}{4}$
- **m** $\frac{2}{3} - \frac{3}{10}$
- **n** $\frac{7}{8} - \frac{3}{5}$
- **o** $\frac{1}{4} - \frac{1}{5}$
- **p** $\frac{13}{2} - \frac{8}{5}$

5 Simplify.

a $2\frac{3}{4} + 2\frac{1}{2}$ b $1\frac{3}{4} - 1\frac{1}{3}$ c $2\frac{7}{8} + 1\frac{3}{5}$ d $3\frac{7}{10} + 2\frac{9}{11}$

e $4\frac{3}{4} + 1\frac{5}{6}$ f $8\frac{2}{5} - 3\frac{1}{2}$ g $7\frac{1}{4} - 2\frac{9}{10}$ h $9\frac{3}{7} - 2\frac{4}{5}$

i $6\frac{3}{5} - 1\frac{11}{13}$ j $2\frac{11}{20} - 1\frac{9}{10}$ k $8 - 2\frac{3}{4}$ l $9\frac{3}{5} - 7\frac{1}{2}$

6 Simplify.

a $\frac{1}{4} \div \frac{1}{4}$ b $\frac{1}{2} \div \frac{1}{4}$ c $\frac{1}{5} \div \frac{2}{7}$ d $\frac{-6}{7} \div \frac{2}{5}$

e $\frac{1}{8} \div \frac{7}{9}$ f $\frac{2}{11} \div \frac{-3}{5}$ g $\frac{5}{9} \div \frac{3}{7}$ h $\frac{-5}{12} \div \frac{-1}{2}$

7 Simplify.

a $\frac{3}{4} \div -2\frac{1}{2}$ b $2\frac{1}{2} \div \frac{2}{5}$ c $3\frac{1}{5} \div 2\frac{1}{2}$ d $1\frac{7}{8} \div 2\frac{3}{4}$

8 Calculate.

a $4 + \frac{2}{3} \times \frac{1}{3}$ b $2\frac{1}{8} - \left(2\frac{1}{5} - \frac{7}{8}\right)$ c $\frac{3}{7} \times \left(\frac{2}{3} + 6 \div \frac{2}{3}\right) + 5 \times \frac{2}{7}$

d $2\frac{7}{8} + \left(8\frac{1}{4} - 6\frac{3}{8}\right)$ e $\frac{5}{6} \times \frac{1}{4} + \frac{5}{8} \times \frac{1}{3}$ f $\left(5 \div \frac{3}{11} - \frac{5}{12}\right) \times \frac{1}{6}$

g $\left(\frac{5}{8} \div \frac{15}{4}\right) - \left(\frac{5}{6} \times \frac{1}{5}\right)$ h $\left(2\frac{2}{3} \div 4 - \frac{3}{10}\right) \times \frac{3}{17}$ i $\left(7 \div \frac{2}{9} - \frac{1}{3}\right) \times \frac{2}{3}$

EXERCISE 6C

1 Nicci buys a 4 kg packet of nuts and raisins.

She notices that $\frac{3}{8}$ of the contents are raisins.

How many kilograms of nuts are there?

2 Josh eats 8 packets of crisps each week. Nick eats $1\frac{3}{4}$ times as many packets.

How many packets do they eat altogether?

3 Kevin is a professional deep-sea diver.

He needs to keep track of how much time he spends underwater to make sure he has enough air left in his tank.

If he spends $9\frac{3}{4}$ minutes diving to a wreck, $12\frac{5}{6}$ minutes exploring the wreck and $3\frac{5}{6}$ minutes examining corals, how much time has he spent in total?

4 $\frac{5}{12}$ of the people at a conference are from Britain, $\frac{3}{16}$ are from India, $\frac{7}{24}$ are from Brazil and the rest are from Malaysia.

a What fraction of the people at the conference are from Malaysia?

b Which country has most people at the conference?

c Which country has the fewest people at the conference?

d The Indian delegation left the conference a day before everyone else. What fraction of the original number of people is left?

5 In a café, $\frac{1}{4}$ of the customers order coffee and $\frac{2}{5}$ order tea. The rest of the customers order juice.

a What fraction of the customers ordered hot drinks?

b What fraction of the customers ordered juice?

6 A litre carton of milk is $\frac{3}{4}$ full. Sandra uses $\frac{1}{3}$ of a litre to make breakfast.

What fraction of a litre is left after breakfast?

7 Mrs Smith spends $\frac{1}{4}$ of her wages on rent and $\frac{2}{5}$ on other expenses. What fraction does she have left?

8 There are $1\frac{3}{4}$ cakes left over after a party. These are shared out equally among 6 people.

What fraction does each person get?

9 If I have $5\frac{2}{3}$ litres of juice, how many cups containing $\frac{2}{15}$ of a litre can I pour?

10 Nico buys 6 trays of chicken pieces for his restaurant.

Each tray contains $2\frac{1}{2}$ kg of chicken. Each chicken meal served uses $\frac{3}{8}$ kg of chicken.

How many chicken meals can he serve?

Section 3: Fractions of quantities

Read and think about these statements.

- I'll pay half of the £50 costs.
- $\frac{1}{3}$ of the 24 apples in this pack are bad.
- I save $\frac{1}{10}$ of my income of £1200 every month.
- I make about 220 calls a month and about $\frac{4}{5}$ of them are for business.

Some of the fractions in the statements are easy to work out.

You know that half of 50 is 25.

$$\frac{1}{2} \times 50 = 25$$

$\frac{1}{3}$ of 24 is 8.

$$\frac{1}{3} \times 24 = 8$$

$\frac{1}{10}$ of 1200 = 120.

$$\frac{1}{10} \times 1200 = 120$$

Find answers at: cambridge.org/ukschools/gcsemaths-studentbookanswers

Tip

When you see the word 'of' in a fraction problem, replace it with × and do the multiplication as you normally would.

Write whole numbers over 1.

This shows that the word 'of' means 'multiply' (×).

Look at the last statement. It isn't so easy to work out $\frac{4}{5}$ of 220 at a glance. Multiplying by $\frac{4}{5}$ gives you the answer.

$$\frac{4}{5} \times \frac{220}{1} = \frac{880}{5} = 176$$

Expressing one quantity as a fraction of another

It is easy to express one quantity as a fraction of another if you remember that the numerator in a fraction tells you how many parts of the whole quantity you are dealing with and the denominator represents the whole quantity. So, a fraction of $\frac{3}{5}$ means you are dealing with 3 of the 5 parts that make up the whole.

Tip

To write a quantity as a fraction of another quantity make sure the two quantities are in the same units and then write them as a fraction and simplify.

WORKED EXAMPLE 5

a What fraction is 20 minutes of 1 hour?

b Express 35 centimetres as a fraction of a metre.

a 20 minutes is the part of the whole, so it is the numerator.

The hour is the whole, so it is the denominator.

20 minutes is part of 60 minutes:
$\frac{20}{60} = \frac{2}{6} = \frac{1}{3}$ of an hour.

> You cannot form a fraction using one unit for the numerator and another for the denominator, so you need to convert the hour to minutes. There are 60 minutes in one hour.

b 35 cm is the part of the whole, so it is the numerator.

The metre is the whole, so it is the denominator.

35 cm is part of 100 cm:
$\frac{35}{100} = \frac{7}{20}$ of a metre.

> Again, you cannot use centimetres and metres in the same fraction, so you convert 1 m to 100 cm.

EXERCISE 6D

1 Calculate.

a $\frac{3}{4}$ of 12 b $\frac{1}{3}$ of 45 c $\frac{2}{9}$ of 36 d $\frac{3}{8}$ of 144

e $\frac{4}{5}$ of 180 f $\frac{1}{3}$ of 96 g $\frac{1}{2}$ of $\frac{3}{4}$ h $\frac{1}{3}$ of $\frac{3}{10}$

i $\frac{4}{9}$ of $\frac{3}{14}$ j $\frac{1}{4}$ of $2\frac{1}{2}$ k $\frac{3}{4}$ of $2\frac{1}{3}$ l $\frac{5}{6}$ of $3\frac{1}{2}$

2 Calculate the following quantities.

a $\frac{3}{4}$ of £28 b $\frac{3}{5}$ of £210 c $\frac{2}{5}$ of £30 d $\frac{2}{3}$ of £18

e $\frac{1}{2}$ of 3 cups of sugar f $\frac{1}{2}$ of 5 cups of flour

3 Calculate the following quantities.

a $\frac{1}{2}$ of $1\frac{1}{2}$ cups of sugar b $\frac{3}{4}$ of $2\frac{1}{3}$ cups of flour c $\frac{2}{3}$ of $1\frac{1}{2}$ cups of sugar

d $\frac{2}{3}$ of 4 hours e $\frac{1}{3}$ of $2\frac{1}{2}$ hours f $\frac{3}{4}$ of 5 hours

g $\frac{1}{3}$ of $\frac{3}{4}$ of an hour h $\frac{2}{3}$ of $3\frac{1}{2}$ minutes i $\frac{3}{15}$ of a minute

4 Express the first quantity as a fraction of the second.

a 12p of every £1
b 35 cm of a 2 m length
c 12 mm of 30 cm
d 45 minutes per 8-hour shift
e 5 minutes per hour
f 150 m of a kilometre
g 45 seconds of 30 minutes
h 575 ml of 4 litres

5 Nick earns £18 000 per year. His friend Samir earns £24 000 per year.

What fraction of Samir's salary does Nick earn?

6 The floor area of a room is 12 m². Pete buys a rug that is 110 cm wide and 160 cm long.

What fraction of the floor area will be covered by this rug?

Checklist of learning and understanding

Equivalent fractions

- Fractions that represent the same amount are called equivalent fractions.
- You can change fractions to their equivalents by multiplying the numerator and denominator by the same value or by dividing the numerator and denominator by the same value (simplifying).

Operations on fractions

- To add or subtract fractions, find equivalent fractions with the same denominator.
- Add or subtract the numerators once the denominators are the same. Do not add or subtract denominators.
- To multiply fractions, multiply numerators by numerators and denominators by denominators.
- To divide fractions, multiply by the reciprocal of the divisor.

Fractions of a quantity

- The word 'of' means multiply.
- A quantity can be written as a fraction of another as long as they are in the same units. Write one quantity as the numerator and the other as the denominator and simplify.

For additional questions on the topics in this chapter, visit GCSE Mathematics Online.

Chapter review

1 Simplify.

a $\dfrac{15}{90}$ **b** $\dfrac{195}{230}$ **c** $4\dfrac{18}{48}$

2 Write each set of fractions in ascending order.

a $\dfrac{8}{9}, \dfrac{4}{5}, \dfrac{5}{6}, \dfrac{3}{7}$ **b** $2\dfrac{2}{5}, \dfrac{23}{7}, 1\dfrac{3}{5}, \dfrac{16}{9}$

3 Evaluate.

a $\dfrac{7}{5} + \dfrac{3}{8}$ **b** $\dfrac{7}{5} \times \dfrac{3}{8}$ **c** $\dfrac{7}{5} \div \dfrac{3}{8}$

d $3\dfrac{1}{7} + 2\dfrac{2}{5}$ **e** $3\dfrac{1}{15} - 1\dfrac{3}{5}$ **f** $\dfrac{1}{7}$ of $3\dfrac{3}{12}$

g $\dfrac{2}{7} \times \dfrac{8}{18} \div 3$ **h** $28 \div \dfrac{3}{4}$ **i** $\dfrac{2}{9}$ of $\dfrac{3}{4}$

4 Simplify.

a $\left(\dfrac{3}{8} \div \dfrac{13}{4}\right) + \left(\dfrac{5}{9} \times \dfrac{3}{5}\right)$

b $2\dfrac{2}{3} \times \left(8 \div \dfrac{4}{7} + \dfrac{7}{8}\right)$

5 a Work out $\dfrac{2}{5} \times \dfrac{3}{8}$

Give your answer in its simplest form. *(2 marks)*

b Work out $\dfrac{3}{8} + \dfrac{1}{4}$ *(2 marks)*

©*Pearson Education Ltd 2012*

6 Express 425 g as a fraction of $2\dfrac{1}{2}$ kg.

7 Sandy has $12\dfrac{1}{2}$ litres of water. How many bottles containing $\dfrac{3}{4}$ litre can she fill?

8 A surveyor has to divide a 15 km² area of land into equal plots each measuring $\dfrac{1}{2}$ km².

How many plots can she make?

7 Decimals

In this chapter you will learn how to ...
- convert decimals to fractions and fractions to decimals.
- order fractions and decimals.
- carry out the four basic operations on decimals without using a calculator.
- solve problems involving decimal quantities.

For more resources relating to this chapter, visit GCSE Mathematics Online.

Using mathematics: real-life applications

Food technologists analyse the contents of different raw and prepared foods to work out what they contain and how much there is of each ingredient. For example, how much water, protein and fat there is in a cut of meat. They use decimal fractions to give the quantities correct to tenths, hundredths or even smaller parts of a gram.

"The laws about labelling food are fairly strict. Manufacturers need to state exactly what is in their product and give exact amounts of different ingredients so I have to measure things very accurately." *(Food technologist)*

Before you start ...

KS3	You need to be able to work confidently with place value.		31.098 0.0398 300.098 0.98308 19.308
		1	Choose the number from the set that has a 3 in the: **a** hundreds position **b** hundredths position **c** tenths position **d** thousandths position **e** tens position.
KS3	Check that you can compare decimal fractions and order them by size.	**2**	Fill in <, = or > between each pair of decimal fractions. **a** 0.65 ☐ 0.7 **b** 0.08 ☐ 0.01 **c** 0.8 ☐ 0.85 **d** 2.87 ☐ 0.99 **e** 4.230 ☐ 4.23
KS3	You need to know the fractional equivalents of some common decimals.	**3**	Make equivalent pairs by matching the decimal in the upper box with the fraction in the lower box.

a 0.25	**b** 0.375	**c** 0.4	**d** 0.75	**e** 0.5	**f** 0.025
$\frac{2}{5}$	$\frac{3}{4}$	$\frac{45}{90}$	$\frac{1}{40}$	$\frac{4}{16}$	$\frac{3}{8}$

7 Decimals

Assess your starting point using the Launchpad

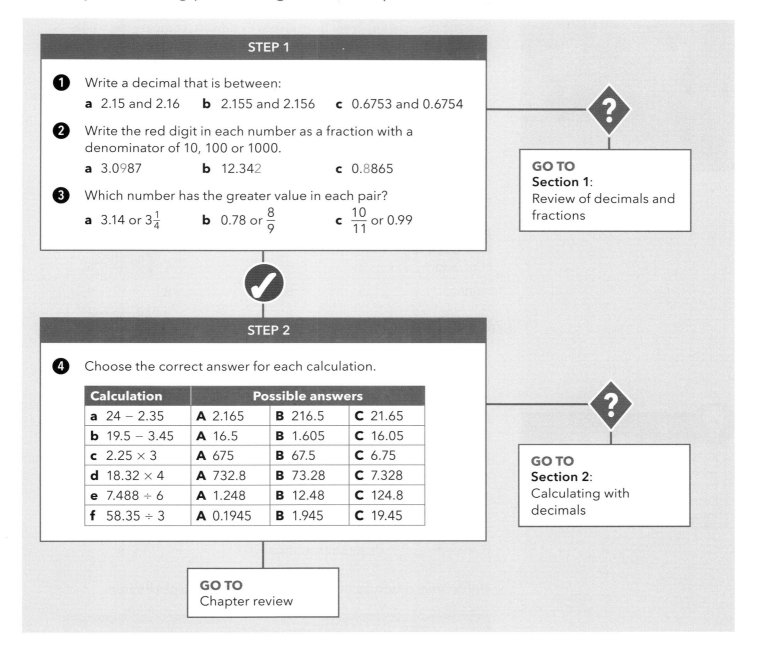

Section 1: Review of decimals and fractions

Five teams completed the men's 4 × 100 m relay final at the 2010 Commonwealth Games in New Delhi.

The times are given as decimals.

The numbers that follow the decimal point indicate parts (decimal fractions) of a second.

There are two digits after the decimal place so the times are given correct to a hundredth of a second.

Team	Time (seconds)
Australia	39.14
Bahamas	39.27
England	38.74
India	38.89
Jamaica	38.79

Find answers at: cambridge.org/ukschools/gcsemaths-studentbookanswers

Comparing decimals

To write the times in order from fastest to slowest, compare the whole number parts of each first. If those are the same, compare the decimal parts.

England, India and Jamaica ran the relay in 38 seconds and a fraction of a second.

This is less time than Australia and the Bahamas who ran the race in 39 seconds and a fraction of a second.

To decide first, second and third place, you need to look at the decimal parts.

Here are the times written in a place value table.

Team	Tens	Ones/units	.	tenths	hundredths
England	3	8	.	7	4
India	3	8	.	8	9
Jamaica	3	8	.	7	9

Start by looking at the tenths.

India has 8 in the tenths place. This is greater than the others, so India came third.

England and Jamaica both have 7 in the tenths place, so compare the hundredths.

England has 4 in the hundredths and Jamaica has 9, this means that England was faster.

The places were: England (1st), Jamaica (2nd) and India (3rd), followed by Australia then the Bahamas.

> **Tip**
> Remember you are comparing winning times, so you are looking for the smallest time as this is the fastest time. The greater the time, the slower the team ran.

Converting decimals to fractions

The decimal times can be written as fractions using place value.

Tens	Ones/units	.	tenths	hundredths
3	8	.	7	4

The England team ran the relay in 38 seconds and $\frac{74}{100}$ of a second.

The fraction part can be simplified further $\frac{74}{100} = \frac{37}{50}$. So, $38.74 = 38\frac{37}{50}$

Any decimal can be converted to a fraction in this way. For example:

$0.6 = \frac{6}{10} = \frac{3}{5}$ $0.25 = \frac{25}{100} = \frac{1}{4}$ $0.075 = \frac{75}{1000} = \frac{3}{40}$

Converting fractions to decimals

There are different methods for converting fractions to decimals. You should choose the method that is easiest for the fraction involved.

Method 1:	Method 2:	Method 3:
Equivalent fractions with denominators of 10, 100, 1000 and so on.	Pen and paper division.	Calculator division.
Express $\frac{61}{125}$ as a decimal. $\frac{61}{125} = \frac{122}{250} = \frac{244}{500}$ $= \frac{488}{1000}$ $\frac{61}{125} = 0.488$ This method works well if the denominator is a factor of 10, 100 or 1000.	Express $\frac{5}{8}$ as a decimal. Work out $5 \div 8$ using division. 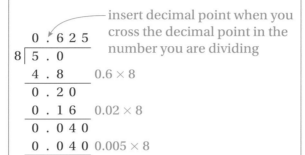 insert decimal point when you cross the decimal point in the number you are dividing	Express $\frac{2}{3}$ as a decimal. Input $2 \div 3$ on your calculator.  The 6s continue forever (they recur). Show this by writing the answer as $0.\dot{6}$

> **Tip**
> Remember that you write a dot above the first and last digit of the recurring numbers if more than one digit recurs.

> **Tip**
> When you have to compare and order ordinary fractions you can convert them all to decimals and compare them easily using place value. This is often quicker than changing them all into equivalent fractions with a common denominator.

EXERCISE 7A

1 Write each of the following decimals as a fraction in its simplest form.
  a 0.6   **b** 0.84   **c** 1.64   **d** 0.385   **e** 0.125
  f 1.08 **g** 0.875 **h** 0.008 **i** 3.064   **j** 0.333

2 Convert the following fractions to decimals without using a calculator.
  a $\frac{3}{5}$ **b** $\frac{3}{4}$ **c** $\frac{18}{25}$ **d** $\frac{19}{20}$ **e** $\frac{34}{50}$
  f $\frac{110}{250}$ **g** $\frac{89}{200}$ **h** $\frac{76}{500}$ **i** $\frac{185}{20}$ **j** $\frac{145}{50}$
  k $\frac{11}{6}$ **l** $\frac{3}{8}$ **m** $\frac{9}{4}$ **n** $\frac{8}{9}$ **o** $\frac{19}{8}$

3 Use a calculator to convert the fractions from $\frac{1}{9}$ to $\frac{8}{9}$ into decimals.

   a What pattern do you notice?
   b What do you call decimals of this nature?
   c Repeat this for the fractions from $\frac{1}{6}$ to $\frac{5}{6}$.
   d Convert $\frac{1}{11}$ and $\frac{2}{11}$ to decimals.
   e Predict what $\frac{3}{11}$ and $\frac{4}{11}$ will be if you convert them to decimals.

   Check your prediction using a calculator.

4 Arrange the following in descending order.

   a 5.2, 5.29, 8.62, 4.92, 4.09
   b 7.42, 0.76, 0.742, 0.421, 3.219
   c 14.3, 14.72, 14.07, 14.89, 14.009
   d 0.23, 0.26, 0.273, 0.287, 0.206
   e 0.403, $\frac{1}{2}$, $\frac{2}{3}$, 0.68, 0.45, $\frac{5}{11}$
   f $\frac{7}{9}$, $\frac{3}{8}$, 0.625, 0.88, 0.718

5 Fill in the boxes using <, = or > to make each statement true.

   a 13.098 ☐ 13.099
   b 0.312 ☐ 0.322
   c $\frac{5}{6}$ ☐ 0.84
   d 0.375 ☐ $\frac{3}{8}$
   e 2.05 ☐ $\frac{205}{1000}$
   f $\frac{3}{5}$ ☐ 0.7
   g $\frac{2}{5}$ ☐ 0.35
   h $\frac{18}{25}$ ☐ 0.67
   i $\frac{1}{3}$ ☐ 0.37

6 Write a decimal fraction that is between each pair of decimals.

   a 3.135 and 3.136
   b 0.6645 and 06646
   c 4.998 and 4.999

7 The lengths of some of the world's longest roller coaster rides are given in the table.

Roller coaster	Length of ride (km)
The Beast (USA)	2.243
California Screaming (USA)	1.851
Formula Rossa (United Arab Emirates)	2.0
Fujiyama (Japan)	2.045
Steel Dragon (Japan)	2.479
The Ultimate (UK)	2.268

   a Which is the longest roller coaster?
   b Which is the shortest?
   c Is the Steel Dragon longer or shorter than $2\frac{1}{2}$ km?
   d Which roller coasters are longer than $2\frac{1}{4}$ km?
   e Write the lengths in order from longest to shortest.

Section 2: Calculating with decimals

You need to be able to add, subtract, multiply and divide decimals without using a calculator.

Estimating and reasoning

When you calculate with decimals it is useful to estimate the answer.

An estimate helps you decide whether your solution is reasonable and whether you have the decimal point in the correct place. There's a big difference between £10 and £0.10!

EXERCISE 7B

Estimate the answers for each of the following problems.

Record the strategies you used to estimate.

1. The masses of some coins are given in the diagram.
   a. What is the approximate difference in mass between the heaviest and lightest coins?
   b. Approximate the combined mass of the four coins.
   c. What is the approximate total mass of a £1 coin and two 10p coins?
   d. Nick has five £1 coins in his pocket. Approximately how much do they weigh altogether?
   e. Xena has a packet of 50p coins that weighs 162 grams.
      Approximately how many coins are in the packet?
   f. Will ten 10p coins weigh more or less than six £1 coins?
   g. Anna has a pile of twenty 2p coins and Ben has a pile of twenty 10p coins. What is the approximate difference in the mass of the two piles?

7.1 g   6.6 g

8.1 g   9.5 g

> **Tip**
>
> The ability to estimate well is an important skill, and you will often rely on estimation rather than working out exact answers.

2. A bottle of medicine contains 0.375 litres and costs £2.55.
   a. Can you buy two bottles for £5?
   b. About how many litres of medicine will you need to fill five bottles?
   c. Estimate how many bottles you can fill with 1 litre of the medicine.
   d. Tanja has to take 15 ml of the medicine twice a day. Approximately how long will the bottle last?
   e. Nina buys two bottles of the medicine and pays with a £20 note. Estimate how much change she should get.
   f. In one day, the pharmacy sold 23 bottles of this medicine. Is that more or less than 5 litres in total?
   g. What is the approximate cost per 100 ml of this medicine?

Find answers at: cambridge.org/ukschools/gcsemaths-studentbookanswers

3 Compare your work with a partner.

a Explain the strategies you used to estimate the answers.

How did you use rounding and approximation?

How did you decide what to do?

b Look at your answers.

Did you get the same estimates? If not, is one estimate closer than the other? Try to explain why.

Adding and subtracting decimals

Add or subtract decimals in columns by lining up the places and the decimal points.

> **WORKED EXAMPLE 1**
>
> Calculate.
>
> a 12.7 + 18.34 + 3.087   b 399.65 − 245.175
>
> a 12.7   b   399.650
>   18.34   − 245.175
>   + 3.087   154.475   Write a 0 as a place holder here.
>   34.127

Multiplying and dividing decimals

Andy made the following notes about multiplying and dividing decimals when he was studying for exams at the end of last year.

- When multiplying or dividing by a power of 10 (10, 100, 1000, and so on):

  Move the digits as many places to the left as the number of zeros when multiplying.

  Move the digits as many places to the right as the number of zeros when dividing.

- When multiplying decimal fractions by decimal fractions:

  Ignore the decimal points and multiply the numbers.

  Place the decimal point in the answer so it has the same number of digits after the decimal point as there were altogether in the multiplication problem.

- When dividing by a decimal:

  Make the divisor a whole number by multiplying the divisor and the dividend by the same power of 10.

  Then divide as normal, keeping the decimal point in the answer directly above the decimal point in the number you are dividing.

Tip

The dividend is the amount you are dividing up; the divisor is the quantity you are dividing it by. If you are asked to divide 18 by 9, then 18 is the dividend and 9 is the divisor.

EXERCISE 7C

Work with a partner.

1 a Read through Andy's summary notes.

b Provide one or two examples for each summary point using actual numbers to show what he means.

c Write your own summary point for dividing a decimal by a whole number.

Include two numerical examples to illustrate your point.

2 Use a place value chart and provide some examples to show why you can multiply and divide decimals in the way Andy summarised.

EXERCISE 7D

1 Estimate first then calculate.

a 0.8 + 0.78
b 12.8 − 11.13
c 0.8 + 0.9
d 15.31 − 1.96
e 2.77 × 8.2
f 9.81 × 3.5

2 Evaluate without a calculator.

a 12.7 + 18.34 + 35.01
b 12.35 + 8.5 + 2.91
c 6.89 − 3.28
d 34.45 − 12.02
e 345.297 − 12.39
f 56 + 8.345 − 34.65
g 27.4 + 9.01 − 12.451

3 Without using a calculator, work out the value of the following.

a 0.786 × 100
b 54.76 × 2000
c 1.234 × 0.65
d 87.87 × 2.34
e 1.83 ÷ 61
f 0.358 ÷ 4
g 5.053 ÷ 0.62
h 31.72 ÷ 0.04

4 The world record for the women's 4 × 100 m relay is 40.82 seconds (USA, 2012) and the Commonwealth Games record is 41.83 seconds (Jamaica, 2014).

a What is the time difference between the World Record and the Commonwealth Games Record?

b In the 2010 Commonwealth Games, the winning time for this relay was 44.19 seconds.

How much slower is this than the 2014 record?

c Each of the four runners in a relay runs 100 m.

Calculate the average time taken for 100 m during the world record winning race.

d Do you think each runner takes the same amount of time? Explain your reasoning.

5 Nadia wants to make a dish that requires 1.5 litres of cream.

She has four 0.385 litre cartons of cream. Does she have enough?

Find answers at: cambridge.org/ukschools/gcsemaths-studentbookanswers

Tip

In Question **6**, mg stands for milligrams; 'milli' means one thousandth. There are one thousand millimetres (mm) in one metre.

In 'kilowatt', 'kilo' means 'one thousand'; 1 kilowatt is 1000 watts.

6 Josh takes a multivitamin tablet every morning.

He calculates that if he takes one tablet every day for a week he will take in 1166.69 mg of vitamin C, 54.6 mg of boron and 257.95 mg of calcium.

Work out how much of each ingredient there is in a tablet.

7 The Chetty household uses about 25.75 kilowatt hours of electricity per day.

Calculate how much they will use in:

a one week   b one (non-leap) year.

Decimals in context

In daily life you use decimals when you deal with money, distances and other measurements.

When you solve problems involving decimals you need to make sure that your answer is both reasonable and sensible in the context.

For example, if you work out the price of an item and you get an answer of £15.987, it makes sense to round it to £15.99 because we don't have coins smaller than 1p (a hundredth of a pound).

Problem-solving framework

Salman saved £20 to go and see an exhibition at the Natural History Museum in London.

His train ticket cost £6.35. The ticket for the exhibition was £8.00. He bought an exhibition booklet for £2.50 and spent £1.55 on a snack at the museum café.

How much money did he have left?

Steps for solving problems	What you would do for this example
Step 1: Read the question carefully to work out what you have to do.	The words 'how much does he have left' tell you that you need to find the change. You need to add up what he spent and then subtract it from the money he started with.
Step 2: What information do you need? Have you got it all?	You need the starting amount and how much he spent. You have that information.
Step 3: Is there any information that you don't need?	You don't need to know where he went or what he spent the money on. That detail is unnecessary.
Step 4: Decide what maths you can do.	You can add up the amounts he spent. You can then subtract the total spending from £20.00.
Step 5: Set out your solution clearly. Check your working and that your answer is reasonable.	6.35 20.00 8.00 −18.40 2.50 1.60 + 1.55 18.40
Step 6: Check that you have answered the question.	Salman has £1.60 left over.

EXERCISE 7E

Use a calculator if you need one to solve these problems.

1 Sandra has £87.50 in her purse. She buys two sweaters that cost £32.99 each.

How much money does she have left?

2 George travels from York to Oxford by car.

His odometer reads 123 456.8 km when he leaves York and 123 642.7 km when he arrives in Oxford.

How far did he travel?

3 If I have 5.67 litres of juice, how many cups containing $\frac{2}{15}$ of a litre can I pour?

4 Find 0.75 of 2400

5 Sheldon places fence posts 0.84 m apart all along his boundary fence.

If the total fence is 59.64 metres long, how many posts will there be?

6 June bought 6.65 litres of petrol at £1.29 per litre. How much did she pay?

7 Toni earns £28 650 per year.
  a How much is this per day? Remember there are 365.25 days in a year.
  b How much is this per five-day week? Give your answer correct to 2 decimal places.
  c If Toni is paid the amount in part **b** every week for 52 weeks of a year, will she earn more or less than the original total? Why?

Checklist of learning and understanding

Decimals and fractions
- You can express decimals as fractions by writing them with a denominator that is a power of ten and then simplifying them. $0.4 = \frac{4}{10} = \frac{2}{5}$
- You can change fractions to decimals by dividing the numerator by the denominator. $\frac{3}{4} = 3 \div 4 = 0.75$
- Changing ordinary fractions to decimals makes it easier to compare their sizes using place value.

Calculations with decimals
- Pen and paper methods are important in the non-calculator exam paper.
- You can use any method as long as you show your working.
- Decimals can be added and subtracted by lining up the places and the decimal points.

Find answers at: cambridge.org/ukschools/gcsemaths-studentbookanswers

- Decimals can be multiplied like whole numbers as long as you insert the decimal point so there are the same number of decimal places in the answer as there were altogether in the numbers being multiplied.
- Decimals can be divided by making the divisor a whole number (multiply both numbers by a power of ten to do this). Then divide normally and insert the decimal point in the answer when you cross the decimal point in the number being divided.

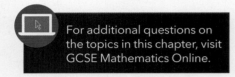

Chapter review

1 Arrange each set of numbers in ascending order.

a 4.2, 4.8, 4.22, 4.97, 4.08

b 2.96, 2.955, $2\frac{46}{50}$, $2\frac{9}{25}$, 2.12

c $\frac{3}{4}$, 0.86, $\frac{4}{5}$, 0.78, $\frac{5}{6}$, 0.91

2 Write these numbers in order of size.

Start with the smallest number. *(2 marks)*

$\frac{1}{4}$ 0.2   40% $\frac{3}{4}$ 0.5

©*Pearson Education Ltd 2013*

3 Convert to decimals and insert <, = or > to compare the fractions.

a $\frac{3}{5} \square \frac{12}{20}$ b $\frac{5}{6} \square \frac{7}{11}$ c $\frac{2}{9} \square \frac{1}{7}$

4 Write each as a fraction in its simplest terms.

a 0.88   b 2.75   c 0.008

5 a Increase $\frac{2}{5}$ by 2.75   b Reduce 91.07 by $\frac{1}{2}$ of 42.8

c Divide 4 by 0.125   d Multiply 0.4 by 0.8

6 a Add 4.726 and 3.09   b Subtract 2.916 from 4.008

c Multiply 8.76 by 100   d Divide 18.07 by 1000

e Multiply 4.12 by 0.7   f Simplify $\frac{32.64}{2.4}$

7 Jarryd and Kate have £16 each. Jarryd spends 0.415 of his money and Kate spends $\frac{7}{20}$ of hers.

Who has more money left? How much more?

8 Powers and roots

In this chapter you will learn how to ...
- use positive and negative powers to represent numbers in index notation.
- calculate with powers and roots.
- apply the rules for multiplying and dividing indices.

 For more resources relating to this chapter, visit GCSE Mathematics Online.

Using mathematics: real-life applications

Interior designers use square units to work out the area of floors to be tiled and walls to be painted. They then work out how much paint to buy and use the size of tiles (also in square units) to work out how many are needed.

 Calculator tip

Make sure you know which buttons to use to evaluate different powers and find different roots of numbers.

"I'm pretty good at estimating. I can usually look at a room and guess the area of the floor and walls quite accurately. Tiles are harder, I do rough sketches on squared paper to help me work out how many tiles of a particular size are needed to cover a floor area." *(Interior designer)*

Before you start ...

Ch 1	You should be able to quickly add and subtract pairs of integers mentally.	**1** Choose the correct sign: $<$, $=$ or $>$. **a** $-3 + -3 \square 4 + 2$ **b** $6 - 7 \square 3 - 4$ **c** $4 - (-5) \square -3 + -6$ **d** $-2 + 6 \square 9 - 5$

Ch 4	You need to be able to find the squares, cubes, square roots and cube roots of numbers.	**2** Choose the correct answer.
		 a The area of a square with sides of 3 cm.
		      A $6\,cm^2$ **B** $9\,cm^2$ **C** $12\,cm^2$
		 b $\sqrt[3]{27}$
		      A 9 **B** 5.2 **C** 3
		 c $\sqrt{810000}$
		      A 9 **B** 90 **C** 900

Ch 6	You need to be able to find the reciprocal of a number or fraction.	**3** Find the reciprocal of each number. Choose from the values in the box. **a** $\frac{3}{4}$ **b** 12 **c** $1\frac{2}{5}$ $\boxed{\frac{12}{1} \quad \frac{4}{3} \quad \frac{1}{12} \quad \frac{7}{5} \quad \frac{5}{7}}$

Find answers at: cambridge.org/ukschools/gcsemaths-studentbookanswers

Assess your starting point using the Launchpad

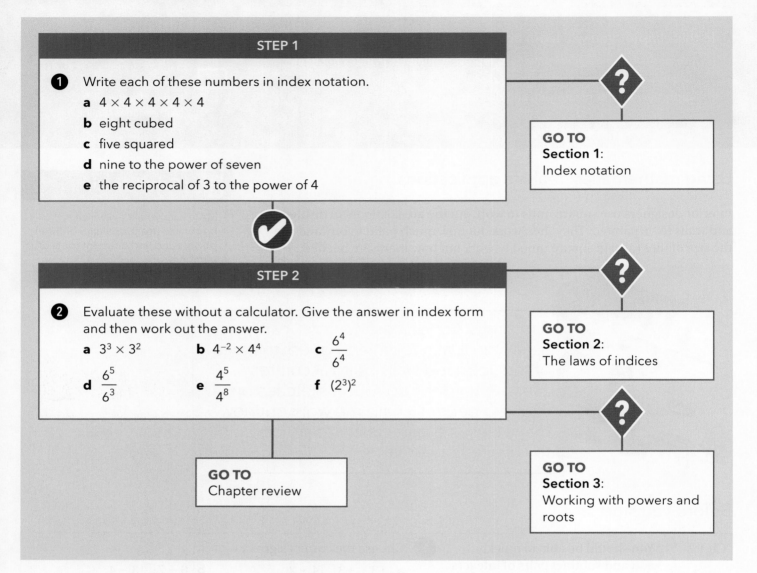

Section 1: Index notation

You can use powers to write repeated multiplications in a shorter form.

For example: $6 \times 6 = 6^2$ and $5 \times 5 \times 5 = 5^3$

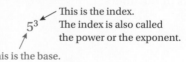

The **index** tells you how many times the base number is used in a multiplication.

When you write a number using an index you are using **index notation**.

7^4 is in index notation.

When you write the multiplication out in full you are using expanded form.

$7 \times 7 \times 7 \times 7$ is in expanded form.

The plural of index is indices.

Key vocabulary

index: a power or exponent indicating how many times a base number is used in a multiplication.

index notation: writing a number as a base and index, for example 2^3.

WORKED EXAMPLE 1

Simplify:

a $3^4 - 2^4$ **b** $3^2 \times 3^3$ **c** $3^5 \div 3^2$

a $3^4 - 2^4$

> You cannot subtract these in index form because the bases are not the same.

$= 3 \times 3 \times 3 \times 3 - 2 \times 2 \times 2 \times 2$

> Write each term in expanded form.

$= 9 \times 9 - 4 \times 4$
$= 81 - 16$
$= 65$

> Multiply mentally. You can do this in pairs.

b $3^2 \times 3^3$
$= 3 \times 3 \times 3 \times 3 \times 3$
$= 9 \times 9 \times 3$
$= 81 \times 3$
$= 243$

> Write each part in expanded notation.

c $3^5 \div 3^2$
$= \dfrac{3 \times 3 \times 3 \times 3 \times 3}{3 \times 3}$
$= 9 \times 3$
$= 27$

Tip
Any number to the power of 1 stays the same number so you don't usually write powers of 1.

Tip
In Section 2 you will learn how to use the laws of indices to simplify expressions in index form without having to expand them.

EXERCISE 8A

1 Write each of the following in index notation. You do not need to work out the value.

- **a** $4 \times 4 \times 4$
- **b** $3 \times 3 \times 3 \times 3 \times 3 \times 3$
- **c** $7 \times 7 \times 7 \times 7$
- **d** $9 \times 9 \times 9$
- **e** $5 \times 5 \times 5 \times 5 \times 5$
- **f** $12 \times 12 \times 12$
- **g** $18 \times 18 \times 18 \times 18 \times 18 \times 18 \times 18$
- **h** $11 \times 11 \times 11 \times 11 \times 11 \times 11 \times 11 \times 11 \times 11$
- **i** 19 to the power of 8
- **j** 23 to the power of 6
- **k** 11 multiplied by itself 14 times
- **l** 9 multiplied by itself 8 times

2 Write in expanded form. Don't work out the answers.

- **a** 3^4
- **b** 9^3
- **c** 4^5
- **d** 8^3
- **e** 5^6
- **f** 3^8
- **g** 23^5
- **h** 51^4
- **i** 72^5
- **j** 203^3
- **k** 121^4
- **l** 100^5

Find answers at: cambridge.org/ukschools/gcsemaths-studentbookanswers

3 Evaluate each expression without using a calculator.

a 2^3　　b 6^2　　c 1^8
d 8^3　　e 10^4　　f 10^6
g $2^3 - 1^5$　　h $1^6 + 7^2$　　i $2^4 \times 2^2$
j $2^4 + 4^2$　　k $2^3 \times 2^4$　　l $3^3 \times 3^3$
m $2^4 \div 2^3$　　n $4^5 \div 4^3$　　o $7^2 \times 10^3$
p 7×10^6　　q $2 \times 10^2 + 3 \times 10^3$　　r $6^2 \times 10^6$

Index notation on your calculator

Most calculators have one key to square a number: $\boxed{x^2}$

Your calculator is also likely to have a key that allows you to enter any other powers quickly and easily. It may be $\boxed{y^x}$ or $\boxed{x^y}$ or $\boxed{a^b}$.

To enter 13^4, you press: $\boxed{1}\,\boxed{3}\,\boxed{y^x}\,\boxed{4}\,\boxed{=}$

You will get a result of 28 561.

EXERCISE 8B

1 Use your calculator to evaluate the following.

a 4^6　　b 12^3　　c 8^5　　d 7^4
e 15^3　　f 10^4　　g 28^2　　h 25^3

2 Use a calculator to find the value of each expression.

a $12^3 - 2^8$　　b $20^4 - 15^2$　　c $15^3 \times 15^2$
d $3^{12} + 3^4$　　e $3^6 + 2^8$　　f $35^3 \div 5^3$

3 Fill in < or > to make each statement true.

a $4^6 \square 6^4$　　b $10^3 \square 3^{10}$　　c $4^9 \square 9^4$
d $15^2 \square 2^{15}$　　e $9^8 \square 8^9$　　f $2^{10} \square 10^2$

Zero and negative indices

Look at this table of powers of 10.

Index notation	Expanded form	Value
10^6	$10 \times 10 \times 10 \times 10 \times 10 \times 10$	1 000 000
10^5	$10 \times 10 \times 10 \times 10 \times 10$	100 000
10^4	$10 \times 10 \times 10 \times 10$	10 000
10^3	$10 \times 10 \times 10$	1 000
10^2	10×10	100
10^1	10	10

In the table each value is $\frac{1}{10}$ of the value above it. (In other words $10^6 \div 10 = 10^5$).

If you continue dividing by 10 you get this pattern for smaller and smaller indices:

Index notation	Expanded form	Value
10^0	$10 \div 10 = 1$	1
10^{-1}	$1 \div 10 = \frac{1}{10}$	$\frac{1}{10}$
10^{-2}	$\frac{1}{10} \div 10 = \frac{1}{100}$	$\frac{1}{100}$
10^{-3}	$\frac{1}{100} \div 10 =$	$\frac{1}{1000}$
10^{-4}	$\frac{1}{1000} \div 10$	$\frac{1}{10000}$

> **Tip**
>
> Remember that we use the reciprocals to change fraction divisions into multiplications.
>
> $\frac{1}{10} \div 10 = \frac{1}{10} \times \frac{1}{10} = \frac{1}{100}$

The pattern in the table shows two very useful features of indices.

> Any number with an index of 0 is equal to 1
> $a^0 = 1$ (except for 0^0 which is undefined).
> So, for example, $5^0 = 1$ and $7^0 = 1$

> Any number with a negative index is equal to its reciprocal with a positive index: $a^{-m} = \frac{1}{a^m}$
>
> So, for example, $4^{-2} = \frac{1}{4^2}$ and $5^{-3} = \frac{1}{5^3}$

EXERCISE 8C

1 Write each of the following using positive indices only.

   a 2^{-1} **b** 3^{-1} **c** 4^{-1}
   d 3^{-2} **e** 4^{-3} **f** 3^{-5}
   g 3^{-4} **h** 6^{-6} **i** 34^{-5}

2 Express the following with negative indices.

   a $\frac{1}{3}$ **b** $\frac{1}{5}$ **c** $\frac{1}{7}$ **d** $\frac{1}{3^2}$
   e $\frac{1}{4^5}$ **f** $\frac{1}{2^6}$ **g** $\frac{1}{7^2}$ **h** $\frac{1}{10^5}$
   i $\frac{1}{2^2}$ **j** $\frac{1}{12^3}$ **k** $\frac{1}{10^4}$ **l** $\frac{1}{3(2)^2}$

3 Fill in = or ≠ in each of these statements.

a $10^{-1} \square \frac{1}{10}$ b $6^0 \square 1$ c $6^{-1} \square \frac{1}{6}$

d $10^{-2} \square \frac{2}{10}$ e $6^{-3} \square \frac{1}{6^3}$ f $10^0 \square 1$

g $6^{-4} \square \frac{1}{6^4}$ h $\frac{1}{10^4} \square 10^{-4}$ i $\frac{1}{6^3} \square \frac{3}{6}$

Section 2: The laws of indices

It can take a long time to write an expression like $x^2 \times 2x^3 \times 5x^{11}$ in expanded form to simplify it.

The laws of indices are a set of rules that allow you to multiply and divide powers without writing them out in expanded form.

Law of indices for multiplication

Consider: $3^4 \times 3^2 = (3 \times 3 \times 3 \times 3) \times (3 \times 3) = 3^6$

Can you see a short-cut?

$3^4 \times 3^2 = 3^{4+2} = 3^6$

You get the same result by adding the indices.

To multiply two numbers in index notation you add the indices.

$a^m \times a^n = a^{m+n}$

Tip

Note that this only works if it is the **same base number**.
So, $2^3 \times 2^5 = 2^8$
but $4^3 \times 5^6$ cannot be simplified by this rule.

This law works for all indices, including negative indices.

For example $2^3 \times 2^{-2} = 2^{3+(-2)} = 2^1 = 2$

Law of indices for division

Consider: $2^5 \div 2^2 = \frac{2 \times 2 \times 2 \times 2 \times 2}{2 \times 2} = 2^3$

You should notice that:

$2^5 \div 2^2 = 2^{5-2} = 2^3$

You get the same result by subtracting the indices.

To divide two numbers in index notation you subtract the indices.

$a^m \div a^n = a^{m-n}$

Tip

$a^m \div a^n$ can be written as $\frac{a^m}{a^n}$

This law works for all indices, including negative indices.

$2^2 \div 2^4 = 2^{2-4} = 2^{-2}$

You can understand how this works by looking at the expanded notation:

$2^2 \div 2^4 = \frac{2 \times 2}{2 \times 2 \times 2 \times 2}$

If you cancel you get $\frac{1}{2 \times 2}$

$\frac{1}{2^2}$ is equal to 2^{-2}

Law of indices for powers of indices

$(3^2)^3$ means 3^2 all to the power of 3 which is $3^2 \times 3^2 \times 3^2$

$3^2 \times 3^2 \times 3^2 = 3^6$ (applying the law of indices for multiplication).

You should notice that:

$(3^2)^3 = 3^{(2 \times 3)} = 3^6$

To find the power of a power you multiply the indices.

$(a^m)^n = a^{mn}$

This law works for all indices, including negative indices.

$(4^3)^{-4} = 4^{(3) \times (-4)} = 4^{-12}$

> **Tip**
>
> The laws of indices also help to show that $a^0 = 1$
>
> $4^3 \div 4^3 = 4^{3-3} = 4^0$
>
> You already know that any number divided by itself is 1, so $4^3 \div 4^3 = 1$
>
> But this is also equal to 4^0, so 4^0 must equal 1.

EXERCISE 8D

1 Simplify. Leave the answers in index notation.

   a $2^4 \times 2^3$
   b $10^2 \times 10^5$
   c $4^3 \times 4^3$
   d 5×5^6
   e $2^4 \times 2^7$
   f $3^2 \times 3^{-4}$
   g $2^{-2} \times 2^5$
   h $3^0 \times 3^2$
   i $2 \times 2^3 \times 2^{-5}$
   j $3^2 \times 3^2 \times 3$
   k $10^2 \times 10^{-3} \times 10^2$
   l $10^0 \times 10^{-2} \times 10^2$

2 Simplify. Leave the answers in index notation.

   a $6^4 \div 6^2$
   b $10^5 \div 10^2$
   c $6^5 \div 6^3$
   d $6^3 \div 6^5$
   e $10^3 \div 10^5$
   f $3^{10} \div 3^0$
   g $3^8 \div 3$
   h $10^4 \div 10^4$
   i $\dfrac{5^4}{5^{-2}}$
   j $\dfrac{10^6}{10^{-4}}$
   k $\dfrac{3^{-2}}{3^{-3}}$
   l $\dfrac{2^0}{2^3}$

3 Simplify each expression. Give the answers in index notation.

   a $(2^2)^3$
   b $(2^3)^3$
   c $(2^4)^2$
   d $(10^2)^2$
   e $(10^2)^3$
   f $(10^4)^2$
   g $(2^4)^{-3}$
   h $(10^{-2})^2$
   i $(10^2)^{-3}$
   j $(3^4)^{-2}$
   k $(2^3)^0$
   l $(2^2 \times 2^3)^2$

4 Say whether each statement is true or false. If it is false, write the correct answer.

   a $3^3 \times 3^5 = 3^8$
   b $3^8 \div 3^2 = 3^4$
   c $10^8 \div 10^2 = 10^6$
   d $(3^3)^2 = 3^6$
   e $121^0 = 1$
   f $4^5 \times 4^2 = 4^7$
   g $3^{10} \div 3^2 = 3^5$
   h $(4^2)^4 = 4^8$
   i $(3^2)^0 = 1$

Section 3: Working with powers and roots

Builders, painters and decorators need to work out areas using square units (powers of 2 and square roots). Volume calculations use cube units (powers of 3 and cube roots). Bankers and accountants who do calculations involving growth rates or decay rates use different powers and roots, and many science formulae rely on being able to work with powers and roots.

Find answers at: cambridge.org/ukschools/gcsemaths-studentbookanswers

Powers of 2, 3, 4 and 5

It is useful to recognise the first few powers of 2, 3, 4 and 5. This can help you to work out their roots as well.

EXERCISE 8E

1. Your teacher will give you a copy of this table to complete.

Base \ Index	−3	−2	−1	0	1	2	3	4	5
2	$2^{-3} = \frac{1}{8}$	$2^{-2} =$	$2^{-1} = \frac{1}{2}$	$2^0 = 1$	$2^1 = 2$	$2^2 = 4$	$2^3 =$	$2^4 = 16$	$2^5 =$
3									
4									
5									

a Use a calculator to work out the missing values in the table.

b Compare the positive and negative values for the same index. What do you notice?

c Compare the powers of 2 with the powers of 4. What do you notice?

d How can you decide quickly that a whole number is *not* a power of 5?

Roots

You know that $5^2 = 25$ and so $\sqrt{25} = 5$

Also, $2^3 = 8$ and so $\sqrt[3]{8} = 2$

Finding the square root of a number involves working out what number multiplied by itself gives you the value under the root sign. For any other roots, the little number in front of the root sign tells you how many times the number has to be multiplied by itself.

Mathematically, finding the root of a number is the inverse of working out the powers of the number.

Look at the table in the exercise. You can see that $2^4 = 16$.

Then $\sqrt[4]{16} = 2$. You know this is correct because $2 \times 2 \times 2 \times 2 = 16$.

Calculator tip

You can find the square root or cube root of a number with your calculator using the 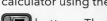 and buttons. The button allows you find any root (fourth root, fifth root, and so on).

EXERCISE 8F

1. Use the table that you completed in Exercise 8E to decide whether each statement is true or false.

a $2^4 = 4^2$
b $2^5 > 3^5$
c $2^0 = 5^0$
d $2^1 = 2^{-1}$
e $3^4 > 4^3$
f $4^4 < 3^4$
g $5^2 = 2^5$
h $3^5 > 5^3$
i $2^{-3} > 3^{-2}$
j $3^{-3} = \frac{1}{27}$
k $5^{-1} = 4^{-1}$
l $2(2^{-1}) = 1$

8 Powers and roots

2 Use your table to work these out without using a calculator.

a $\sqrt{25}$
b $\sqrt[3]{8}$
c $\sqrt[4]{256}$
d $\sqrt[3]{125}$
e $\sqrt[5]{243}$
f $\sqrt[3]{64}$
g $\sqrt[3]{8} + \sqrt[4]{625}$
h $\sqrt{2500}$
i $\sqrt[5]{32} + \sqrt[4]{81}$
j $\sqrt[3]{27\,000}$
k $\sqrt[4]{160\,000}$
l $\sqrt[5]{3125} \times \sqrt[4]{625}$

Solving problems involving powers and roots

Knowing that squaring or cubing numbers and finding their roots are inverse operations can help you to solve problems involving area and volume.

WORKED EXAMPLE 2

Mr Jones wants to pave a courtyard 4 m long and 4 m wide with square paving slabs. The label on the box of paving slabs says that each slab covers an area of 0.25 m².

a What are the dimensions (length and breadth) of the slabs?

b How many slabs will he need for his courtyard?

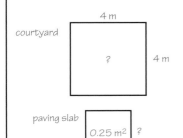

In problems like this it helps to make a sketch and to label it with the information you are given.

a Slab length = L; L^2 = 0.25 m²
Length of each side = $\sqrt{0.25}$ m
= 0.5 m

Once you've made sense of the problem, present your solution clearly.

b Courtyard area = 4 × 4 = 16 m²
Area of one paving slab = 0.25 m²
16 ÷ 0.25 = 64 paving slabs

EXERCISE 8G

1 Find the lengths of the sides of each of these square areas.

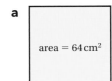

a area = 64 cm²
b area = 0.09 cm²
c area = 0.16 m²
d area = 1600 mm²

2 Jane has 900 square mosaic tiles. Is it possible to arrange them to make a perfect square?

Find answers at: cambridge.org/ukschools/gcsemaths-studentbookanswers

3. Sandy has a square piece of plastic with sides of 140 cm. Is this big enough to cover a square table with an area of 1.69 m²?

4. Mr Khan wants to tile a square floor of length 3.5 m. The square tiles he plans to use each have an area of 1024 cm².
   a. What is the area of the floor?
   b. What is the length of one side of each tile?
   c. How many tiles will he need to tile the floor? Show your working.
   d. Tiles come in boxes of eight. How many boxes are needed?
   e. Mr Khan knows, from experience, that he needs to buy one more tile for each full box in case of breakages. How many extra boxes of tiles should he buy?
   f. The tiles cost £23.50 per box. How much will it cost Mr Khan to buy the tiles he needs (including the extra).

5. Nisha and Sandy want to buy pizzas. The pizzas come in small or large size.

   The small pizza has a diameter of 16 cm and the large one has a diameter of 32 cm. Nisha says that two small pizzas will give the same amount of pizza as one large one. Sandy disagrees.

   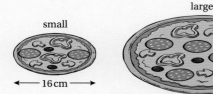

   a. Use the formula Area = $3.14 \times (\frac{1}{2} \times \text{diameter})^2$ to work out the area of each pizza correct to 2 decimal places. Is Nisha correct?
   b. What happens to the area of the pizza when you double the diameter?

Tip

Volume of a cube = $L \times L \times L$

6. Work out the length of the sides of each of these cubes.

   a 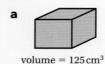 volume = 125 cm³

   b 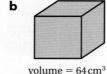 volume = 64 cm³

   c 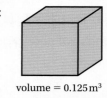 volume = 0.125 m³

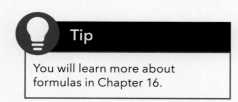

7. Nick uses small boxes like the one on the left to store electronic components. Each box is 4 cm long and 4 cm wide with a volume of 24 cm³. How deep are the boxes?

8. Shamila received an inheritance of £2500. She wants to invest it for ten years in an account that offers 5% growth but she wants to know how much money she will have after the ten-year period. Her auntie works in a bank and she tells Shamila that there is a formula to work this out quickly:

   value of future investment = original amount × $(1.05)^{10}$

   a. Work out how much money Shamila will have in 10 years.
   b. How much will she have if she decides to spend £500 and put the rest of the money into this investment?

Tip

You will learn more about formulas in Chapter 16.

9 Pam took a mortgage of £55 000 to buy a flat. The bank manager showed her this formula for working out how much she will repay over a 20-year period.

total amount paid = mortgage amount × $(1.04)^{20}$

a Work out how much her mortgage will cost if she takes 20 years to repay it.

b The power of 20 in the formula represents the number of years over which it is repaid. Work out what the total amount paid would be if Pam paid her mortgage off in 15 years.

c How much would she save by paying over the shorter period?

> **Tip**
>
> You will meet more problems like these last two in Chapter 32.

Checklist of learning and understanding

Index notation

- Numbers can be expressed as powers of their factors using index notation.
  - $2 \times 2 \times 2 \times 2$ can be written in index notation as 2^4.
  - The 4 is the index (also called the power or the exponent) and it tells us how many times the base (2) must be multiplied by itself.
- Any number to the power of 0 is equal to 1: $a^0 = 1$
- A negative index is the reciprocal of a positive index: $a^{-m} = \dfrac{1}{a^m}$

Laws of indices

- Multiplication: $a^m \times a^n = a^{m+n}$
- Division: $\dfrac{a^m}{a^n} = a^{m-n}$
- Raising to a power: $(a^m)^n = a^{mn}$

Working with powers and roots

- Finding the root of a number is the inverse of raising the number to a power.

  For example, $4^2 = 16$ and $\sqrt{16} = 4$
- Powers and roots are useful for solving problems related to area, volume and future value of investments.

> For additional questions on the topics in this chapter, visit GCSE Mathematics Online.

Chapter review

1 Write each number in index form.

a $8 \times 8 \times 8 \times 8 \times 8$ **b** three cubed

c nine squared **d** 14 to the power of five

2 Write each expression in expanded form and work out the answer.

a 4^4 **b** 9^3 **c** $(4^3 - 3^3) \times 13^2$

Find answers at: cambridge.org/ukschools/gcsemaths-studentbookanswers

3
   a Work out 3^4 *(1 mark)*

   b Write down the cube root of 64   *(1 mark)*

©*Pearson Education Ltd 2012*

4 Put each set of expressions in order from smallest to greatest.

   a $3^4, \sqrt{81}, 10^2, 4^3, 2 \times \sqrt{121}$ **b** $4^5, 5^4, 10^3, 96^0, 3^2, 20^2$

5 Write these numbers with positive indices.

   a 3^{-3} **b** 2^{-10} **c** 5^{-2}

6 Use the laws of indices to simplify each expression and write it as a single power of 4.

   a $4^2 \times 4^2$ **b** $4^6 \times 4^{-3}$ **c** $4^7 \div 4^3$

   d $4^3 \div 4^5$ **e** $(4^3)^2$ **f** $(4^{-2})^2$

7 Evaluate. Check your answers with a calculator.

   a $\sqrt{121}$ **b** $\sqrt{0.25}$ **c** $\sqrt[3]{125}$

   d $\sqrt[5]{32}$ **e** $\sqrt[4]{81}$ **f** $\sqrt{\dfrac{1}{4}}$

8 Find the length of each side of a cube of volume $0.027\,\text{m}^3$.

9 $\dfrac{V}{30} = \sqrt{h}$. Find V when $h = 25$

10 $P - y = x^2$. Find P when $x = 2$ and $y = 8$

11 Electricians use the formula $V = \sqrt{PR}$ to work out voltage (V) when P is the power in watts and R is the resistance in ohms. Calculate the voltage when $P = 2000$ and $R = 24.2$

9 Rounding, estimation and accuracy

In this chapter you will learn how to ...
- approximate values by rounding them to different degrees of accuracy.
- recognise the difference between rounding and truncating.
- use approximations to estimate and check the results of calculations.
- understand and apply limits of accuracy in numbers and measurements.

For more resources relating to this chapter, visit GCSE Mathematics Online.

Using mathematics: real-life applications

When you read that 34 000 people attended a festival, the actual number is likely to be slightly less or slightly more than that. When you roughly estimate what you spent over the weekend, look at an object and guess it is about $2\frac{1}{2}$ m long or say things like, "I live about 15 kilometres from school", you are estimating and using approximate values.

"I round off the prices to the nearest pound and keep a mental running total of the costs of things I put in my trolley so I know that I am not over-spending." *(Consumer)*

Before you start ...

Chs 1, 4	You should be able to use rounding to quickly estimate the answers to calculations.	**1** Use rounded values to estimate and decide whether each answer is correct without doing the calculation. **a** 312 − 56 = 256 **b** 479 × 17 = 3142 **c** 350 + 351 − 96 = 798
Ch 7	You should be able to calculate with decimals and estimate to decide whether an answer is reasonable.	**2** Say whether each statement is true or false. **a** $5.8 \times 6.72 \approx 42$ **b** $3.789 + 234.6 \approx 4 + 230$ **c** $0.00432 + 3.55 \approx 4$ **d** $4 \times \pi \approx 12$
Ch 7	You need to be able to work confidently with decimals and place value.	**3** Write the number halfway between: **a** 3.0 and 5.0 **b** 3.5 and 3.6 **c** 0.02 and 0.07

Find answers at: cambridge.org/ukschools/gcsemaths-studentbookanswers

Assess your starting point using the Launchpad

STEP 1

1 Round each value to the degree of accuracy specified.
  a 86 to the nearest 10.
  b 1565 to the nearest 1000.
  c 134.1234 to 2 decimal places.
  d 19.999 to 1 decimal place.
  e 1235.26 to 1 significant figure.
  f 234 650 034 to 3 significant figures.

2 The length of a metal component is found to be 0.937 cm. What is its length to the nearest millimetre?

3 The cost of an international call on an itemised phone bill is given as £5.159 32.
  a What is this amount truncated to the nearest penny?
  b What is this amount rounded to 2 decimal places?

GO TO Section 1: Approximate values

STEP 2

4 Estimate the cost of 12 packets of seeds at £1.36 each.

5 Approximately how many litres of petrol can you get for £20 if each litre costs £1.89?

6 Use rounding to find an approximate answer to each calculation.

  a $\dfrac{784 + 572}{109}$ b $(2.099)^2$ c $\dfrac{3.803 + 7.52}{3.29}$

7 A piece of wire is 10 m long, correct to the nearest metre. Complete the following statement to show the longest and shortest possible lengths that this wire could be.

$$\square \leqslant 10\ \text{m} < \square$$

GO TO Section 2: Approximation and estimation

GO TO Section 3: Limits of accuracy

GO TO Chapter review

Section 1: Approximate values

If someone asks you what you spend on mobile phone calls each week you are more likely to answer, "about £10" than to say, "exactly £10 and 24 pence". In the same way, the weather service will report that temperatures reached a record high of 37° rather than saying 36.895 79°.

Approximation allows you to give numbers in a more convenient form by writing them in a simpler, but less accurate way.

Rounded values

Rounding is used in calculations where a precise answer is not required because of the size of the numbers involved or the purpose for which you need the answer.

Numbers can be rounded to:

- the nearest whole number or place value (tens, hundreds or thousands)
- a particular number of decimal places
- a particular number of significant figures.

To round, or approximate, a number correct to a given place value, find the specified place value and look at the next digit to the right. If this is 5 or greater, you round up. If it is less than 5 you round down.

This rule applies to whole numbers and decimals whether you are rounding to given place values or to a given number of significant figures.

Rounding whole numbers

Consider the number 456 789. It can be rounded to different degrees of accuracy.

456 789 rounded to the nearest ten	456 790
456 789 rounded to the nearest hundred	456 800
456 789 rounded to the nearest thousand	457 000

In a test or exam you might be told to round numbers to a given **degree of accuracy**.

Key vocabulary

rounding: writing a number with zeros in the place of some digits.

Tip

You might need to write in some zeros so that the digits of your answer have the correct place value.

Key vocabulary

degree of accuracy: the number of places to which you round a number.

EXERCISE 9A

1 Choose the correct approximation of each animal's weight.

Weight of an animal			Rounded to	Identify the correct one	
a	cow	635 kg	nearest 10 kg	**A** 630 kg	**B** 640 kg
b	horse	526 kg	nearest 100 kg	**A** 500 kg	**B** 600 kg
c	sheep	96 kg	nearest 10 kg	**A** 90 kg	**B** 100 kg
d	dog	32 kg	nearest 10 kg	**A** 30 kg	**B** 40 kg
e	cat	5.2 kg	nearest 10 kg	**A** 0 kg	**B** 10 kg

Find answers at: cambridge.org/ukschools/gcsemaths-studentbookanswers

2
a Round each value to the nearest whole number.
   i 54.8   ii 10.6   iii 9.4   iv 12.3
b Round each value to the nearest 10.
   i 26   ii 57.5   iii 111.1   iv 35 814
c Round each value to the nearest 100.
   i 458   ii 5732   iii 2389   iv 35 814
d Round each value to the nearest 1000.
   i 2590   ii 176   iii 35 814   iv 66 876
e Round the following to the nearest hundred thousand.
   i 123 456   ii 1 234 567   iii 12 354 642   iv 123 456 789
f Round the following to the nearest million.
   i 545 000   ii 555 000   iii 14 354 642   iv 546 267 789

3
a A food bill is £27.60. How much is this to the nearest pound?
b There are 27 students in a class. What is this to the nearest ten students?
c I have £175 saved up. What is this to the nearest £100?
d You need 167 cm of material to make a kite. Approximately how many metres do you need?
e Sue says that the population of the United Kingdom is 63 793 234 which is 63.7 million to the nearest hundred thousand people. Is she correct?

Rounding decimals

In calculations you will often get answers with many more decimal places than you need. You will usually be told to give your answers to a specific number of decimal places.

> **Tip**
>
> The same rules for rounding whole numbers apply to decimals, but you also leave off any digits after the required number of decimal places.

WORKED EXAMPLE 1

Round:

a 54.149 to 1 decimal place   b 0.8751 to 2 decimal places
c 0.100 24 to 3 decimal places.

a 54.149
  54.1
> There is a 1 in the first decimal place. The next digit is 4, so round down.
>
> Leave the 1 unchanged and take off the digits to the right of it.

b 0.8751
  0.88
> There is a 7 in the second decimal place. The next digit is 5.
>
> Round 7 up to 8 and take off the digits to the right of it.

Continues on next page …

c 0.100 24
   0.100

> There is a 0 in the third decimal place. The next digit is 2.
>
> Leave the 0 unchanged and take off the digits to the right of it.

In part **c**, the answer 0.100 is the same as 0.1, but you write the two zeros to show that the number is rounded to 3 decimal places.

Suitable levels of accuracy

Some questions will ask you to round answers to a suitable degree of accuracy. You will have to decide what to round to. This will depend on the degree of accuracy required by the situation or problem.

If the question involves quantities that can only be whole numbers, you round to the nearest whole number. It does not make sense to talk about 5.45 bricks or 9.15 tins of paint.

If the question involves money, you always round to two decimal places. An answer of £4.568 96 would be rounded to £4.57.

In mathematical or scientific calculations you usually work to a higher degree of accuracy than when you describe quantities in real life. If you are working with small values you might round them to tenths, if you are working with large numbers you might round to the nearest ten thousand or million.

When you calculate with decimals or significant figures, you generally round answers to no more than the number of places in the original values.

Tip

If you need 9.15 tins of paint, you can't round down to 9 tins. You need 10 tins. With only 9 tins, you might not have enough paint to finish the job.

EXERCISE 9B

1 Round each number to:

  i 1 decimal place **ii** 2 decimal places **iii** 3 decimal places.

  a 4.526 38   **b** 25.256 37   **c** 125.617 38
  d 0.537 921   **e** 32.3972

2 Write each value correct to 2 decimal places.

  a 19.869 03   **b** 302.0428   **c** 0.292
  d 0.205 28   **e** 21 245.8449   **f** 0.0039
  g 0.0972   **h** 0.999 999 9   **i** 99.997

3 Round each value to a suitable level of accuracy. Explain your decisions.

  a A large dog weighs 24.4872 kg.

  b To calculate a circumference I use the value
     π = 3.141 592 653 589 793 238 46…

  c Dan's car can travel 13.7895 km per 1.000 098 7 litres of petrol.

  d My share of a phone bill is £14.098 76

4 Bricks cost 83p each.

   a Ian wants 45 bricks, how much should this cost?

   b Ian pays £40 and tells the merchant to keep the change. How much has he paid per brick? (Give your answer to a suitable accuracy.)

5 A motorist travels 379 miles on 43 litres of petrol. How many miles per litre is this? (Give your answer to a suitable accuracy.)

Significant figures

In measurement and scientific calculations values are normally given to a level of accuracy that is certain to be correct. For example, a measure of 10.0 ml means the person measuring is certain that the measurement is 10 point something and not 9 point something or 11 point something. This measurement is given to three significant figures.

When you need to work with values with many digits or decimal places it is useful to **round to significant figures (s.f.)**.

The first **significant figure** in a number is the first non-zero digit when you read the number from left to right. All digits that follow are significant.

For example:

120 000 000   208.130   1.000 87   0.000 560 3

1st s.f.   1st s.f.   1st s.f.   1st s.f.

⑫0 000 000   ⑫08.130   ①.000 87   0.000 ⑤60 3

significant   significant significant significant
figures   figures   figures   figures

Tip

One significant figure does not mean that you will have only one digit in the answer. 12 756 is 10 000 correct to 1 significant figure, not '1' on its own.

Key vocabulary

significant figure (s.f.): a position in a number used to decide the level of accuracy. The first significant figure is the first non-zero digit when reading a number from the left.

round to significant figures: round to a specified level of accuracy from the first significant figure.

WORKED EXAMPLE 2

Write each figure correct to the given number of significant figures (s.f.).

a 308 000 000 (2 s.f.)   **b** 476.372 (4 s.f.)   **c** 2531.8 (2 s.f.)   **d** 0.004 36 (1 s.f.)

a

30|8000000
1st significant figure ↗ ↑ digit to the right of the specified place
2nd significant figure

> Read the number from left to right. Mark the first non-zero digit.

30|8000000 ≈ 310 000 000 (2 s.f.)
       |>5
1st s.f. round   round up   state the
          to 2 s.f.   level of
                                 accuracy

Include the zeros. 31 is not the same as 310 000 000

b

476.3|72 ≈ 476.4 (4 s.f.)
      |>5
1st s.f. round   round up   state the
          to 4 s.f.   level of
                               accuracy

Ignore digits after the rounding point if they are decimals

> Count the required number of significant figures from there to find the rounding place.

Continues on next page …

c 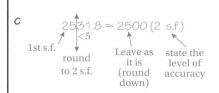 Replace digits before the decimal point with zeros. Ignore digits after the decimal point.

Look at the digit to the right of this. If it is 5 or more round up; if it is less than 5 leave the digit unchanged.

d 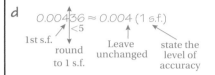 Ignore digits to the right.

Use 0 as a place holder to fill any gaps between the rounding place and the decimal point (if there is one). Leave off any digits past your rounding place if they are after the decimal point.

EXERCISE 9C

1 a Round each value to 1 significant figure.
   i 789   **ii** 3874   **iii** 69 356   **iv** 0.0456

b Write correct to 2 significant figures.
   i 789   **ii** 3145   **iii** 0.003 325   **iv** 0.000 749 9

c Express each number correct to 3 significant figures.
   i 789   **ii** 46 712   **iii** 0.004 214   **iv** 753 413

d Round each value to 2 significant figures.
   i 37.673   **ii** −4127   **iii** 3.0392   **iv** 1 999 000

e Write correct to 3 significant figures.
   i 37.673   **ii** −4127   **iii** 3.0392   **iv** 1 999 000

2 Explain why it is more useful to round a value such as 0.000 134 567 to 2 significant figures than to 2 decimal places when you need to work with it.

3 a $\pi \approx 3.141\,592\,6$. What is this to 3 significant figures?

b The density of a gas is $1.234\,\text{kg/m}^3$. What is this to 2 significant figures?

c The speed of light is 299 792 458 m/s. What is this to 2 significant figures?

d The strength of gravity at the Earth's surface is $9.806\,65\,\text{m/s}^2$. What is this to 3 significant figures?

Tip

Rounding numbers to a given number of decimal places means that you start the rounding at the decimal point.

Rounding numbers to significant figures means you start at the first significant figure, which can be before or after the decimal point.

Truncation

Truncated means 'cut off'.

A decimal can be truncated by cutting off all the digits past a given point without rounding.

Key vocabulary

truncation: cutting off all digits after a certain point without rounding.

Tip

When you calculate with rounded or truncated values your results will not be completely accurate. You will learn more about this in *Section 3*.

Did you know?

Truncation is used in statistics. A truncated mean is an average worked out by discarding (cutting off) very high or very low values in the data.

Key vocabulary

estimate: an approximate answer or rough calculation.

You will not generally use **truncation** to approximate values for calculation. However, you need to be aware that some calculator displays give a truncated rather than rounded value.

You can see this if you enter 2 ÷ 3. The display might show 0.6666666666

You know that $\frac{2}{3}$ can be expressed as the recurring decimal $0.\dot{6}$. So the calculator is showing a truncated value.

If you enter $\frac{20}{3}$ your screen will either show 6.6666666666 or 6.6666666667.

The first value is truncated, the second is rounded to 10 decimal places.

EXERCISE 9D

1 Truncate each number after the second decimal place.
  a 37.673   **b** −4.1275   **c** 3.0392   **d** 0.997

2 Truncate each number after the third significant figure.
  a 4.52638   **b** 25.25637   **c** 125.61738
  d 0.537921   **e** 32.397   **f** 200.6127

3 Consider splitting a bill of £20 equally between three people. What would each person pay? What method of approximation is most useful for deciding?

Section 2: Approximation and estimation

Approximate values are useful for estimating or checking the results of calculations without using a calculator.

An **estimate** is a very useful tool for checking whether your answer is sensible. If your estimate and your actual answer are not similar, then you may have made a mistake in your calculation.

"I estimate measurements and prices to give customers a fairly accurate quote telling them what a building job will cost." *(Builder)*

For estimating an approximate answer you generally round the original figures to 1 significant figure.

9 Rounding, estimation and accuracy

WORKED EXAMPLE 3

Estimate the value of $\dfrac{8.3 \times 536}{2.254 \times 9.612}$.

$$\dfrac{8.3 \times 536}{2.254 \times 9.612} \approx \dfrac{8 \times 500}{2 \times 10} = \dfrac{4000}{20} = 200$$

Start by rounding each number to 1 significant figure.

Tip

≈ means approximately equal to.
= means exactly equal to.

If you work out this problem using a calculator, the answer is actually 205.34 (to 2 decimal places).

Comparing this with your estimate tells you that your calculated answer is reasonable.

WORK IT OUT 9.1

A group of four friends are travelling to a festival. They intend to split the cost of everything between them.

The costs are: tent hire, £86.50; travel, 140 mile round trip with petrol costing roughly 20p per mile and camping entry tickets at £44 per night for three nights.

They decide to estimate how much they will each have to pay.

Which is the best estimate?

Why is it better than the others?

Estimate A	Estimate B	Estimate C
The cost of tent hire is roughly £88, which is £22 per person split between four.	The tent hire is roughly £100 which is £25 per person.	The tent hire is roughly £80, which is £20 each.
Petrol costs roughly £28 (140 × 0.2) which is £7 per person.	The petrol is roughly £150 × 0.2, which is £30, so split four ways is £7.50.	Petrol is about £100 × 0.2 = £20 so £5 per person.
The camping ticket costs are £44 × three nights, which is roughly £120 which is £30 per person.	The camping costs about £50 × 3 = £150, which is about £40 each.	Camping cost is about £40 × 3 = £120 for three nights, so about £30 each.
So total cost per person is £22 + £7 + £30 = £59.	Total cost is approximately £25 + £7.50 + £40 = £72.50.	Total cost is approximately £20 + £5 + £30 = £55.

EXERCISE 9E

1 Estimate the following by rounding each number to 1 significant figure.

a 111.11×3.6 b 378×1.07 c 0.99×16.7 d -13.6×0.48

e $\pi \times (5.3)^2$ f 4.8×12.5 g $\dfrac{192}{17.2}$ h $\dfrac{58.38}{0.5185}$

Tip

Always re-write the calculation with your rounded values before working out your estimate.

Find answers at: cambridge.org/ukschools/gcsemaths-studentbookanswers

2 Which calculation would provide the best estimate (A, B or C)?

  a 186×9.832 **A** 200×10 **B** 190×9 **C** 190×10

  b $15.76 \div 7.6$ **A** $15 \div 7$ **B** $16 \div 8$ **C** $16 \div 7$

3 Estimate the following by rounding each number to 1 significant figure.

  a $\dfrac{82.65 \times 0.4654}{42.4 \times 2.53}$ **b** $\dfrac{16.96 + 3.123}{16.96 - 6.432}$ **c** $\dfrac{879 \div 43.6}{2.36 \times 0.23}$ **d** $\dfrac{976.9 \div 492.9}{21.6 \div 43.87}$

4 Estimate the following.

  a $\sqrt{\dfrac{3.2 \times 4.05}{0.39 \times 0.29}}$ **b** $\sqrt{\dfrac{4.1 \times 11.9}{7.9 \times 0.25}}$

5 Shafiek runs a cross country race at an average speed of 6.25 m/s.

  a Estimate how far he will have run after 6 minutes.

  b Estimate how long it takes him to cover 1467 m. Assume he runs at a steady speed.

6 Look at the calculator display answers for each calculation. Use your estimation skills to say whether the answer is sensible or not without actually doing the calculation.

  a $3 \times \pi \times 5^2$ `125.6637061`   **b** 5×8.9 `445`

  c 50×8.9 `445`   **d** 3×192.5 `57.75`

  e $\dfrac{\sqrt{86}}{2.8 \times 16.18}$ `0.204697565`   **f** $0.0253 \div 0.45$ `56.222222222`

7 A rectangle has an area of 54.67 cm². The base of the rectangle is 7.9 cm long.

  a When a student tries to find the height on his calculator, he gets a result of 69 202 531. This is clearly wrong. Give the height correct to 2 significant figures.

  b Estimate the perimeter of the rectangle to the nearest cm.

Section 3: Limits of accuracy

All recorded measurements are given to a certain level of accuracy. Even with very accurate measuring instruments, quantities such as mass, length and capacity cannot be measured exactly.

When you are given a measurement you assume it is accurate except for the last digit. However, the rules of rounding mean the measurement has to fall within certain limits.

A piece of wood that is 47 cm to the nearest centimetre, could be anything from 46.5 cm up to, but not including, 47.5 cm long.

You can see this on the number line.

If the length was less than 46.5, it would have been rounded down to 46.

If it was 47.5, it would have been rounded up to 48.

If we let l represent the length of the piece of wood, the possible measurements can be expressed as $46.5 \leqslant l < 47.5$

This is called inequality notation and it means the length is greater than or equal to 46.5 and smaller than 47.5.

> **Tip**
>
> It is helpful to draw a number line to work out the largest and smallest possible values.

WORKED EXAMPLE 4

Use inequality notation to write down the possible values for each.

a 10 cm correct to the nearest cm.

b 22.5 kg to 1 decimal place.

c 128 000 correct to 3 significant figures.

a Let the length be x.
$9.5 \leqslant x < 10.5$

b Let the mass be m.
$22.45 \leqslant m < 22.55$

c Let the value be x.
$127\,500 \leqslant x < 128\,500$

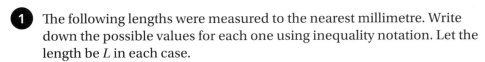

> **Tip**
>
> $\leqslant$ means less than or equal to
> $\geqslant$ means greater than or equal to
> $<$ means less than
> $>$ means greater than

For any measurement correct to a given level of accuracy, the exact values lie in a range half a unit below and half a unit above the measurement.

EXERCISE 9F

1 The following lengths were measured to the nearest millimetre. Write down the possible values for each one using inequality notation. Let the length be L in each case.

    a 4.9 cm   **b** 12.520 m   **c** 43.0 cm   **d** 29 mm

2 **a** There are 36 litres of petrol in a car's tank, to the nearest litre. What is the least and greatest possible volume of petrol in the tank?

    b A length of wood is 1.40 m to the nearest cm. Is it possible for the wood to be 137 cm long?

    c The weight of a stone is 43.4 kg to the nearest tenth of a kg. What is the least and greatest weight it could be?

3 Verna is a jeweller. She buys 9 carat gold for £10.66 per gram and platinum for £33.46 per gram.

a Use this information to complete the table.

Mass of a piece of jewellery	Maximum value of gold (to nearest pence)	Maximum value of platinum (to nearest pence)
18 grams to nearest gram		
18 grams to nearest 0.1 gram		
18 grams to nearest 0.01 gram		

b Use your data to explain why jewellers tend to use scales that are accurate to a hundredth of a gram or more to weigh the metal they use to make jewellery.

4 Graham has counted the marbles in a bag and says there are 100, to the nearest 10. What is the least and most marbles there could be in the bag?

5 The number of students at a school is 450 to the nearest 10. What is the most that this could be?

Checklist of learning and understanding

Approximate values

- When you round to a specified place, if the number following is 5 or above, then the original number goes up to the next number. If it is not more than 5, then the number stays the same.
- Rounding to a given number of significant figures specifies the number of digits, starting from the first non-zero digit, that are used to express a number.
- Truncating a decimal number, means removing all digits after a specified number of decimal places and expressing the number that remains without rounding.

Estimation

- Complex calculations can be estimated without using a calculator by using approximations of each term in the calculation to make a simple calculation.
- Visual estimation techniques can also be used to approximate sizes and measurements.

Level of accuracy

- Measurement is really approximation within a range of values.
  A measurement can be expressed using inequality notation ($\leqslant$, $<$).

Chapter review

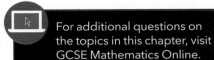
For additional questions on the topics in this chapter, visit GCSE Mathematics Online.

1 Are the following true or false?

	Original number	Approximation	Answer
a	123.456	rounded to 2 decimal places	123.456
b	123.456	rounded to 1 significant figure	120.000
c	123.456	rounded to 1 decimal place	123.5
d	123.456	rounded to 4 significant figures	123.5
e	123 456.789	truncated to 1 decimal place	123 456.8

2 Elizabeth says she is 24 years old to the nearest year. What is the youngest age she could be and the oldest age she could be?

3 The population of a town is given as 425 000 to the nearest 1000. What is the largest the population could be?

4 The weight of the large box is 357 grams, to the nearest gram.
  a What is the minimum possible weight of the box?   *(1 mark)*
  b What is the maximum possible weight of the box?   *(1 mark)*

©*Pearson Education Ltd 2011*

5 Given that a camping field measures 45 m by 30 m to the nearest metre, what is the maximum perimeter of the field needed for fencing requirements?

6 If there are 140 people at a party, to the nearest 10 people, what is the maximum number of people that could be there?

7 If a length of rope has been rounded to the nearest centimetre and it measures 90 cm what is the possible range of values for its length?

Find answers at: cambridge.org/ukschools/gcsemaths-studentbookanswers

10 Mensuration

In this chapter you will learn how to ...

- work with and convert standard units of measurement.
- use and convert compound units of measurement.
- work with map scales and bearings.
- construct and use scale diagrams to solve problems.

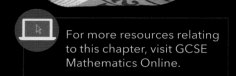

For more resources relating to this chapter, visit GCSE Mathematics Online.

Using mathematics: real-life applications

Measurement has practical applications in many jobs, and also in everyday activities. Being able to read and work with measurements is important when you make or alter clothes, work out what materials you need to build things, and weigh ingredients to make a recipe.

"I use accurate measurements to work out the scale when I draw maps. The people who use maps need to understand the scale so that they can make sense of map distances." *(Cartographer)*

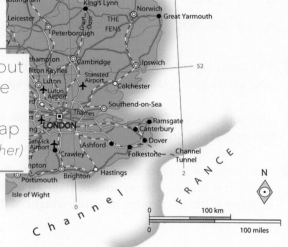

Before you start ...

Ch 7	You must be able to multiply and divide using multiples of 10.	**1** Work out. **a** 1000×10 **b** $340 \div 100$ **c** $16 \div 1000$	
Ch 5	You should be able to substitute numbers into a simple formula.	**2** Use the formula to work out the pay of each person: pay = hours worked × rate of pay **a** Amelia: 20 hours worked at £7 per hour. **b** Billy: 15 hours worked at £6 per hour. **c** Catrin: 40 hours worked at £5.50 per hour.	
KS3	You should be able to solve problems involving direct proportion.	**3** Six pencils cost 90p. Work out the cost of: **a** 12 pencils **b** 1 pencil **c** 4 pencils.	

Assess your starting point using the Launchpad

STEP 1

1 Convert.
  a 11 569 grams into kilograms
  b $4\frac{1}{2}$ hours into seconds
  c 123 456 pence into pounds (£)
  d 5 cm² into m²

GO TO Section 1: Standard units of measurement

STEP 2

2 A car travels 16 kilometres in 20 minutes.
  a What is the average speed of the car in kilometres per hour (km/h)?
  b Express this speed in metres per second (m/s).

3 An object is travelling at a speed of 25 m/s. Express this as a speed in km/h.

GO TO Section 2: Compound units of measurement

STEP 3

4 A helicopter is drawn using a scale of 1 : 100.
On the scale drawing the length of the helicopter blade is 8 cm.
How long is the actual blade?

5 A helicopter takes off from a point X and flies due north for 30 km to reach point Y.
It then flies on a bearing of 150° for 15 km to reach point Z.
  a Use a scale of 1 cm to represent 10 km to make a scale drawing showing this journey.
  b Use your diagram to find the bearing from X to Z.
  c By measuring your diagram, find the actual distance directly between X and Z.

GO TO Section 3: Maps, scale drawings and bearings

GO TO Chapter review

Find answers at: cambridge.org/ukschools/gcsemaths-studentbookanswers

Section 1: Standard units of measurement

In the metric system, units of measurement are divided into sub-units with prefixes such as milli-, centi-, deci-, deca-, hecto- and kilo-.

The same prefixes are used for length, mass and capacity.

Each sub-unit is 10 times bigger than the one before it. Centimetres are 10 times bigger than millimetres, decimetres are 10 times bigger than centimetres and so on.

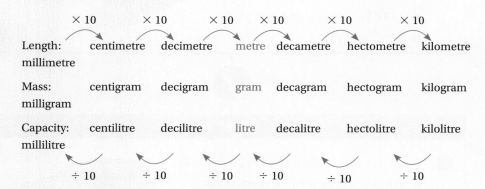

Tip

When you convert from a smaller to a larger unit there will be fewer of the larger units, so you divide by a power of 10.

When you convert from a larger to a smaller unit there will be more of the smaller units, so you multiply by a power of 10.

Converting between units

To convert between units in the metric system you need to multiply or divide by powers of 10.

This diagram shows how to convert between centimetres and metres.

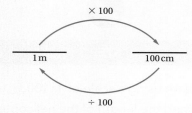

Tip

Not all the prefixes are in common use.

For example, length is usually only measured using mm, cm, m and km.

The **conversion factor** is 100 because you are changing across two sub-units.

Each sub-unit is 10 times greater or smaller than the one next to it, so here you have to multiply or divide by $10^2 = 100$.

You will use the following conversions often, so it is useful to try to remember them.

1 centimetre (cm) = 10 millimetres (mm)

1 metre (m) = 100 centimetres (cm)

1 kilometre (km) = 1000 metres (m)

1 kilogram (kg) = 1000 grams (g)

1 tonne (t) = 1000 kilograms (kg)

1 litre (l) = 1000 millilitres (ml)

1 litre (l) = 1000 cubic centimetres (cm^3)

1 cubic centimetre (cm^3) = 1 millilitre (ml)

Key vocabulary

conversion factor: the number that you multiply or divide by to convert one measure into another smaller or larger unit.

Tip

In the UK, the imperial unit the mile is used for measuring long distances.
1 mile ≈ 1.6 km

WORK IT OUT 10.1

In a sponsored swim the total number of lengths swum is 94. Each length is 25 metres.

How many kilometres were swum in total?

Which of the answers below is correct?

Option A	Option B	Option C
Total number of metres $= 25 \times 94 = 2350$ metres	Total number of metres $= 25 \times 94 = 2350$ metres	Total number of metres $= 25 \times 94 = 2350$ metres
Conversion: 100 metres = 1 km	Conversion: 100 metres = 1 km	Conversion: 1000 metres = 1 km
$2350 \div 100 = 23.5$ km swum in total.	$2350 \times 100 = 235\,000$ km swum in total.	$2350 \div 1000 = 2.35$ km swum in total.

Converting areas and volumes

Area is measured in square units such as mm^2 (square millimetres), cm^2, m^2 or km^2 so any conversion factor also has to be squared.

For example, to convert cm to mm you would multiply by 10, but to convert cm^2 to mm^2 you would need to multiply by 10^2 which is 100.

The two squares on the right have the same area.

The conversion factor from m^2 to cm^2, and vice versa, is 10 000.

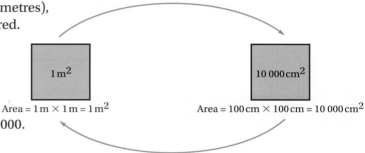

WORKED EXAMPLE 1

Convert each measure to the units given.

a $10\,m^2$ to cm^2 **b** $8.6\,km^2$ to m^2 **c** $3500\,mm^2$ to cm^2

a $10\,m^2$ to cm^2
$= 10 \times 10\,000 = 100\,000\,cm^2$ ← Conversion factor = 10 000

b $8.6\,km^2$ to m^2
$= 8.6 \times 1\,000\,000$
$= 8\,600\,000\,m^2$ ← Conversion factor = 1 000 000

c $3500\,mm^2$ to cm^2
$= 3500 \div 100 = 35\,cm^2$ ← Conversion factor = 100

Tip

$1\,cm^2 = 100\,mm^2$
 (Conversion factor = 100)
$1\,m^2 = 10\,000\,cm^2$
 (Conversion factor = 10 000)
$1\,km^2 = 1\,000\,000\,m^2$
 (Conversion factor = 1 000 000)

Volume is measured in cubic units such as mm^3 (cubic millimetres), cm^3 or m^3 so any conversion factor also has to be cubed.

Again, to convert cm to mm you would multiply by 10, but to convert cm^3 to mm^3 you would need to multiply by 10^3 which is 1000.

The two cubes on the right have the same volume.

The conversion factor from m^3 to cm^3, and vice versa, is 1 000 000.

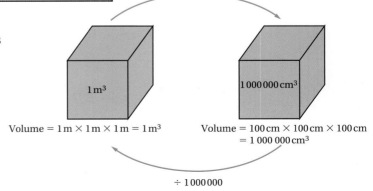

Find answers at: cambridge.org/ukschools/gcsemaths-studentbookanswers

Did you know?

The USA, Liberia and Myanmar are the only three countries in the world that do not officially use the metric system of measurement. All other countries have adopted the metric system.

 Tip

$1\,cm^3 = 1000\,mm^3$ (Conversion factor = 1000)
$1\,m^3 = 1\,000\,000\,cm^3$ (Conversion factor = 1 000 000)
$1\,litre = 1000\,cm^3$ (Conversion factor = 1000)

WORKED EXAMPLE 2

Convert each measurement to the unit given.

a $6.3\,m^3$ to cm^3 **b** $96\,500\,000\,cm^3$ to m^3 **c** $750\,cm^3$ to mm^3

a $6.3\,m^3$ to cm^3
$= 6.3 \times 1\,000\,000 = 6\,300\,000\,cm^3$ — Conversion factor = 1 000 000

b $96\,500\,000\,cm^3$ to m^3
$= 96\,500\,000 \div 1\,000\,000 = 96.5\,m^3$ — Conversion factor = 1 000 000

c $750\,cm^3$ to mm^3
$= 750 \times 1000 = 750\,000\,mm^3$ — Conversion factor = 1000

 Tip

You will use measurement conversions when you deal with scale drawings and maps in Section 3 and when you deal with perimeter, area and volume in Chapters 11, 12 and 17.

This is the plan of a room.

The real length of each section of wall is given but the units of measurement are different.

You can make mistakes if you try to calculate with measurements in different units.

Convert the measurements to the same unit before doing any calculations.

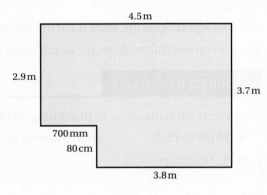

WORKED EXAMPLE 3

A builder is to put a wallpaper border around the ceiling of the room drawn above. How many metres of wallpaper border does the builder need?

$80\,cm \div 100 = 0.8\,m$
$700\,mm \div 1000 = 0.7\,m$ — The answer needs to be given in metres so it makes sense to work in metres.

$2.9 + 4.5 + 3.7 + 3.8 + 0.8 + 0.7 = 16.4\,m$ — Add the given lengths.

The builder needs 16.4 m of wallpaper border.

EXERCISE 10A

1 Match each measurement with one from the box.

    a 10 000 mm   **b** 10 000 ml   **c** 10 kg

    d 0.01 kg   **e** 0.1 cm

10 l	10 g	10 m	1 mm	10 000 g

2 Convert the following lengths and masses into the given units to complete the following.

    a 2.5 km = ☐ m   **b** 85 cm = ☐ mm   **c** 34 m = ☐ mm

    d 1.55 m = ☐ mm   **e** 0.07 m = ☐ cm   **f** 5.4 kg = ☐ g

    g 0.9 kg = ☐ g   **h** 102 g = ☐ kg   **i** 14.5 g = ☐ kg

3 Add the following capacities. Give your answers in the units indicated in brackets.

    a 3.5 l + 5 l (ml)   **b** 2.3 l + 450 ml (l)   **c** 20 l + 4.5 l + 652 ml (l)

Tip

Think units!

It is important to consider the units in which you are working. Always look back to the original question to determine the units. With a formula, whatever units you put in must match the units that come out.

4 Mandy wants to use concrete slabs to form a border round a rectangular garden. The garden is 480 cm wide and 7.2 metres long.

    a Draw a diagram to represent the garden.

    b Calculate the perimeter of the garden in metres.

    c If the concrete slabs are square with sides of 120 cm, how many will Mandy need to form the border?

    d The slabs cost £4.55 each. Work out how much it will cost Mandy to buy the slabs she needs.

    e What is the cost per metre for the concrete border?

5 Convert each of the following into the required units.

    a Total weight in kg of 3 bags of flour, each of mass 1200 g.

    b The length in cm of a 7.763 m long whale.

    c 3567 kg of lead into tonnes.

    d Area of 5 m^2 into mm^2.

    e 96.35 m^3 of sand into cm^3.

    f 345 cm^3 of water into litres.

6 An area of 250 000 cm^2 needs to be painted. A pot of paint can cover an area of 10 m^2. How many pots of paint are needed?

Time

Tip

1 year = 365 days
        (366 in a leap year)

1 day = 24 hours = 1440 minutes

1 hour = 60 minutes
      = 3600 seconds

1 minute = 60 seconds

The units of time we use on a daily basis are not decimal units.

To convert from weeks to days, you would need to multiply by 7 as there are 7 days in a week.

To convert from seconds to hours, you would need to divide by 60 to get minutes and then by 60 again to get hours. (You can also do this in one step by dividing by 3600.)

Find answers at: cambridge.org/ukschools/gcsemaths-studentbookanswers

Time is sometimes given in decimal form, for example, 4.8 hours.

You can convert these times back to ordinary units in different ways.

There are 60 minutes in an hour, so

4.8×60 minutes = 288 minutes = 4 hours 48 minutes.

Or, you can think of this as 4 hours and 0.8 hours.

$0.8 \times 60 = 48$, so the time is 4 hours and 48 minutes.

In athletics, and other timed sporting events, the times are often given using decimal fractions of a second.

For example, in August 2009, Usain Bolt ran the 100 m in the World Record time of 9.580 seconds.

In the metric system, 1 second = 1000 milliseconds.

9.580 is time recorded exact to $\frac{1}{1000}$ of a second.

This is 9 seconds and 580 milliseconds. You cannot convert it in any other way.

12-Hour and 24-hour times

The 12-hour time system uses am to show times from midnight to noon and pm for time from noon till midnight.

The 24-hour time system shows the times 00:00 to 23:59. Midnight is 00:00.

Calculator tip

Modern scientific calculators have a mode that you can use to do sexagesimal calculation (hours, minutes and seconds). Different models work in different ways, so read the manual or check online to see how your calculator works.

Tip

Add 12 to write a pm time using the 24-hour clock, for example
10.35pm + 12 = 22:35

Subtract 12 to write a time between 13:00 and 23:59 using the 12-hour clock, for example 15:40 − 12 = 3.40pm

Problem-solving framework

Mr Smith is in Moscow.

He needs to return to London for an urgent meeting in the morning.

The flight from Moscow to London takes 3.6 hours. The local time in Moscow is 3 hours ahead of the UK.

The flight is scheduled for take-off at 18:55 local time, and on arrival it will take 45 minutes to pass through customs and exit the airport.

Mr Smith will stop to buy a coffee, sandwich and newspaper for the train.

Trains for central London leave at 5, 27 and 46 minutes past the hour, and the journey will take 29 minutes.

There is a 7-minute walk from the train station to the meeting location.

What is the earliest time that Mr Smith can expect to arrive at his meeting? Give your answer using the 12-hour system of time.

Continues on next page …

Tip

When you work with time, treat hours and minutes separately. If you carry over from hours to minutes, remember you are carrying 60 minutes.

Steps for solving problems	What you would do for this example
Step 1: What have you got to do?	First work out the time of arrival in London and then work out how long it takes from there to the meeting location.
Step 2: What information do you need?	Flight departure time: 18:55 (local time) Time difference between London and Moscow: 3 hours Flight time: 3.6 hours Time to pass through customs: 45 minutes Train departure times: 5 past, 27 minutes past, 46 minutes past the hour Length of train journey: 29 minutes Walk time: 7 minutes
Step 3: What information don't you need?	Assume time to buy a coffee, sandwich and newspaper is negligible (as no time for this is provided).
Step 4: What maths can you do?	18:55 minus 3 hours = 15:55 (London time) Convert 3.6 from a decimal to time in hours and minutes: 3 hours and 0.6×60 = 3 hours 36 minutes Arrival time in London: 15:55 + 3 hours 36 minutes = 19:31 Add on time in customs: 19:31 + 45 minutes = 20:16 Next possible train is 20:27 Time at end of train journey: 20:27 + 29 minutes = 20:56 Arrival time at venue following walk: 20:56 + 7 minutes = 21:03 21:03 − 12 = 9.03 pm or three minutes past nine in the evening
Step 5: Have you used all the information? At this point you should check to make sure you have calculated what was asked of you.	Flight departure time ✓ Train departure times ✓ Time difference ✓ Length of train journey ✓ Flight time ✓ Walk time ✓ Time through customs ✓
Step 6: Is it correct?	Estimate to check $3\frac{1}{2}$ hours flying + 45 min at airport + 10 min wait + 30 min train + 7 min walk = about 5 hours Take off the time difference leaves 2 hours Leave 7 pm + 2 hours = 9 pm

Money

£1 = 100p so £x = 100x pence and x pence = £$\dfrac{x}{100}$

The rate at which one currency is converted to another is called an **exchange rate**. For example, £1 = €1.26 and €1 = £0.794 or 79.4p (at 2014 rates).

WORKED EXAMPLE 4

If the exchange rate for the pound to the euro is £1 = €1.26, convert:

a £400 to euros   b €400 to pounds.

a £400 = 400 × 1.26 = €504

> When the exchange rate is given as 1 unit of A = x units of B, you can convert A to B by multiplying by x.

b €400 = $\dfrac{400}{1.26}$ = £317.46

> Currency B can be converted to currency A by dividing by x.

Key vocabulary

exchange rate: the value of one currency used to convert that currency to an equivalent value in another currency.

Tip

Exchange rates can change over short periods of time due to economic and political factors. So, unlike conversions for distance, time and so on, you shouldn't memorise a particular exchange rate.

EXERCISE 10B

1. The starting pistol for a road race is fired at 12:15:30.
   The first runner crosses the finishing line at 14:07:22.
   What was the winning time for the race?

2. Sandra is exactly 16 years old. Calculate her age in:
   a weeks   b days   c hours   d seconds.

3. A boat leaves port at 14:35 and arrives at its destination $6\frac{1}{2}$ hours later.
   At what time does the boat arrive?

4. Ashwin and Luke plan to meet up in London. Ashwin's journey will take 2 hours and 27 minutes. Luke's journey will take 1.3 hours.
   What time must each set off if they are to meet at 11:15?

5. The table gives the value of the pound against four other currencies in July 2014.

British pound (£)	Euro (€)	US dollar ($)	Australian dollar (AS$)	Indian rupee (INR)
1	1.26	1.70	1.80	102.28

   a Calculate the value of each of the other currencies in pounds at this rate.
   b Convert £125 to US dollars.
   c How many Indian rupees would you get if you converted £45 at this rate?
   d Dilshaad has 8000 Indian rupees. What is this worth in pounds at this rate?

Section 2: Compound units of measurement

Compound measures involve more than one unit of measurement.

- Rates of pay (such as pounds per hour) involve units of money and time.
- Unit pricing (such as pence per gram) involves units of money and mass (or capacity or volume).

Compound measures usually express how many of the first unit correspond with 1 of the second unit. A rate of pay of £8.20/hour shows how many pounds you earn for 1 hour of work.

You can simplify compound measures by multiplying or dividing both units by the same factor.

WORKED EXAMPLE 5

40 litres of petrol cost £52.

a What is the cost in £/litre?

b Convert the cost in £/litre to pence per millilitre.

a $\dfrac{52}{40} = £1.30/\text{litre}$

The compound measure £/litre tells you that pounds are the first unit.

b Convert pounds to pence:
$£1.30 = 130\,p$
Convert litres to millilitres:
$1\,l = 1000\,ml$

You want a compound measure comparing pence and millilitres.

$130\,p$ per $1000\,ml$

Compare the two quantities

$130 \div 1000 = 0.13$
$1000 \div 1000 = 1$

You want a rate per **one** millilitre, so divide both quantities by 1000.

So $130\,p$ per $1000\,ml = 0.13\,p/\text{millilitre}$

Tip

A forward slash symbol / is often used instead of the word 'per'. So £8.30/hour means the same as £8.30 per hour.

Speed

Speed compares the distance travelled to the time taken. The units of speed depend on the situation.

A car or train's speed is often given in km/h or mph (kilometres or miles per hour).

An athlete's running speed might be given in m/s (metres per second).

You need to know the formula for calculating speed:

$$\text{Average speed} = \dfrac{\text{distance travelled}}{\text{time taken}}$$

Did you know?

Speed isn't just a measure of how fast something is travelling. Run rates per over in cricket and the number of words you can type per minute are both examples of speed.

Learn this formula

$$\text{Average speed} = \dfrac{\text{distance travelled}}{\text{time taken}}$$

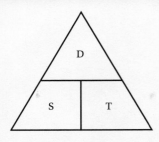

The speed is an average speed because a journey might involve faster and slower speeds over the given period. You can see the different speeds at different points in a car journey by looking at the speedometer. The speed shown on the speedometer is the speed at that particular time.

The triangle (left) shows the relationship between speed, distance and time.

Using the triangle:

- To find distance: cover D with your finger.
  The position of S next to T tells you to multiply speed by time.
  Distance = Speed × Time

- To find time: cover T with your finger. The position of D over S tells you to divide distance by speed. Time = Distance ÷ Speed

- To find speed: cover S with your finger. The position of D over T tells you to divide distance by time. Speed = Distance ÷ Time

The units of speed given in a problem usually let you know what units to use.

For example, if the problem talks about km/h, then express distances in kilometres and time in hours to calculate the speed.

Tip

Unless the unit is specified in the question, you can use any suitable unit you like. If you change the unit from the one in which the question is written, you must state the units you are using.

WORK IT OUT 10.2

A car travels 330 miles in $5\frac{1}{2}$ hours.

What is the average speed of the car in miles per hour (mph)?

Which of the answers below is correct?

Option A	Option B	Option C
Speed = Distance ÷ Time	Speed = Distance × Time	Speed = Distance − Time
D = 330 miles	D = 330 miles	D = 330 miles
T = $5\frac{1}{2}$ hours = 5.5 hours	T = $5\frac{1}{2}$ hours = 5.5 hours	T = $5\frac{1}{2}$ hours = 5.5 hours
S = 330 ÷ 5.5 = 60 mph	S = 330 × 5.5 = 1815 mph	S = 330 − 5.5 = 324.5 mph

To convert speeds from one set of units to another, you need to work systematically and take care with the units.

WORKED EXAMPLE 6

An athlete runs at an average speed of 10.16 m/s. Is this faster or slower than an average speed of 40 km/h?

Convert 10.16 m/s so that both speeds are in km/h.

× 60 to get metres per minute
  10.16 × 60 = 609.6 m/min   *Convert seconds to hours.*

× 60 to get metres per hour
  609.6 × 60 = 36 576 m/hour

÷ 1000 to get kilometres per hour   *Convert metres to kilometres.*
  $\frac{36576}{1000}$ = 36.576 km/h

So the athlete's speed is slower than 40 km per hour.

EXERCISE 10C

1 Joe is 18 years old and works for a minimum wage of £5.03 per hour.
If he works for 14 hours, how much will he earn?

2 Henry earns £8.75 per hour.
Sian earned £220.40 for working 22 hours.
How much more does Sian earn per hour?

3 A bricklayer lays 680 bricks in 4 hours.
How many does she lay per minute, to the nearest brick?

4 Bernie cycles 168 km in 8 hours. What is his average speed?

5 A car travels 528 km at an average speed of 88 km/h. Work out the time taken.

6 Usain Bolt set an Olympic Record over 100 m at the London Olympics in 2012 with a time of 9.63 seconds.

  a Express this speed in m/s.   **b** How fast is this in km/h?

Density and pressure

Density is the ratio between the mass and the volume of an object.

The triangle (right) helps you see the relationship between density, mass and volume.

  Mass = Density × Volume

  Volume = Mass ÷ Density

The units of density are grams per cubic centimetre (g/cm³) or kilograms per cubic metre (kg/m³).

Express the mass and the volume in the units given when you solve problems involving density.

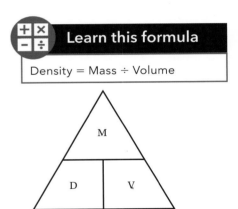

Learn this formula

Density = Mass ÷ Volume

WORKED EXAMPLE 7

A gold bar has a volume of 725 cm³ and a mass of 14.5 kg. What is the density of the gold bar in g/cm³?

M = 14.5 kg = 14.5 × 1000 = 14 500 g   Units of density are g/cm³ so mass must be in g.

Density = Mass ÷ Volume   Write the formula you are using.

Density = 14 500 ÷ 725
       = 20 g/cm³

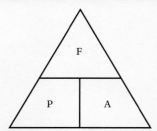

Pressure is defined by the formula:

Pressure = Force ÷ Area

The units given for force and area give you compound units for the pressure.

For example, if force is measured in Newtons and area in mm^2, the compound unit of pressure would be N/mm^2 (Newtons per mm^2).

WORKED EXAMPLE 8

A brick exerts a force of 5N.

Calculate the pressure exerted on the ground when the brick is in each of the following positions.

A On flat surface   **B** On short edge   **C** On long edge

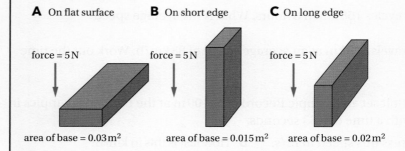

force = 5 N   force = 5 N   force = 5 N

area of base = $0.03 m^2$ area of base = $0.015 m^2$ area of base = $0.02 m^2$

A Pressure = $\dfrac{5 N}{0.03 m^2}$ = $167 N/m^2$

B Pressure = $\dfrac{5 N}{0.015 m^2}$ = $333 N/m^2$

C Pressure = $\dfrac{5 N}{0.02 m^2}$ = $250 N/m^2$

EXERCISE 10D

1 The mass of $1 cm^3$ of different substances is shown in the diagram.

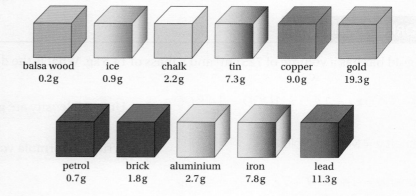

balsa wood 0.2 g   ice 0.9 g   chalk 2.2 g   tin 7.3 g   copper 9.0 g   gold 19.3 g

petrol 0.7 g   brick 1.8 g   aluminium 2.7 g   iron 7.8 g   lead 11.3 g

a Calculate the density of each substance in g/cm^3.

b Express each density in g/m^3 and then convert each to kg/m^3.

2 A cube of material with side length 30 mm has a mass of 0.0642 kg.
Calculate the density of the material in g/cm³.

3 Calculate the volume of a piece of wood with a mass of 0.1 kg and a density of 0.8 g/cm³.

4 Two metal blocks both exert a force of 18 N.
Block A is a cube with sides 100 cm long. Block B is a cuboid with a base of area 6 m² in contact with the floor.
Calculate the pressure exerted by each block in N/m².

5 A car exerts a force of 6000 N on the road.
Each of the four wheels has an area of 0.025 m² in contact with the road.
What pressure does the car exert on the road?

Section 3: Maps, scale drawings and bearings

A **scale drawing** is a diagram in which measurements are either reduced or enlarged by a **scale factor**.

The scale tells you how much the dimensions are reduced or enlarged.

Maps are scaled representations of areas of the real world.

The scale of a map describes the relationship between lengths in real life and lengths on the map.

The scale allows you to convert distances that you measure on the map to real-life distances.

Map scales are shown in different ways.

- **Bar or line scales.** The divided line or bar shows you distances on the map but the number on the bar tells you what the real distances are (usually in km).
- **Ratio scales.** You will often see scale given as a ratio, for example: 1 : 25 000. This means that one unit measured on the map is equivalent to 25 000 of the same units in real life. So, 1 cm on the map represents 25 000 cm = 0.25 km in real life.

Bar or line scales are useful for finding small distances. You measure the distance and then compare it with the line scale.

On the line scale below, each block is 1 cm long, and 1 cm on the map represents 1 km in real life.

To find a distance in real life:

- Measure the map distance using a piece of paper.
- Make pencil marks on the paper to record the distance.
- Compare your marked distance with the line scale.
- Read off the real distance.

> **Key vocabulary**
>
> **scale factor**: a number that scales a quantity up or down.

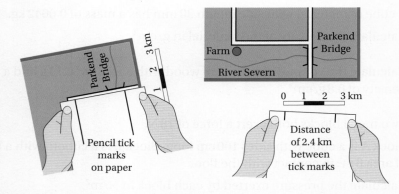

For bigger distances it is more efficient to use the ratio scale. You can convert any distance on a map to a real distance using the following formula.

Distance on the map × Scale = Distance on the ground

WORKED EXAMPLE 9

A scale drawing has a scale of 1 : 10.

a Calculate the real-life length when a length on the drawing is 5 cm.
b Calculate the length on the drawing when the real-life length is 30 cm.

a 5 cm on the drawing = 5 × 10 = 50 cm in real life
b 30 cm in real life = 30 ÷ 10 = 3 cm on the drawing.

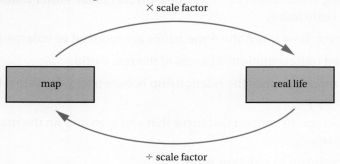

Take care with the units. When you multiply by the scale your answer will be in the same units that you used to measure on the map.

If the question asks for different units, you will need to convert the measurement to get the units you need.

WORK IT OUT 10.3

The distance between two towns on a 1 : 25 000 map is 3.4 cm.

How many kilometres apart are these towns in reality?

Which student has worked out the correct answer?

What have the other two done incorrectly?

Student A	Student B	Student C
$\frac{1}{25\,000} \times \frac{3.4}{1}$ $= 0.000\,136$ cm $= 1.36$ km	Map distance = 34 mm Scale = 1 : 25 000 Real distance 34 × 25 000 = 850 000 mm = 8.5 km	3.4 × 25 000 = 85 000 The distance is 85 000 cm ÷ 100 = 850 m ÷ 1000 = 0.85 km

EXERCISE 10E

1 Here are three line scales. Work out what distance is represented by 1 cm in each case.

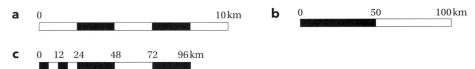

2 Work out the real distance (in kilometres) that a map distance of 45 mm would represent at each scale.

  a 1 : 120   **b** 1 : 1200   **c** 1 : 12 000   **d** : 120 000
  e 1 : 1 200 000   **f** 1 : 12 000 000   **g** 1 : 120 000 000

3 Andrew says that a map drawn to a scale of 1 : 15 000 is a larger scale map than one drawn to 1 : 150 000.

Is he correct? How do you know?

4 The map to the right shows three towns: Coltown, Bracwich and Nostley.

The map has a scale of 1 : 75 000.

Work out the actual distance directly from Coltown to Bracwich.

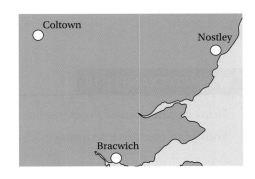

5 The red line on the map (below right) shows the flight path of a plane flying from Edinburgh to London. The flight took 55 minutes.

  a Calculate the distance flown in kilometres.

  b What was the plane's average speed on this flight?

6 A set of toy furniture is manufactured using a scale of 1 : 50.

Work out:

  a the height of a cupboard if the model is 5 cm high

  b the width of the model bed if the actual bed is 1.5 m wide

  c the length of a table if the model is 2.7 cm long.

7 A model of an F15 fighter jet has a scale of 1 : 32.

The real aircraft is 12.5 m long.

What is the length of the model?

8 A map of Scotland on an A4 sheet of paper has a scale of 1 : 2 000 000.

  a The map distance from Inverness to Glasgow is 90 mm.
  What is the real distance between these places?

  b The actual distance by road from Aberdeen to Dundee is 96.5 km.
  How long would this road be on the map?

1 : 10 000 000

Tip

You will use scale factors again in Chapter 28 when you deal with similar triangles and with enlargements of shapes.

Constructing scale drawings

To make a scaled drawing or simple map you need to:
- Find out or measure the real lengths involved.
- Decide what size your drawing is going to be so you can work out a scale.
- Choose an appropriate scale (if you are not given one to use). Use ratio to work this out.

  scale = length on drawing : length in real life.
- Use the scale to convert the real lengths to ones you need for the scaled drawing.

WORKED EXAMPLE 10

Draw a scale plan of a rectangular park that is 115 m long and 85 m wide.

Your plan must fit into a space 7.5 cm long and 5 cm wide.

A scale drawing 6 cm long will fit into the given space.
Don't try to draw to the edge of the paper. Choose a convenient length.

Scale = length on drawing : length in real life
= 6 cm : 115 m
Work out the scale.

= 6 cm : 11 500 cm
Convert the metres to centimetres.

= 1 : 1917
Divide both sides of the ratio by 6 to get 1 on the left.

≈ 1 : 2000
1 : 1917 is a clumsy scale. Most scales are rounded.

Scaled length = 115 m ÷ 2000 = 0.0575 m = 5.75 cm
Scaled width = 85 m ÷ 2000 = 0.0425 m = 4.25 cm
Use the scale to convert the real distances.

4.25 cm is less than the width of the paper, so your choice of about 6 cm for the length was a good choice.

Scale 1 : 2000

Draw a 5 cm by 7.5 cm frame.

Draw an accurate rectangle 4.25 cm by 5.75 cm within the frame.

Remember to write the scale you used on the diagram.

10 Mensuration

EXERCISE 10F

1 The floor of a school hall is 40 m long and 20 m wide.

Draw scaled diagrams to show what it would look like at each of these scales.

a 1 : 500 **b** 1 : 1000

2 Measure the dimensions of your desk in centimetres.

Work out a suitable scale and draw a scaled diagram of your desk, including anything on it.

3 Jules drew this rough plan of a classroom block at her school.

She wrote the actual measurements on the plan.

Use the dimensions on the plan to draw a scaled diagram of this classroom block that fits into the width of your exercise book.

Indicate windows and doors as shown on the plan.

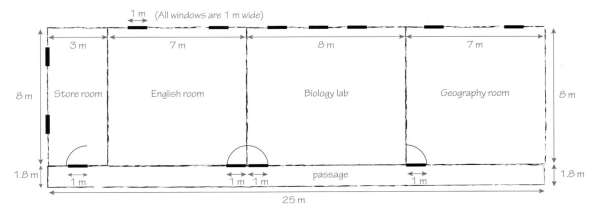

4 A plan of a house is to be drawn at a scale of 1 : 80.

a What should the scaled dimensions of the kitchen be if the real dimensions are 4900 mm by 3800 mm?

b Calculate the scaled length of the sink unit if it is 1.2 metres long in reality.

Bearings

Compass directions can be given using cardinal points as shown on the compass rose.

More accurate directions can be given using degrees or bearings.

Bearings are measured in degrees from 0° (north) around in a clockwise direction to 360° (which is back at north).

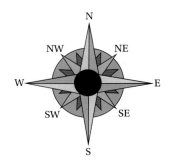

 Tip

You may need to create your own north line if it is not on the diagram.

 Find answers at: cambridge.org/ukschools/gcsemaths-studentbookanswers

Bearings are written as three-figure numbers. East is 90° clockwise from north and, as a bearing, is written as 090°.

WORKED EXAMPLE 11

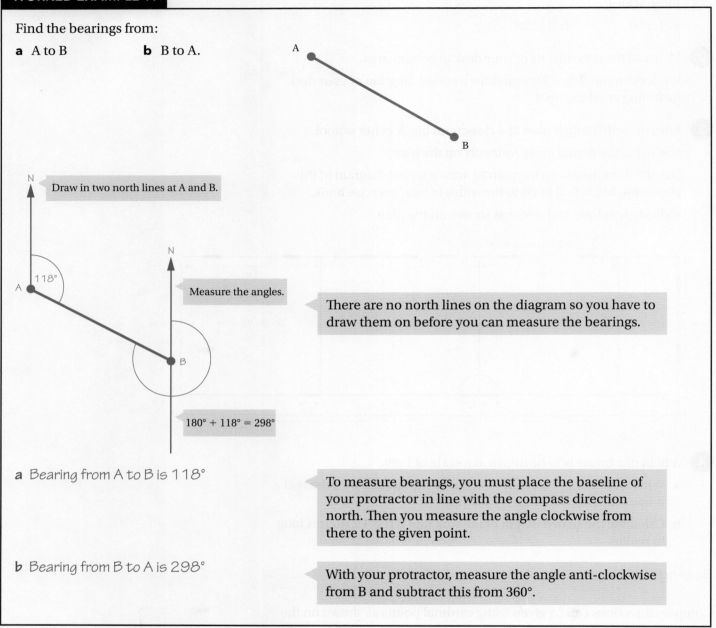

a Bearing from A to B is 118°

b Bearing from B to A is 298°

EXERCISE 10G

1 Write the three-figure bearing that corresponds with these directions:
   a due south   b north-east   c west.

2 Use a protractor to measure each of the following bearings on the diagram. Give your answer to the nearest 5°.

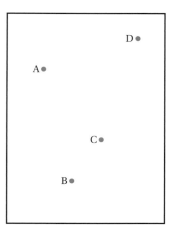

a A to B **b** B to A **c** A to C
d B to D **e** D to A **f** D to B

3 Beville is 140 km west and 45 km north of Lake Salina.

Draw a scale drawing with a scale of 1 cm to 20 km and use it to find:

a the bearing from Lake Salina to Beville

b the bearing from Beville to Lake Salina

c the shortest distance between the two places in kilometres.

Checklist of learning and understanding

Standard units of measurement

- You can convert between metric units of length, mass and capacity by multiplying or dividing by powers of ten.
- To convert squared units, you need to square the conversion factors.
- To convert cubed units, you need to cube the conversion factors.
- Units of time are not metric, so you need to use the number of parts in the sub-units when you convert units of time.

Compound units of measurement

- Compound units of measurement involve more than one unit.
- Rates such as £/hour or cost per kilogram are compound units.
- Speed = $\dfrac{\text{distance}}{\text{time}}$
- Density = $\dfrac{\text{mass}}{\text{volume}}$
- Pressure = $\dfrac{\text{force}}{\text{area}}$

Find answers at: cambridge.org/ukschools/gcsemaths-studentbookanswers

Maps, scales and bearings

- The scale of a map or diagram describes how much smaller (or bigger) the lengths on the diagram are compared to the original lengths.
- Real length = map length × scale
- The scale can be written as, length on diagram : real length
- Bearings are accurate directions given in degrees from 0° to 360°. 0° corresponds with north.
- Bearings are measured clockwise from 0° and written using three figures.

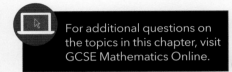

Chapter review

1. **a** How many minutes are there between 8.50 pm and 10.05 pm? *(1 mark)*

   b i Write 15:25 using the 12-hour clock.

   ii Write 9.15 pm using the 24-hour clock. *(2 marks)*

   Lucy and Saad went to a cafe on the same day.

   Lucy was in the cafe from 10.15 am to 10.45 am.

   Saad was in the cafe from 10.25 am to 11.05 am.

   c Work out the number of minutes that Lucy and Saad were in the cafe at the same time. *(2 marks)*

   ©Pearson Education Ltd 2011

2. Match each statement to the correct number in the box below.

5	475	182.5	259 200

   a The number of seconds in 3 days.

   b The number of kilometres travelled in $2\frac{1}{2}$ hours by a car travelling at 73 km/h.

   c The distance in km in real life of a length of 5 cm on a map with a scale of 1 : 100 000.

   d The number of litres in 475 000 millilitres.

3. True or false?

   a Tony's fish tank contains 72 litres of water.

   He says this is 72 000 millilitres.

   b The school is 15 km from the bus stop. The bus travels at 40 km/h.

   Molly says it will take her half an hour to get to school.

   c The distance from Liverpool to Manchester is about 55 km.

   The scale of a map is 1 : 250 000.

   The distance on the map would be 5.2 cm.

4 Convert 60 000 cm² into m².

5 How many mm² are there in 2 m²?

6 A car is travelling at an average speed of 80 km/h for one hour on a bearing of 120°.

   a Use a scale of 1 cm to 20 km to show this journey.

   b After 45 km, the driver stopped for petrol. Mark this spot on the diagram. If this was after 40 minutes, calculate his speed for that part of the journey.

7 The density of an object is 8 kg/m³.

Work out the mass of 25 m³ of the object.

8 A cyclist travels due east from point A for 10 km to reach point B.

She then travels 6 km on a bearing of 125° from B to reach point C.

   a Use a scale of 1 cm to 2 km to represent her journey on a scale diagram.

   b Find the bearing from C to A.

   c Find the direct distance from C to A in kilometres.

   d If it takes the cyclist $1\frac{1}{2}$ hours to cycle back using the direct route from C to A, find her average speed:

   i in km/h   **ii** in m/s.

Find answers at: cambridge.org/ukschools/gcsemaths-studentbookanswers

11 Perimeter

In this chapter you will learn how to …
- calculate the perimeter of simple shapes such as rectangles and triangles.
- find the perimeter of composite shapes.
- calculate the circumference of a circle.
- calculate the perimeter of composite shapes including circles or parts of circles.

For more resources relating to this chapter, visit GCSE Mathematics Online.

Using mathematics: real-life applications

Working out the amount of fencing needed for a field, or the number of tiles needed to edge a swimming pool, or the number of perimeter security cameras needed to secure an area all require the calculation of a perimeter.

> "Security cameras in a car park are only effective if they can see all round the perimeter of the car park."
>
> *(Security camera technician)*

Before you start …

Ch 2	You must be able to recognise and name some common polygons.	**1**	Match each name to the correct shape. octagon pentagon hexagon  a b c
Ch 10	You must be able to convert between basic metric units.	**2**	Convert these units. **a** 5 km into m **b** 12 km into cm **c** 8500 mm into m **d** 4.8 m to mm
KS3	You need to know and use the correct names of circle parts.	**3**	True or false? **a** The diameter is twice the length of the radius. **b** The diameter is always shorter than the radius. **c** The angles at the centre of a circle add up to 180°.
Ch 5	You should be able to expand brackets and factorise expressions.	**4**	Which is the correct expansion of $3(x + 2)$? **a** $3x + 2$ **b** $3x + 5$ **c** $3x + 6$ **5** Factorise $3\pi + 6$.
KS3	You should be able to change the subject of a simple formula.	**6**	Make l the subject of the formula $P = 2(l + w)$. **7** Make r the subject of the formula $P = \pi r + 2r$.

Assess your starting point using the Launchpad

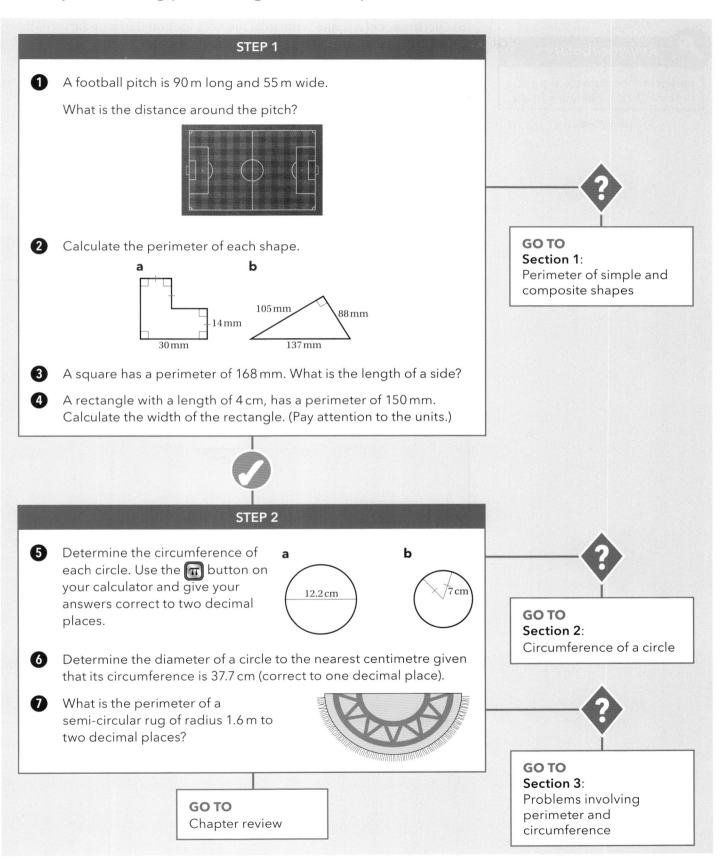

Key vocabulary

perimeter: the distance around the boundaries (sides) of a shape.

Section 1: Perimeter of simple and composite shapes

The **perimeter** of a shape is the total distance around the boundaries of the shape. To calculate the perimeter of a shape:

- determine the lengths of all the sides, making sure they are in the same units
- add the lengths together.

Tip

The perimeter must include **all** the lengths. You can use what you know about the properties of shapes to deduce missing lengths.

WORKED EXAMPLE 1

Find the perimeter (P) of this T-shaped piece of cardboard.

All angles are right angles and all dimensions are in centimetres.

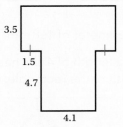

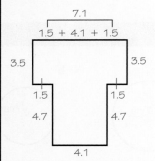

$P = 7.1 + 3.5 + 1.5 + 4.7 + 4.1 + 4.7 + 1.5 + 3.5$
$= 30.6$

Start by working out the lengths of the missing sides.

Add up the side lengths.

Using formulae to find perimeter

The properties of different polygons can be used to derive formulae for calculating the perimeter without adding up all the sides.

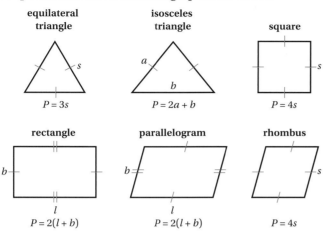

> **Tip**
> You might remember some of these formulae from KS3. You do not need to memorise them, but you must be able to apply and manipulate them.

For **regular polygons** you only need the length of one side to find the perimeter.

Multiply the side length by the number of sides.

Each side length of this regular hexagon is 6 cm so the perimeter is: $6 \times 6 = 36$ cm.

If each side length was a cm, the expression for the perimeter would be $6 \times a$ or $6a$.

> **Tip**
> Writing a general expression is a way of showing the quantity of something when the actual numbers are not known. Numbers can be substituted into the expression to find the quantity required.

WORKED EXAMPLE 2

Write an expression for the the perimeter of this shape in terms of x.

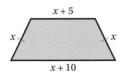

Perimeter $= x + (x + 5) + x + (x + 10)$ ◁ Start at one side and move systematically round the shape.

$= x + x + 5 + x + x + 10$ ◁ Remove the brackets and gather like terms.
$= 4x + 15$

> **Tip**
> Simplifying expressions by collecting like terms was covered in Chapter 5.

EXERCISE 11A

1 What is the perimeter of an equilateral triangle of side length 10 cm?

2 A yard is fully enclosed by a fence of lengths 12 m, 4.7 m, 354 cm, 972 cm. What is the perimeter of the yard?

Find answers at: cambridge.org/ukschools/gcsemaths-studentbookanswers

3) Write expressions for the perimeter of these shapes.

a An equilateral triangle, side length a.

b A rectangle, width x and length y.

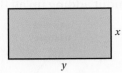

c A regular octagon, side length z.

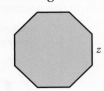

4) Work out the perimeter of this shape.
Each side length is 10 cm.

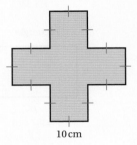

5) A tessellated hexagon, where each hexagon fits exactly with the next, creates a pattern for a patchwork quilt.

If the side length of each hexagon is 8 cm, what is the perimeter of this shape?

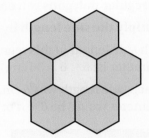

6) Each side length of this tiling pattern is 15 cm.
Work out its perimeter.

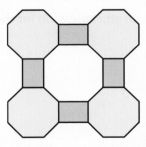

7) A field with dimensions shown on the diagram is to be fenced to protect the vegetation growing there. The fence consists of upright posts and four strands of wire (as shown).

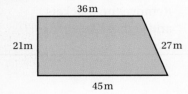

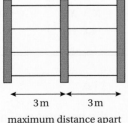

3 m   3 m
maximum distance apart
(from middle of posts)

a Find the total length of wire needed for the fencing (the wire runs through the fence posts).

b The fence posts are placed a maximum distance of 3 m apart (measured from the middle of the post to the middle of the next post). Work out the minimum number of fence posts needed for the whole perimeter fence. Make sure to include one fence post at each corner. (Assume there is no gate, only a stile to get over the fence.)

c Calculate the cost of fencing if each post costs £2.39 and the wire costs £1.78 per metre.

Finding lengths when the perimeter is known

You can calculate missing side lengths of shapes if you know the perimeter and have enough information about the shape. For example, if you know a regular hexagon has a perimeter of 72 cm, you can deduce that each of the six equal sides are 12 cm long because 72 cm ÷ 6 = 12 cm.

WORKED EXAMPLE 3

What is the length of a rectangle of perimeter 20 cm and width 5.5 cm?

$P = 2(L + W)$
$P = 20, W = 5.5$

Think about the information you have.

$20 = 2L + 2(5.5)$
$20 = 2L + 11$
$20 - 11 = 2L$
$9 = 2L$
$L = 4.5 \text{ cm}$

Substitute the values into the formula and solve for L.

WORKED EXAMPLE 4

Calculate the length of each side of a rhombus of perimeter 1.8 m.

$P = 4s$

Write down the formula.

$s = \dfrac{P}{4}$

$s = \dfrac{1.8}{4}$

$s = 0.45 \text{ m}$

Change the subject of the formula and substitute the values you know to find s.

Tip

You will do more work with changing the subject in Chapter 16.

Perimeter of composite shapes

Composite shapes are formed by combining shapes or by removing parts of a shape.

This shape is made up of a rectangle and a triangle.

You add up the side lengths around the outside boundary of the shape to find the perimeter. The perimeter is: $4 + 3 + 3 + 4 + 3 = 17$ cm

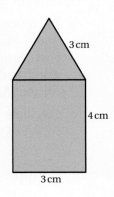

The rectangle below has a parallelogram cut out of it. The perimeter of the shape must include all of the boundaries, so, in examples like these, you must include the outer and inner boundaries of the shape.

$P = 2(20 + 10) + 2(10 + 5)$
$ = 2(30) + 2(15)$
$ = 60 + 30$
$ = 90$ cm

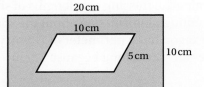

WORK IT OUT 11.1

This shape was made by combining five identical squares with sides of 6.5 cm with four identical equilateral triangles.

Which calculation will result in the correct perimeter? Why?

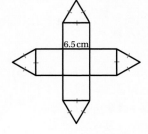

Option A	Option B	Option C
$P = 16 \times 6.5$	$P = 20 \times 6.5$	$P = 24 \times 6.5$
$= 104$ cm	$= 130$ cm	$= 156$ cm

EXERCISE 11B

1. Work out the lengths of the sides in each shape.

   a A rectangle has a perimeter of 242 mm and a longer side length of 77 mm.

   b A parallelogram has a perimeter of 200 mm. One side length is 55 mm.

   c A rhombus has a perimeter of 14 cm.

   d A square has a perimeter 47.2 cm.

2 Calculate the perimeter of each shape.

a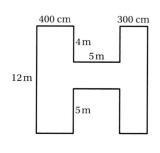

b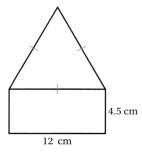

c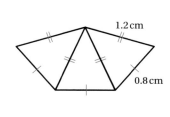

d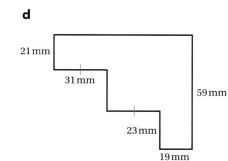

e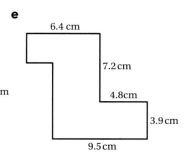

3 Petra is a mosaic artist. She has the following mosaic tiles:

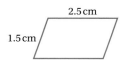 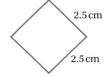 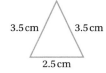 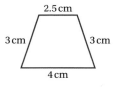

She arranges the tiles to make these shapes.

a   b  c   d

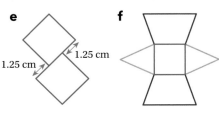

Use the dimensions above to find the perimeter of each shape.

4 The end of a maze consists of a regular pentagon with side length 5 m.
If a child runs round this three times, how far will they run in total?

5 A rectangular allotment has a perimeter of 25 metres.
If the length of one side is 6.8 metres, how wide is the allotment?

Find answers at: cambridge.org/ukschools/gcsemaths-studentbookanswers

Section 2: Circumference of a circle

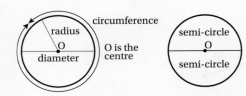

Make sure you remember these names for different parts of a circle.

The perimeter of a circle is called its **circumference (C)**.

The radius (r) is the distance from the centre to the circumference.

The diameter (d) is a line through the centre of a circle. The diameter is twice the radius ($2r$).

The Ancient Greeks discovered that when you measured the circumference of a circle and divided it by its diameter you got a constant ratio of approximately 3.142.

$\pi = \dfrac{C}{d}$ This ratio is called pi and the symbol π is used to represent it.

This ratio can be rearranged to give two formulae for finding the circumference of any circle.

Pi has no exact decimal or fractional value (it is an irrational number). When you do calculations involving pi you might need to give rounded or approximate answers.

Your calculator can work with more precise values of π but you might be given an approximate value of π to use in a problem context.

Learn this formula

$C = \pi d$ and $C = 2\pi r$

WORKED EXAMPLE 5

a Calculate the circumference of a circle with diameter 7 cm.

Give your answer to the nearest cm.

b Calculate the circumference of a circle with radius 4.5 m.

Leave your answer as a multiple of π.

a $C = \pi d$

Use the version of the formula with d, as you are given the diameter.

$= \pi \times 7$
$= 21.991\,149 \approx 22 \text{ cm}$

Substitute $d = 7$ and use the $\boxed{\pi}$ button on your calculator.

b $C = 2\pi r$

Use the version of the formula with r, as you are given the radius.

$= 2 \times \pi \times 4.5$
$= 9\pi \text{ m}$

Substitute $r = 4.5$. **Do not** put in a value for π.

11 Perimeter

Problem solving

Problem-solving framework

Racing wheelchairs can travel at speeds of up to 45 mph and can cost up to £20 000.

The rear wheels of a chair have a diameter of 70 cm.

The hand wheels have a radius of 20 cm.

a Find the difference in circumference between the two wheels.
b Find the number of revolutions that the rear wheels will make over a 100 m race.

Steps for solving problems	What you would do for this example
Step 1: What have you got to do?	Find the difference in circumference between the two wheels, and find the number of revolutions (turns) that the rear wheels will make over 100 m.
Step 2: What information do you need?	Diameter (d) of rear wheel = 70 cm Radius (r) of hand wheel = 20 cm Length of race = 100 m Circumference formula: $C = \pi d$ or $2\pi r$
Step 3: What information don't you need?	The speed of the wheelchair and its cost are irrelevant.
Step 4: What maths can you do?	Circumference of rear wheel = $\pi d = \pi \times 70 = 219.9$ cm (to 1 decimal place) Circumference of hand wheel = $2\pi r = 2 \times \pi \times 20 = 125.7$ cm (to 1 decimal place) Difference in circumference = $219.9 - 125.7 = 94.2$ cm Change the units so that they are the same: $100\,\text{m} = 100 \times 100\,\text{cm} = 10\,000\,\text{cm}$ Number of revolutions made by rear wheel = $10\,000 \div 219.9 = 45.5$ (to 1 decimal place)
Step 5: Have you used all the information? At this point you should check to make sure you have calculated what was asked of you.	All information used ✓ Calculated part **a** ✓ Calculated part **b** ✓
Step 6: Is it correct?	Used diameter for rear wheel ✓ Used radius for hand wheel ✓ Converted cm into m correctly ✓

Find answers at: cambridge.org/ukschools/gcsemaths-studentbookanswers

Knowing the circumference of a circle means that you can calculate the diameter and/or radius.

WORKED EXAMPLE 6

A circle has a circumference of 200 cm.

Calculate the diameter of the circle to 1 decimal place.

$C = \pi d$ — Use the version of the formula with d.

You *could* re-arrange the formula here to give $d = \dfrac{C}{\pi}$

$200 = \pi \times d$ — Substitute $C = 200$.
$200 \div \pi = d$

$d = 63.7 \text{ cm}$

> **Tip**
>
> If you are asked to find r and use $C = \pi d$, remember to divide d by 2 to get r.

EXERCISE 11C

1 Use the key on your calculator to calculate the circumference of each circle.

Give your answer correct to two decimal places if necessary.

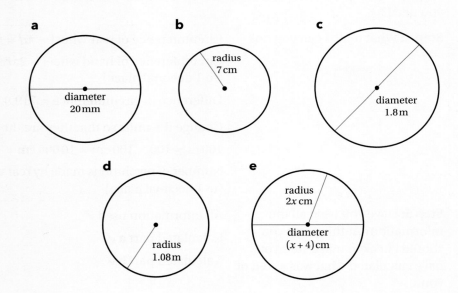

a — diameter 20 mm
b — radius 7 cm
c — diameter 1.8 m
d — radius 1.08 m
e — radius $2x$ cm, diameter $(x + 4)$ cm

2 A car rim has a diameter of 42 cm. Calculate the circumference of the rim.

3 A plastic toy (left), called a slinky spring, consists of 36 coils of plastic.

If the diameter of one coil is 55 mm, what length of plastic is needed to make the spring?

4. Nate has a square piece of metal with sides of 8.5 cm. He wants to cut out a round disc with a radius of at least 4 cm from the square.
   a Draw a rough sketch to show this shape.
   b Calculate the circumference of the disc if the radius is 4 cm.
   c When he cuts the disc out, Nate finds the diameter is actually 8.3 cm. What is the circumference of this disc?
   d What is the perimeter of the piece of metal left after cutting out a disc of:
      i radius 4 cm   ii diameter 8.3 cm?

5. Find the diameter, correct to 2 decimal places, of a circle of circumference:
   a 20 mm   b 15.2 cm.

6. Find the radius of a round CD to the nearest mm, if its circumference is 36.33 cm.

7. What is the smallest possible square plate that you can use for a round cake of circumference 77 cm?

8. The minute hand of a clock is 75 mm long.
   How far will the tip of the hand travel in one hour? Give your answer correct to the nearest centimetre.

9. A circular boating lake has a diameter of 8 m. What is its circumference? Give your answer in terms of π.

10. The radius of a circular stained glass window is 60 cm. In terms of π, what is its circumference?

Sectors of a circle

A sector is a 'slice' of a circle. The perimeter of a sector is formed by two radii and a section of the circumference called an arc.

Perimeter of a sector = radius + radius + arc length.

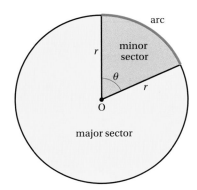

> **Tip**
>
> The term circumference is only used to describe the distance around a whole circle. For the distance around the boundary of a shape that is part of a circle, you use the term perimeter.

To find the perimeter of a sector you have to work out the length of the arc.

The angle at the centre of a circle is a fraction of 360°. You can express this as $\frac{x}{360}$.

The arc length is a fraction of the circumference. To find the length of the arc, use the size of the angle and the circumference of the circle.

$$\text{Arc length} = \frac{x}{360} \times \pi d \quad \text{or} \quad \text{Arc length} = \frac{x}{360} \times 2\pi r$$

fraction   circumference   fraction   circumference

WORKED EXAMPLE 7

Find the length of the arc in each of these circle sectors.

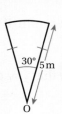

a 30°, 5 m

b 65°, 4 cm

a Arc length $= \frac{30}{360} \times 2\pi r$

                $= \frac{1}{12} \times 2 \times \pi \times 5$

                $= 2.62$ m

Use $2\pi r$ here as you have been given the radius.

Correct to 2 decimal places.

b $360° - 65° = 295°$

The angle inside the sector is not given.

Work it out using angles round a point.

Arc length $= \frac{295}{360} \times 2 \times \pi \times 4$

            $= 20.59$ cm

Correct to 2 decimal places.

The semi-circle and quarter-circle (quadrant) are special cases.

- In a semi-circle, the angle at the centre is 180° and the arc length is half the circumference.
- In a quarter-circle, the angle at the centre is 90° and the arc length is a quarter of the circumference.

EXERCISE 11D

1 Find *l* in each of the following circles.

a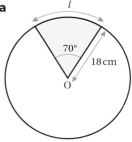

b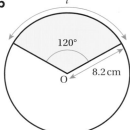

c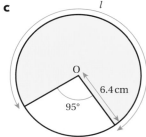

d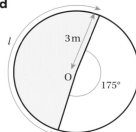

2 Find the perimeter of each shape.

a

b

c

d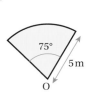

e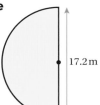

f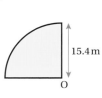

Section 3: Problems involving perimeter and circumference

In this section you are going to combine what you have learned about perimeter and circumference to solve problems involving composite shapes.

To work with irregular and composite shapes you divide them up into known shapes to make it easier to do the calculations.

An athletics track can be divided up into a rectangle and two semi-circles, one at either end. By splitting the shape into known shapes, you can find the perimeter of the track.

Two semi-circles make one circle, so

perimeter = circumference of circle + the lengths of the two straight sides.

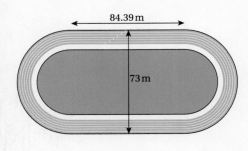

WORKED EXAMPLE 8

The diagram shows a baseball field which is $\frac{1}{4}$ of a circle.

Calculate the perimeter of the field correct to the nearest metre.

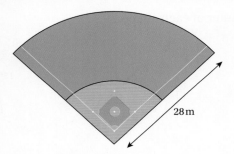

Circumference of whole circle
$= 2\pi r = 2 \times \pi \times 28$

Arc length of $\frac{1}{4}$ circle $= \frac{C}{4}$

$= \frac{2 \times \pi \times 28}{4}$

$= 43.98\,m$

Perimeter = $2r$ + arc length
$= 2 \times 28 + 43.98$
$= 56 + 43.98$
$= 99.98\,m$

$P = 100\,m$ correct to the nearest metre.

> Remember, a quarter-circle is a special case, with the angle at the centre being 90°.

WORK IT OUT 11.2

This is a plan of a children's play area.

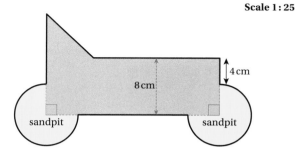

Scale 1 : 25

Edging is to be placed around the curved edges of the yellow sandpits.

Using $\pi = 3.14$, what is the total length of edging required?

Which of the answers below is the correct answer?

What mistakes have been made in the other workings?

Option A	Option B	Option C
Circumference of a circle $= \pi d = 2\pi r$ $r = 4$ cm	Circumference of a circle $= \pi d = 2\pi r$ $r = 4$ cm	Circumference of a circle $= 2\pi d = \pi r$ $r = 4$ cm
$C = 2 \times 3.14 \times 4$ $= 3.14 \times 8$ $= 25.12$ cm	$C = 2 \times 3.14 \times 4$ $= 3.14 \times 8$ $= 25.12$ cm	$C = 2 \times 3.14 \times 8$ $= 50.24$ cm
Scale 1 : 25 So, actual circumference of the circle $= 25.12 \times 25$ $= 628$ cm $= 6.28$ m	Only $\frac{3}{4}$ of the sandpit needs edging: $\frac{3}{4} \times 25.12$ cm $= 18.84$ cm	Only $\frac{3}{4}$ of the sandpit needs edging: $\frac{3}{4} \times 50.24$ cm $= 37.68$ cm
Two sandpits so total edging required $= 6.28 \times 2 = 12.56$ m	Two sandpits: 18.84 cm $\times$ 2 = 37.68 cm of edging Scale = 1 : 25 So actual amount of edging required: 37.68 cm $\times$ 25 = 942 cm $= 9.42$ m	Two sandpits: 37.68 cm $\times$ 2 = 75.36 cm of edging Scale = 1 : 25 So actual amount of edging required: 75.36 cm $\times$ 25 = 1884 cm $= 18.84$ m

Tip

Perimeter can be worked out by measuring lengths, or by using lengths from a scale diagram.

Find answers at: cambridge.org/ukschools/gcsemaths-studentbookanswers

EXERCISE 11E

The diagram shows the shape and dimensions of different throwing event field areas in international competitions. Use the information on the diagram to answer questions **1** to **4**.

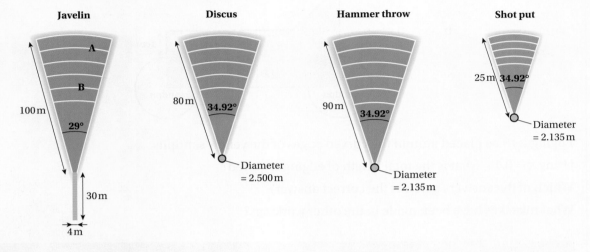

1 Calculate the length of the white line painted around the outside of:

  a the discus area
  b the hammer throw area.

2 Competitors in the discus, shot put and hammer throw have to remain inside a marked circle while the equipment is in their hands (before they throw it).

  a Which sport has the largest marked circle?
  b What is the circumference of the circle in the discus throwing cage?
  c In shot put and hammer throw, a raised edge is built around the circumference of the starting circle. If this edge is 10 cm wide, calculate its inner and outer circumference.

3 Calculate the perimeter of the event space for javelin.

4 The curved measurement lines on each event space are 10 m apart.
Using the javelin field, calculate the length of the lines marked A and B.

5 The average radius of the Earth is 6378.1 km at the Equator.
Calculate the circumference of the Earth, assuming the Earth is a sphere. Give your answer correct to 2 decimal places.

6 A large pizza has a circumference of 94 cm.
What is the side length of the smallest cardboard box it will fit into?

7 Find the perimeter of the symmetrical logo (left).
Use a ruler and protractor to measure and find the dimensions you need.

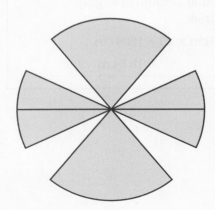

Checklist of learning and understanding

Perimeter
- Perimeter is the total distance around the boundaries of a shape.
- You can calculate perimeter by adding the lengths of the sides or by applying a formula based on the properties of the shape.

Circumference
- The perimeter of a circle is called its circumference.
- $C = \pi d$ or $C = 2\pi r$
- A sector is a part of a circle between two radii. You can find the arc length of a sector by working out what fraction of a circle the sector represents.

Chapter review

For additional questions on the topics in this chapter, visit GCSE Mathematics Online.

1 The perimeters of the two shapes on the right are equal.

What is the side length of the square?

2 The perimeter of a regular pentagon is 90 cm.

Work out the length of each side.

3 A rectangular vegetable plot has a perimeter of 50 metres.

If the width is 6.5 m, what is the length of the plot?

4 An irrigator in a field can water a circular area of radius 14.5 m.

What is the circumference of the area that can be irrigated? Use $\pi = 3.14$.

5 Susan has a round cake.

The cake has a diameter of 20 cm.

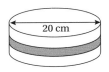

Diagram **NOT** accurately drawn

Susan wants to put a ribbon round the cake.

What is the least length of ribbon she can use?   *(3 marks)*

©*Pearson Education Ltd 2012*

6. A pizza has a diameter of 28 cm. It is placed in a cardboard box. Calculate the perimeter of the smallest box that it will fit into.

7. Calculate the perimeter of the shape shown. Give your answer correct to 1 decimal place.

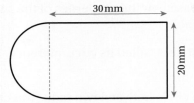

8. The diagram (below) shows some staging for a concert.

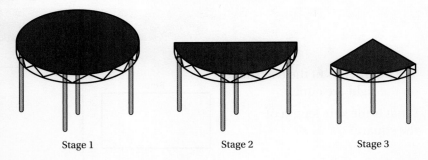

Stage 1   Stage 2   Stage 3

The main stage is a circle of diameter 6 m. The smaller stages can be made by splitting up a main stage into smaller pieces, with stage 2 being half the size of stage 1, and stage 3 being half the size of stage 2. The curved edges of the staging pieces have a patterned edge.

a Work out the total length of the patterned edging on the stage sections shown in the diagram.

A rubber safety strip is to be applied along all the edges of the staging.

b Work out the length of strip required for the three stages sections shown in the diagram.

12 Area

In this chapter you will learn how to ...
- apply formulae to find the area of different shapes, including circles and parts of circles.
- use appropriate formulae to calculate the area of composite shapes.

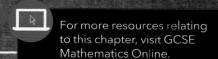

For more resources relating to this chapter, visit GCSE Mathematics Online.

Using mathematics: real-life applications

Ordering the right quantity of turf for a sports field, preparing detailed floor plans, and determining how much fertiliser is needed to treat a field crop all require knowledge and calculation of areas.

 Did you know?

The area of farmland is sometimes given in acres. An acre was traditionally the area of farmland that could be ploughed in one day by oxen. In the metric system, the acre was replaced by the hectare.
1 hectare = $10\,000\,m^2$
1 hectare = 2.471 054 acres

"Fertiliser application rates are normally given in kilograms per hectare. One hectare is an area of 100 m × 100 m or $10\,000\,m^2$. Applying too much or too little fertiliser to an area can have disastrous results for the crops."

(Farmer)

Before you start ...

Ch 2	You should remember the properties of quadrilaterals.	**1** Use the marked properties to name the quadrilaterals correctly. 
Ch 4	You should be familiar with square numbers and square roots.	**2** Calculate. **a** 5^2 **b** 2×10^2 **c** $3^2 + 4^2$ **3** Find the number which is squared to give each of these. **a** 144 **b** 10 000 **c** 0.25
Ch 10	You need to be able to convert between square units of measurement.	**4** Complete these. **a** $5\,m^2 = \square\,cm^2$ **b** $\square\,cm^2 = 87\,000\,mm^2$ **c** $4\,km^2 = \square\,m^2$

Find answers at: cambridge.org/ukschools/gcsemaths-studentbookanswers

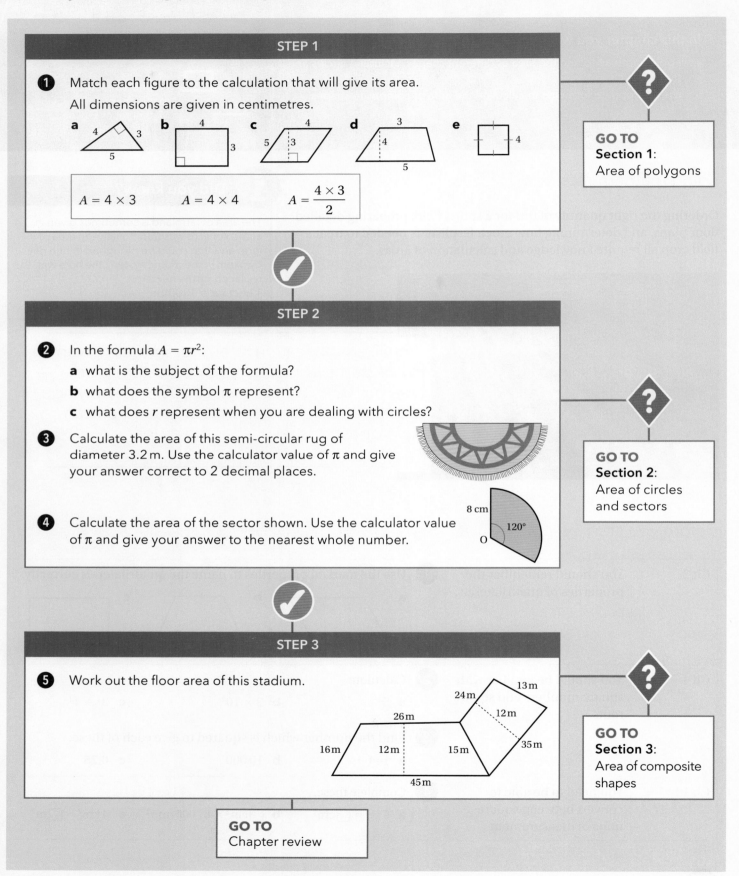

Section 1: Area of polygons

The **area** of a plane shape is the amount of space it takes up. You can think of area of a plane, or flat, shape as the number of squares (square units) that will fit inside the boundary of the shape.

Area is always given in square units. Common units are mm² (square millimetres), cm² (square centimetres), m² and km².

Area of rectangles and squares

The formula for finding the area of a rectangle is:

area = length × width

$A = lw$

A square is a special case of a rectangle, the length and width are equal, so the formula is:

area of the square = $l \times l = l^2$

$A = l^2$

You can change the subject of area formulae to find unknown lengths if you know the area and the other lengths.

> **Tip**
> Base *(b)* and height *(h)* may be used instead of length and width, so you might see this area formula written as $A = bh$.

> **Tip**
> Sometimes the side length is written as *s* (for side) instead of *l*. In this case the formula for the area of a square is written as $A = s^2$.

> **Tip**
> You will learn more on how to change the subject of a formula in Chapter 16.

WORKED EXAMPLE 1

A rectangle of area 45 cm² has one side 9 cm long. How long is the other side?

$A = l \times w$ — Write down the formula you are going to use.

$45 = 9 \times w$ — Substitute the given values. You don't know which of *l* or *w* you have been given in this case so you can replace whichever one you want.

$\dfrac{45}{9} = w$ — Divide each side by 9 to get *w* on its own.

$5 = w$

The other side is 5 cm long.

Area of a triangle

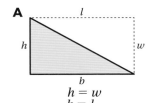
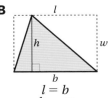

Figure A shows a right-angled triangle. Figure B shows a scalene triangle. In both diagrams the area of the shaded and unshaded parts are equal.

Area of the rectangle = $l \times w$

So, the area of the triangle = $\dfrac{1}{2} \times l \times w$

But, l is equal to the base of the triangle and w is equal to the height of the triangle, so the area of the triangle $= \frac{1}{2} \times$ length of its base $\times$ its height.
This gives a formula for the area of any triangle.

You can use any side of a triangle as the base. The height must be perpendicular to the side that you are using as the base. The perpendicular height can be inside the triangle, one of the sides of the triangle or outside the triangle.

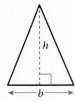 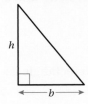 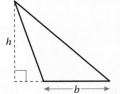

Learn this formula

Area of a triangle
$= \frac{1}{2} \times$ base $\times$ perpendicular height
$= \frac{1}{2} \times b \times h$

WORKED EXAMPLE 2

Calculate the area of each triangle.

a 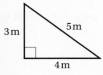   b 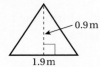   c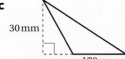

a Area $= \frac{1}{2} \times b \times h$

            $= \frac{1}{2} \times 4 \times 3$

            $= \frac{1}{2} \times 12$

            $= 6\,m^2$

Write down the formula you are going to use.

In a right-angled triangle the two shorter (perpendicular) sides can be used as the base and height.

Remember to include the units in the answer.

b Area $= \frac{1}{2} \times b \times h$

            $= \frac{1}{2} \times 1.9 \times 0.9$

            $= \frac{1}{2} \times 1.71$

            $= 0.855\,m^2$

Use the side marked 1.9 as the base because the height is perpendicular to it.

c Area $= \frac{bh}{2}$

          $= \frac{(170 \times 30)}{2}$

          $= \frac{5100}{2}$

          $= 2550\,mm^2$

This is the same formula expressed differently. Multiplying by $\frac{1}{2}$ is the same as dividing by 2.

When you extend a base line so that you can find a perpendicular height outside the triangle, it does not change the base length you use in your calculations.

Tip

The order of the multiplication is not important:
$\frac{1}{2} \times 4 \times 3 = \frac{1}{2} \times 12 = 6$
$\frac{1}{2} \times 4 \times 3 = 2 \times 3 = 6$
$0.5 \times 3 \times 4 = 1.5 \times 4 = 6$

Did you know?

Two triangles that look completely different might still have the same area.

These two triangles are equal in area.

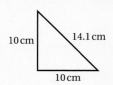

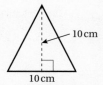

EXERCISE 12A

1 How might each of the following people use area calculations in their jobs? For each one, try to give examples of the shapes they would work with most often and the formulae they might need.

  a House painter

  b Mosaic artist

  c Gardener

  d Dressmaker

  e Carpet layer

2 Calculate the area of each triangle.

  a 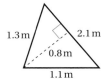   **b**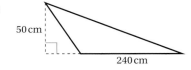

3 The area of a triangle is $36\,m^2$ and its perpendicular height is $6\,m$. Work out the length of its base.

4 Work out the total area of this kite.

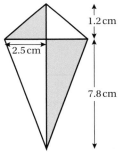

5 A triangle has a base of $3.6\,m$ and a height of $50\,cm$. What is its area in metres squared?

6 A triangle of area $0.125\,m^2$ has a base $25\,cm$ long. What is its height in centimetres?

7 Calculate the total area of the sails on this small boat (right). The larger sail is $2.7\,m$ tall. The smaller sail extends $\frac{2}{3}$ of the way up the larger sail.

Area of a parallelogram

The base of this parallelogram is b and the **perpendicular height** is h.

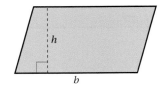

Learn this formula

Area of a parallelogram
= base × perpendicular height
= $b \times h$

GCSE Mathematics for Edexcel (Foundation)

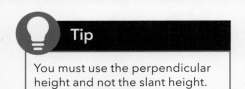

Tip

You must use the perpendicular height and not the slant height.

By removing a triangle from one end and joining it to the other end, you can form a rectangle.

The parallelogram and rectangle have the same area.

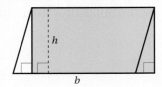

WORKED EXAMPLE 3

Calculate the area of the parallelogram.

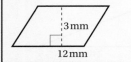

Area = $b \times h$
     = $12 \times 3 = 36 \, mm^2$

WORK IT OUT 12.1

A parallelogram is made by combining a rectangle and two right-angled triangles like this.

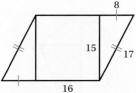

Which of these options will give the correct area?

Which dimensions are incorrect in the other two? Why?

Option A	Option B	Option C
$A = bh$	$A = bh$	$A = bh$
$= 16 \times 17$	$= 24 \times 17$	$= 24 \times 15$

Area of a trapezium

A **trapezium** has parallel sides a and b, and perpendicular height h.

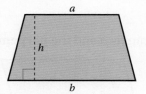

If you join two identical trapezia, you can form a parallelogram.

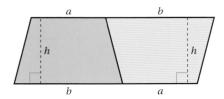

The parallelogram has a base of length $a + b$, and perpendicular height h.

The area of a parallelogram = base × perpendicular height.

So, area = $(a + b) \times h$

But, each trapezium is half the area of the parallelogram.

So, area of a trapezium = $\frac{1}{2} \times (a + b) \times h$, where a and b are the lengths of the parallel sides.

> **Tip**
>
> You might see the trapezium area formula written as $\frac{(a + b)h}{2}$.

WORKED EXAMPLE 4

Calculate the area of the trapezium.

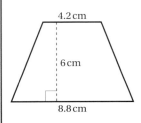

$$\text{Area} = \frac{1}{2} \times (a + b) \times h$$
$$= \frac{1}{2} \times (4.2 + 8.8) \times 6$$
$$= \frac{1}{2} \times 13 \times 6$$
$$= 39 \text{ cm}^2$$

WORK IT OUT 12.2

A solar farm is being built in a field.

Each solar panel measures 98 cm by 150 cm.

If 500 panels can fit onto the field, what is the area of panels being used, in m²?

Which of the answers below is correct?

What errors were made in each of the others?

Answer A	Answer B	Answer C
Area of one panel:	Area of one panel:	Area of one panel:
$98 \times 150 = 14\,700 \text{ cm}^2$	$\frac{1}{2} \times (98 \times 150) = 7350 \text{ cm}^2$	$98 \times 150 = 14\,700 \text{ cm}^2$
$1 \text{ m}^2 = 10\,000 \text{ cm}^2$	$1 \text{ m}^2 = 10\,000 \text{ cm}^2$	$1 \text{ m}^2 = 100 \text{ cm}^2$
1 panel = $14\,700 \div 10\,000 = 1.47 \text{ m}^2$	1 panel = $7350 \div 10\,000 = 0.735 \text{ m}^2$	1 panel = $14\,700 \div 100 = 147 \text{ m}^2$
500 panels = $1.47 \text{ m}^2 \times 500$ = 735 m^2	500 panels = $0.735 \text{ m}^2 \times 500$ = 367.5 m^2	500 panels = $147 \text{ m}^2 \times 500$ = $73\,500 \text{ m}^2$

EXERCISE 12B

1 Calculate the area of each parallelogram.

a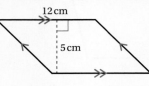

b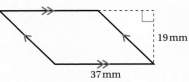

c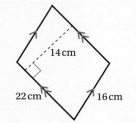

d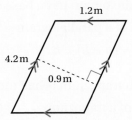

2 Calculate the area of each trapezium.

a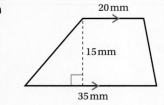

b, d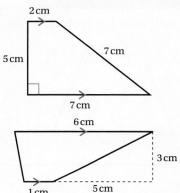

c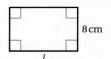

3 The area of this rectangle is 96 cm². What is its length?

4 The area of a parallelogram is 40 m². If the perpendicular height is 10 cm, what is the length of the base?

5 The area and one other measurement are given for each shape. Use the given information to find the unknown length in each figure.

a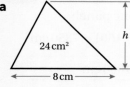

b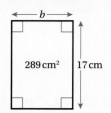

c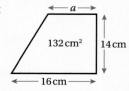

d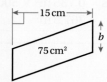

e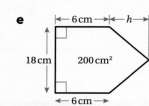

6 Amira works in a community development programme that helps people to grow organic vegetables. The area of land available in one community is shown on the plan.

a Calculate the area of the available land.

b To prepare the soil, the community has to lay down 25 kg of soil and 10 kg of compost per square metre of land. Work out how much soil and compost they will need.

c There is a gate 2 m wide along the 22 m boundary. The rest of the land needs to be fenced. Work out the total amount of fencing needed.

Section 2: Area of circles and sectors

Tip

Use the π key of your calculator to find the area and circumference of circles unless you are given an approximate value to use.

Leave the value you get on the display for the next step and only round off to the required number of places when you have a final value.

Remember always to calculate r^2 before you multiply by π. The rules of BODMAS apply.

In Chapter 11 you worked with the diameter, radius and circumference of circles and sectors of circles. You also used approximate and calculator values of π in circle calculations. You will meet these terms again in this section.

The area of a circle is calculated using the formula:

area, $A = \pi r^2$

where r = radius.

If you are given the diameter, d, you can still use this formula by remembering that:

$r = \frac{1}{2}d$

Learn this formula

Area of a circle, $A = \pi r^2$ where r = radius

WORKED EXAMPLE 5

Calculate the area of this circle correct to 2 decimal places.

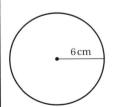

$A = \pi r^2$

$= \pi \times 6 \times 6 = 113.0973355 \text{ cm}^2$

$= 113.10 \text{ cm}^2$ (to 2 decimal places)

When you know the area of a circle you can find the length of a radius (or the diameter).

Tip

If you are asked to find the diameter (d) remember it is twice the radius ($2 \times r$).

WORKED EXAMPLE 6

Calculate the radius of a circle with area $50\,\text{cm}^2$.

$A = \pi r^2$

$50 = \pi \times r^2$

$r^2 = \dfrac{50}{\pi}$

$r = \sqrt{\dfrac{50}{\pi}} = 3.989422804$

$ = 3.99\,\text{cm}$ (to 2 decimal places)

Area of a sector

Tip

A semi-circle is half a circle, so its area is half the area of a circle. $\left(\dfrac{\pi r^2}{2}\right)$

A quarter-circle is one quarter of a circle, so its area is one quarter of the area of a circle. $\left(\dfrac{\pi r^2}{4}\right)$

A sector is a fraction of the area of the whole circle. To find the area of a sector you need to know what fraction the sector is of the circle.

You find this by dividing the sector angle by 360 in the same way that you did to find the arc length in Chapter 11.

WORKED EXAMPLE 7

Calculate the area of the sector shown.

Area of the sector $= \dfrac{\theta}{360} \times \pi r^2$

$= \dfrac{135}{360} \times \pi \times r^2$

$= 0.375 \times \pi \times 14^2$

$= 230.90706$

$= 230.91\,\text{m}^2$ (to 2 decimal places)

Did you know?

A circle can also be split into major and minor segments by a chord.

EXERCISE 12C

Use the calculator value of π and give your final answers correct to 2 decimal places.

1 Find the area of each circle.

a 9 cm   b 12.8 cm   c 14 cm   d 21.3 cm

2 Calculate the area of each sector.

a 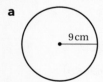 53°

b 18 mm, 105°, 2 cm

c 4 m, 28°

d 122°, 19 mm

3 What is the difference in area between these two sectors?

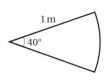

4 A pizza has a diameter of 14 cm.
  a Calculate the area of the pizza.
  b Estimate the area of a round plate that the pizza can fit onto with about 1 cm space around the edge.

5 A pair of sunglasses has circular lenses each 5.4 cm in diameter.
  a What is the total area of the tinted surface of the lenses?

6 The area for discus at an international event has the dimensions shown in the diagram (right).
  a Calculate the area of the grass in the landing zone.
  b What is the area of the starting circle in the throwing cage?

7 A circular disc has a circumference of 75.398 mm. Show clearly how you could use this information to find the area of the disc.

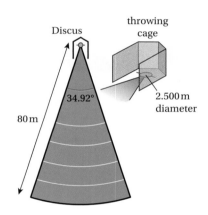

Section 3: Area of composite shapes

Addition of parts

- Divide the figure into smaller known shapes whose area can be found directly.
- Calculate the area of each part separately.
- Add the areas of all the parts to find the total area.

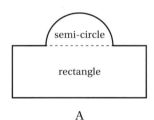 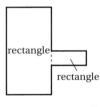 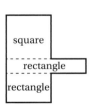

A   B

Figure A can be divided into a rectangle and a semi-circle.

Figure B can be divided in different ways. The first way requires fewer calculations.

Subtraction of parts

When one figure is 'cut out' of another you have to find the area of the larger figure and subtract the area of the cut-out figure.

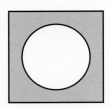

The shaded area = area of square − area of circle.

In some cases, you can find the area by viewing the figure as part of a larger known shape and subtracting the parts that have been 'removed'.

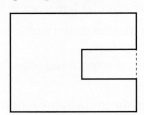

In this shape you can work out the area of the large rectangle (made by drawing the dotted line) and subtract the smaller cut-out rectangle from it. (You could also find this area by addition, but you would need to do more calculations.)

> **Tip**
> Some problems can be solved either way. Think carefully about which method will be more efficient before you decide which one to use.

> **Tip**
> Copy diagrams into your exercise book so you can draw on them and mark dimensions. This helps you keep track of your working.

WORKED EXAMPLE 8

Calculate the area of this shape.

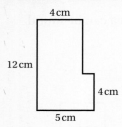

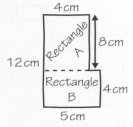

Split the shape into:
Rectangle A: 4 cm × 8 cm
Rectangle B: 4 cm × 5 cm

Area of rectangle A
 = bh = 4 × 8 = 32 cm²
Area of rectangle B
 = bh = 4 × 5 = 20 cm²

Calculate the area of each smaller rectangle.

Make sure you label the working so that someone else knows which calculation goes with each part.

Area of shape = 32 + 20 = 52 cm²

> **Tip**
> This shape can also be split into:
> Rectangle A: 12 cm × 4 cm
> Rectangle B: 4 cm × 1 cm

EXERCISE 12D

1 Find the total area of each of these shapes. Show all your working.

a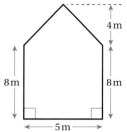

b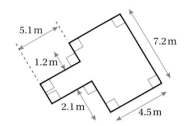

c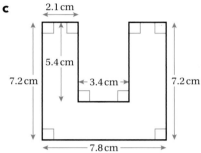

d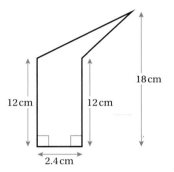

e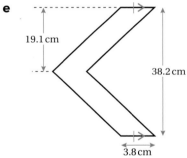

f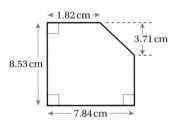

g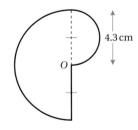

h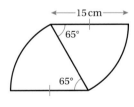

2 Find the area of the shaded part of each figure.

a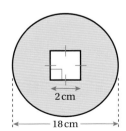

b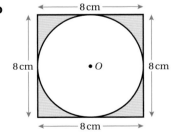

c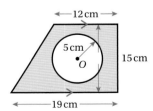

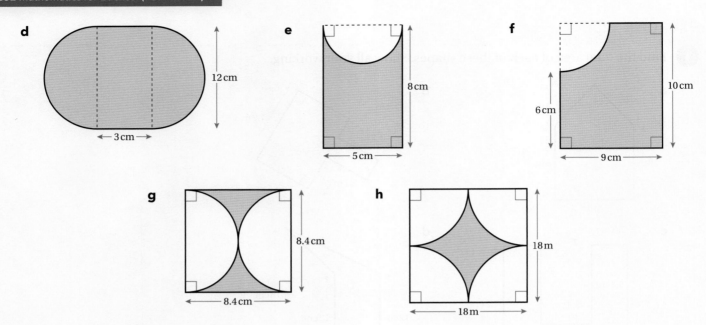

Problem solving

Perimeter, circumference and area are often combined with other calculations in problem-solving situations.

WORKED EXAMPLE 9

This face of a house is to be painted.

If each pot of paint can cover an area of $12\,m^2$, how many pots of paint will be needed?

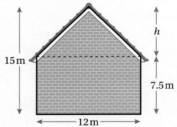

Area of rectangle = $l \times w$
$= 12 \times 7.5 = 90\,m^2$

Area of triangle = $\frac{1}{2} \times b \times h$

Height, h, of triangle = $15 - 7.5 = 7.5\,m$

Area = $\frac{1}{2} \times 12 \times 7.5$
$= 45\,m^2$

Area to be painted = $90 + 45 = 135\,m^2$
Number of pots of paint
$= 135 \div 12 = 11.25$
So 12 pots of paint are needed.

Split the shape into a rectangle and a triangle.
Work out the area of each smaller shape.

Calculate the number of pots needed.

WORKED EXAMPLE 10

A garden pond has a radius of 5 m.

The pond needs netting across the top, and some edging all around the water's edge.

Using 3.14 as the value for π, calculate the quantities of:

a netting and **b** edging required.

The pond lies in a rectangular area, of width 15 m and length 20 m, which is to be made into a lawn.

Grass seed is required for the lawn. Packets of grass seed cover 25 m² and cost £3.75 each.

c What is the cost of seeding the lawn area?

a Area of pond = πr^2 = 3.14 × 5 × 5 = 78.5 m²
A minimum of 78.5 m² of netting is required.

> You are told to use the approximate value of 3.14 for π here.

b Circumference of pond = $2\pi r$ = 2 × 3.14 × 5 = 31.4 m
31.4 m of edging is required for the pond.

c Rectangular lawn area = $l \times w$ = 20 × 15 = 300 m²
Area for seeding = 300 − 78.5 (area of pond)
= 221.5 m²

> Use the 'subtraction of parts' to find the area of lawn.
>
> You already have the circular area from part **a**.

221.5 ÷ 25 = 8.86
So, 9 packets of grass seed will be needed.

> Remember, you have to buy whole numbers of packets.

Cost of seed = 9 × 3.75 = £33.75

EXERCISE 12E

1 How many rectangular tiles 20 cm wide by 30 cm tall would you need to tile the area of wall shown?

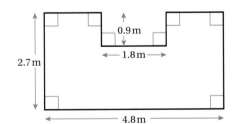

2 The net of a cylinder is shown.

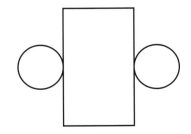

The rectangle has dimensions 10 cm × 16 cm, and each circle has radius 2.55 cm. Using 3.14 as an approximate value of π, calculate the total surface area of the cylinder.

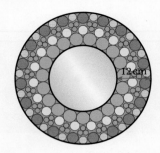

3 The diagram (left) shows a circular mirror that has a decorative metal surround.

The mirror has an edge which is 12 cm wide. The width of the edge is $\frac{4}{5}$ of the radius of the whole mirror.

Calculate the area that the mirror and its surround would cover on a wall. Use 3.14 as an approximate value of π.

4 A piece of fondant icing is rolled out into the shape of a square. A circle with radius 11 cm is cut out from the square.

Given that the largest circle possible is cut out, find the difference between the area of the circular icing and the area of the square.

5 A semi-circular sandpit sits at the end of a lawn. A cover is to be placed over the sandpit.

What is the area that needs to be covered?

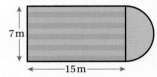

6 A circular photo frame has a plastic surround. The width of the plastic surround is 8 cm. The diameter of the complete photo frame is 28 cm.

Calculate the area available for the photo.

7 The shapes below have the same area. What is the side length of the square?

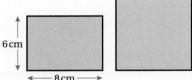

8 Guy is paid £0.15 for every square metre of grass he cuts.

How much would he be paid for cutting the grass in this garden?

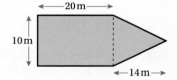

9 A stained glass window is a semi-circle with radius 30 cm. Calculate:

a the perimeter of the window.

b the area of glass.

 Checklist of learning and understanding

Area of polygons

- Area is the amount of space occupied by a plane shape. Area is always given in square units as it describes 2D shapes. You find the area by multiplying the two dimensions.

- Area can be calculated using formulae.

Rectangle	Square	Triangle	Parallelogram	Trapezium
$A = lw$	$A = l^2$	$A = \frac{1}{2}bh$	$A = bh$	$A = \frac{1}{2}(a+b)h$

Area of circles

- Area of a circle = πr^2.
- Area of a circle sector is a fraction of the area of the whole circle.
- Area of a sector = $\frac{\theta}{360} \times \pi r^2$.

Composite shapes

- The area of composite shapes can be found by splitting them into known shapes.
- Areas of smaller shapes can be calculated separately and added to find the total area.
- The area of a given shape can be subtracted from a known shape to find the total area.

Chapter review

For additional questions on the topics in this chapter, visit GCSE Mathematics Online.

1 A rectangular vegetable plot has an area of 100 m².

If the width is 6.5 m, what is the length of the plot?

2 The net of a square-based pyramid is shown. The square base has side length 5 cm. The triangles have a perpendicular height of 4.3 cm.

Calculate the total surface area of the square-based pyramid.

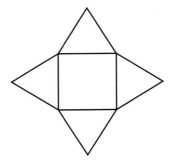

3 A square flower bed, with side length 3 m, sits exactly in the middle of a rectangular lawn of side length 5 m and width 450 cm.

Calculate the area of grass.

Find answers at: cambridge.org/ukschools/gcsemaths-studentbookanswers

4 Mr Weaver's garden is in the shape of a rectangle.

In the garden there is a patio in the shape of a rectangle and two ponds in the shape of circles with diameter 3.8 m.

The rest of the garden is grass.

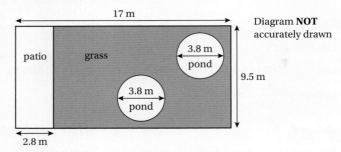

Diagram **NOT** accurately drawn

Mr Weaver is going to spread fertiliser over all the grass.

One box of fertiliser will cover 25 m² of grass.

How many boxes of fertiliser does Mr Weaver need?

You must show your working.

(5 marks)

©Pearson Education Ltd 2012

5 An irrigator in a field can water a circular area of crops within a radius of 14.5 m.

What is the total area of field that can be irrigated?

6 The diagram shows a shape made from a square and a semi-circle. Calculate the area of the shape. Give your answer in cm².

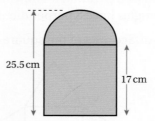

7 Calculate the area of the shape below.

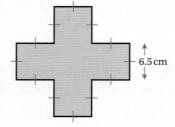

8 A jeweller charges by the centimetre for gold rings. If the charge is £28.50 per cm for a 9 carat ring, what is the cost of a ring to fit a finger that is 2.5 cm wide?

13 Further algebra

In this chapter you will learn how to …
- expand the product of two binomial expressions and use the difference of two squares identity.
- factorise quadratic expressions of the form $x^2 + bx + c$.
- solve problems involving quadratic expressions.

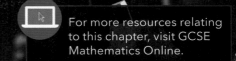

For more resources relating to this chapter, visit GCSE Mathematics Online.

Using mathematics: real-life applications

Situations that involve motion, including acceleration, stopping distance, velocity and distance travelled (displacement) can be modelled using quadratic expressions and formulae.

"At the site of a crash, I measure the length of the tyre skid marks and apply an equation to work out the speed at which vehicles were moving before the accident."

(Police road accident investigator)

Before you start …

Ch 5	Check that you can simplify expressions.	**1**	Simplify. $3x + 6y + 2xy - 2x$
Ch 5	Make sure you can multiply out brackets.	**2**	Expand. $2x(x - y)$
Ch 5	Make sure you can factorise binomial expressions.	**3**	Factorise fully. $27xy - 9x$
Ch 5	You should be able to recognise an identity.	**4**	Is the identity symbol used correctly in each example? Explain why or why not. **a** $3(2a^2 - 4) \equiv 6a^2 - 12$ **b** $2x + 4 \equiv 7x - 8$
Ch 5	You should be able to express situations using algebra.	**5**	3 is added to a number and this new number is multiplied by 6 more than another number. Which expression represents this situation? **A** $(3 + y)x + 6$ **B** $(3 + y)(x + 6)$
Ch 1, 5	Check that you can use the rules for multiplying negative and positive quantities.	**6**	Simplify. **a** 6×-5 **b** -3×-7 **c** $-2a \times b$ **d** $-y \times -y$ **e** $-2a \times -5a$

Find answers at: cambridge.org/ukschools/gcsemaths-studentbookanswers

Assess your starting point using the Launchpad

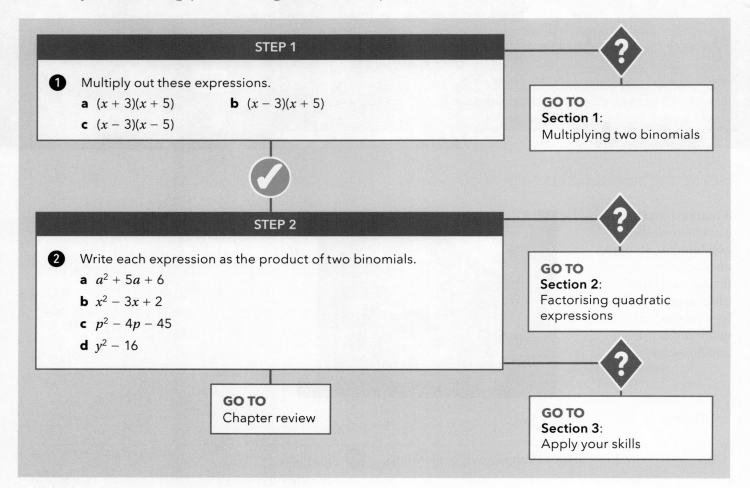

Section 1: Multiplying two binomials

Key vocabulary

binomial: an expression consisting of two terms.

binomial product: the product of two binomial expressions; for example, $(x + 2)(x + 3)$.

Tip

Breadth is another word for width.

A **binomial** is an expression that contains two terms.

For example, $x + 2$ or $2x^2 - 5y^3$

A **binomial product** is the product of two binomials.

For example, $(x + 2)(3x^2 + 4)$

The area of a rectangle is useful for showing how to multiply two binomials.

Consider a large rectangle of length $(a + 2)$ metres and breadth $(b + 3)$ metres.

	a	2
b	ab	$2b$
3	$3a$	6

The area of the whole rectangle is its length × breadth. This is the binomial product $(a + 2)(b + 3)$.

The area of the whole rectangle must also be equal to the sum of the four smaller rectangles.

This means $(a + 2)(b + 3) = ab + 3a + 2b + 6$

To multiply two brackets together each term in the first bracket must be multiplied by each term in the second bracket.

You can use arrows to keep track of your multiplication.

$(a + 2)(b + 3) \quad ab + 3a + 2b + 6$

You can also use a grid to make sure you have multiplied all the terms.

WORKED EXAMPLE 1

Expand and simplify, if possible.

$(x + 3)(x + 5)$.

×	x	3
x	x^2	$3x$
5	$5x$	15

Put the terms of the first binomial on the top row and of the second in the first column.

$(x + 3)(x + 5) = x^2 + 3x + 5x + 15$

Write out the terms in the smaller boxes to show your expansion.

$= x^2 + 8x + 15$

$5x$ and $3x$ are like terms, so add them.

Tip

When the product contains like terms you add these to simplify the expression.

The expression $x^2 + 8x + 15$ is an example of a **quadratic expression**. The highest power of x in the expression is x squared (x^2).

WORKED EXAMPLE 2

Use substitution to determine whether $(a + b)^2 = a^2 + b^2$.

Let $a = 1$ and $b = 2$

Choose small values to make your calculations as simple as possible.

$(a + b)^2 = (1 + 2)^2 = (3)^2 = 9$
$a^2 + b^2 = 1^2 + 2^2 = 1 + 4 = 5$

Calculate the value for each side.

$9 \neq 5$ so the expressions are not identical.

Check to see if the values are the same.

Tip

Finding a binomial product is the same process we could apply to a long multiplication calculation. For example 14×27 can be rewritten as the product of two binomials:

$(10 + 4) \times (20 \times 7)$

We can illustrate this in a grid (the same method we used in Worked example 1)

×	10	4
20	200	80
7	70	28

Total represented in the
grid = 200 + 80 + 70 + 28
 = 378

Key vocabulary

quadratic expression: an expression in which the highest power of x is x^2.

WORKED EXAMPLE 3

Expand and simplify.

a $(x - 2)(x + 9)$ b $(x - 4)(x - 7)$

a $(x - 2)(x + 9) = x^2 + 9x - 2x - 18$
$= x^2 + 7x - 18$

Notice that you get a positive and a negative 'like' term here.

b $(x - 4)(x - 7) = x^2 - 7x - 4x + 28$
$= x^2 - 11x + 28$

Notice that the 'like' terms are both negative here.

Tip

Pay careful attention to the signs when you multiply binomials.

A negative quantity multiplied by a positive quantity gives a negative product.

A negative quantity multiplied by another negative quantity gives a positive product.

WORK IT OUT 13.1

These are the results three students got when they were asked to expand $(x - 5)(x + 6)$.

What did each student do wrong?

What is the correct answer?

Student A:	Student B:	Student C:
$x^2 + 11x + 30$	$2x - 30$	$x^2 + x + 1$

Squaring a binomial

You know that a^2 is short for writing $a \times a$.

So, for a binomial, $(x + 8)^2 = (x + 8)(x + 8)$

WORKED EXAMPLE 4

Expand $(x + 8)^2$

$(x + 8)^2 = (x + 8)(x + 8)$ *Write as a product of two brackets.*

$= x^2 + 8x + 8x + 64$ *Write down the result of multiplying out.*

$= x^2 + 16x + 64$ *Simplify by gathering any like terms.*

EXERCISE 13A

1 Expand and simplify.
  a $(x+2)(x+5)$ **b** $(x-2)(x-5)$ **c** $(x+2)(x-5)$
  d $(x-2)(x+5)$ **e** $(x+3)(x-4)$ **f** $(x+y)(x+y)$

2 Find these products and simplify.
  a $(x-5)(x-1)$ **b** $(a-7)(a-4)$ **c** $(m+4)(m-5)$
  d $(p-6)(p+4)$ **e** $(x-7)(x+6)$ **f** $(x+11)(x-3)$
  g $(x-11)(x-7)$ **h** $(x+8)(x-3)$ **i** $(x-12)(x-6)$

3 Expand and simplify.
  a $(2x+4)(3x+3)$ **b** $(3x+4)(5x+2)$ **c** $(2x-5)(3x+1)$
  d $(4y-3)(5y+1)$ **e** $(3a-5)(2a-1)$ **f** $(2b-5)(b-3)$
  g $(2y-3)(3y-5)$ **h** $(2x+4)(2x-6)$ **i** $(5x-3)(4x-1)$

4 Expand the squares of these binomials.
  a $(x+5)^2$ **b** $(x-5)^2$

5 Expand the squares of each of these binomials.
  a $(x+2)^2$ **b** $(x+7)^2$ **c** $(x-3)^2$
  d $(x-9)^2$ **e** $(2x+1)^2$ **f** $(1-3x)^2$

6 If $A = 3x + 2$ and $B = 2x - 1$ determine:
  a AB **b** $A^2 + B^2$ **c** $(A-B)(A+B)$

7 Find the area, in terms of x, of each of these squares.

  a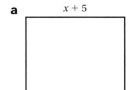
  $x+5$

  b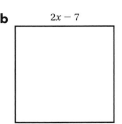
  $2x-7$

  c Calculate the area of each square when $x = 8$.

The difference of two squares identity

Exercise 13B is an investigation. Work through it to find a shortcut for expanding binomials in the form of $(a+b)(a-b)$.

EXERCISE 13B

1 Expand each of the following binomials.
   a $(x+1)(x-1)$
   b $(a+2)(a-2)$
   c $(2x-1)(2x+1)$
   d $(x-2y)(x+2y)$

2 How many terms are there in your answers? Can you explain why this happens?

3 What is special about each term in the binomials you expanded?

4 Write down a rule that you can use to quickly find the answer to any similar expansion.

5 Copy this expansion and fill in the gaps.
$(x+y)(x-y) = x^2 + \square - xy - \square$
$= \square - \square$

6 The example in question **5** shows that $(x+y)(x-y) \equiv x^2 - y^2$

This is called the **difference of two squares identity**.

   a Which are the two squares?
   b How can you recognise when a binomial expansion is a difference of two squares?
   c Is $(3y+2x)(2x-3y)$ a difference of squares? Explain why or why not.

> **Tip**
>
> Do you remember how to factorise an expression by taking out a common factor?
>
> Revise that section in Chapter 5 if you are not sure.

Section 2: Factorising quadratic expressions

In Chapter 5 you learned that factorising an expression is the opposite of expanding it.

$2(x+7)$ is the factorised form of $2x+14$

Expanding a binomial such as $(x+3)(x+4)$ gives you a quadratic expression.

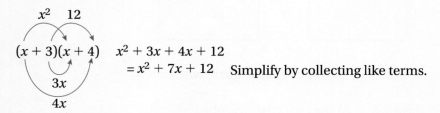

$(x+3)(x+4) \quad x^2 + 3x + 4x + 12$
$\qquad\qquad\qquad = x^2 + 7x + 12 \quad$ Simplify by collecting like terms.

Factorising the quadratic expression means writing it as the product of its factors.

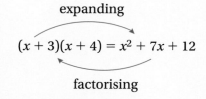

$(x+3)(x+4) = x^2 + 7x + 12$

$x^2 + bx + c$ is a quadratic expression.

In this expression x is the variable, b is the **coefficient** of x and c is a **constant**.

WORK IT OUT 13.2

Two students were asked to identify the coefficient of x^2 in the expression $x^2 + bx + c$

The first student said the coefficient was 1 and the second student said that x^2 has no coefficient.

Which student is correct?

How would you explain to the other student why their answer is wrong?

Key vocabulary

coefficient: the number in front of a variable in a mathematical expression. In the term $5x^2$, 5 is the coefficient and x^2 is the variable.

constant: in algebra, a constant is a value that does not change. It is usually a number. It could be a letter with a fixed value, like π.

In general:

$(x + a)(x + b) = x^2 + (a + b)x + ab$

So, the coefficient of the middle term is the sum of a and b in the original binomials, or $(a + b)$.

The third term is the product of a and b or (ab).

This pattern can be used to develop a strategy for factorising quadratic expressions.

Problem-solving framework

Factorise $x^2 + 7x + 12$.

Steps for factorising a quadratic expression	What you would do for this example
Step 1: Write down the expression you have been asked to factorise.	$x^2 + 7x + 12$
Step 2: Write two pairs of brackets, writing x in each bracket. You can put x in each bracket because to get x^2 you need to multiply x by x.	$x^2 + 7x + 12$ $= (x + \Box)(x + \Box)$
Step 3: Now look at the value of the constant. What are the factor pairs of this constant?	Factor pairs are: 1×12 2×6 3×4
Step 4: Which factor pair, when added, will give the coefficient of the middle term?	$3 + 4 = 7$ $3 \times 4 = 12$ So, $x^2 + 7x + 12 = (x + 3)(x + 4)$
Step 5: Check your answer.	Expand the brackets to check your answer: $(x + 3)(x + 4) = x^2 + 3x + 4x + 12$ $\qquad\qquad\qquad\;\; = x^2 + 7x + 12$ Yes, this is the expression you started with.

Find answers at: cambridge.org/ukschools/gcsemaths-studentbookanswers

Quadratic expressions that have negative terms need a bit more care.

WORKED EXAMPLE 5

Factorise $x^2 - 7x + 12$.

$x^2 - 7x + 12 = (x - \square)(x - \square)$

> Start by writing two pairs of brackets with x to the left of each.
>
> Put negative signs in **both** brackets because the constant is $+12$ but the coefficient of x this time is a negative number, -7.

-1×-12
-2×-6
-3×-4

> Write the factor pairs of the constant term.
>
> The product of two negative numbers is positive so the factor pairs must be negative.

$-3 + (-4) = -7$

> Determine which factor pair, when added, will give the coefficient of the middle term.

$x^2 - 7x + 12 = (x - 3)(x - 4)$

> Complete your solution. (You can expand the brackets to check that it is correct.)

The example below shows you how to work systematically to factorise a quadratic expression which has two negative terms.

WORKED EXAMPLE 6

Factorise $x^2 - 4x - 12$

$x^2 - 4x - 12 = (x + \square)(x - \square)$

> Start by writing two pairs of brackets with x to the left of each.
>
> The constant is negative.
>
> To get a negative product you have to multiply a negative number by a positive number. This means one bracket will have a negative sign and the other will have a positive sign.

1×12
2×6
3×4

> Write the factor pairs of the constant term.
>
> Which of these pairs has a difference of 4 (the value of the number in front of the x term)?

$-2 + 6 = 4$
$-6 + 2 = -4$

> There is a difference of 4 between 2 and 6.
>
> So this is the pair you need.
>
> Notice that the middle term is negative, so the larger number of the factor pair needs to be negative.

$x^2 - 4x - 12 = (x + 2)(x - 6)$

> Complete your solution. (Expand the brackets to check that it is correct.)

Work out the signs before you factorise by looking at the signs in the expression.

If the **constant is positive**, the brackets will have the same sign.

- If the middle term (the x term) is positive, both brackets will have positive signs.
- If the middle term is negative, both brackets will have negative signs.

If the **constant is negative**, the brackets will have different signs.

- When the signs are different, the middle term is the difference between the two factors.
- The largest number in the factor pair will have the same sign as the middle term in the expression.

When you factorise any expression, the first step should be to check for, and remove, common factors.

WORKED EXAMPLE 7

Factorise $4x^2 - 12x - 40$

$4x^2 - 12x - 40$
$= 4(x^2 - 3x - 10)$ — Take out the common factor of 4.
$= 4(x - 5)(x + 2)$ — Factorise the quadratic.

EXERCISE 13C

1 Find two numbers that meet each set of conditions.

  a Have a sum of 5 and a product of 6.
  b Add to give 8 and multiply to give 7.
  c Have a product of −8 and a sum of 2.
  d Multiply to give 24 and add to give −10.
  e Produce −24 when multiplied, and sum to 5.
  f Have a sum of 3 and a product of −18.

2 Factorise each of the following.

  a $x^2 + 14x + 24$ **b** $x^2 + 3x + 2$ **c** $x^2 + 7x + 12$
  d $x^2 + 12x + 35$ **e** $x^2 + 12x + 27$ **f** $x^2 + 7x + 6$
  g $x^2 + 11x + 30$ **h** $x^2 + 10x + 16$ **i** $x^2 + 11x + 10$
  j $x^2 + 8x + 7$ **k** $x^2 + 24x + 80$ **l** $x^2 + 13x + 42$

3 Factorise each of the following.

  a $x^2 - 8x + 12$ **b** $x^2 - 9x + 20$ **c** $x^2 - 7x + 12$
  d $x^2 - 6x + 8$ **e** $x^2 - 12x + 32$ **f** $x^2 - 14x + 49$
  g $x^2 - 8x - 20$ **h** $x^2 - 7x - 18$ **i** $x^2 - 4x - 32$
  j $x^2 + x - 6$ **k** $x^2 + 8x - 33$ **l** $x^2 + 10x - 24$

4 Factorise fully.

  a $2x^2 + 6x + 4$ **b** $6x^2 - 24x + 18$ **c** $5x^2 - 5x - 10$
  d $2x^2 + 14x + 20$ **e** $2x^2 + 4x - 6$ **f** $3x^2 - 30x - 33$

Find answers at: cambridge.org/ukschools/gcsemaths-studentbookanswers

Factorising the difference of two squares

In Exercise 13B, you saw that multiplying out binomials in the form $(a + b)(a - b)$ gives a product that is the difference of two squares.

$$(a - b)(a + b) = a^2 + ab - ab - b^2$$
$$= a^2 - b^2 \text{ (simplifying)}$$

The reverse of this expansion is factorising the difference of two squares.

WORKED EXAMPLE 8

Factorise:

a $x^2 - 4$ **b** $x^2 - 36$ **c** $4a^2 - 9$ **d** $100 - y^2$

a $x^2 - 4 = x^2 - (2)^2$ *Express both terms as squares.*

$= (x + 2)(x - 2)$ *Apply the identity: $a^2 - b^2 \equiv (a + b)(a - b)$.*

b $x^2 - 36 = x^2 - (6)^2$
$= (x + 6)(x - 6)$

c $4a^2 - 9 = (2a)^2 - (3)^2$
$= (2a + 3)(2a - 3)$

d $100 - y^2 = (10)^2 - y^2$
$= (10 + y)(10 - y)$

Tip

You can also think of factorising a difference of squares as taking the square root of each term and writing these in brackets, one with a negative sign and one with a positive sign. The order in which you write down the brackets doesn't matter.
$(a - b)(a + b) = (a + b)(a - b)$

Mathematically a difference of squares such as $x^2 - 4$ is a special case of the quadratic expression $x^2 + bx + c$.

In a difference of squares, the coefficient of x is 0, so there is no x term ($x \times 0 = 0$) and the constant (c) is a negative number.

Using the difference of two squares in number problems

You can use the difference of squares to subtract square numbers such as $86^2 - 14^2$ without working out the square values, which can be very large numbers.

Tip

This method can also be used to find one of the shorter sides in a right-angled triangle. You will meet this in Chapter 30.

WORKED EXAMPLE 9

Find the value of $86^2 - 14^2$

$86^2 - 14^2 = (86 + 14)(86 - 14)$ *Write the subtraction as the product of its factors.*

$= 100 \times 72$ *Add and subtract the values in each bracket.*

$= 7200$ *Find the product.*

EXERCISE 13D

1 Factorise each of the following.

a $x^2 - 36$ b $p^2 - 81$ c $w^2 - 16$
d $p^2 - 36q^2$ e $144s^2 - c^2$ f $64h^2 - 49g^2$

2 Using $(a - b)(a + b) = a^2 - b^2$, evaluate the following.

a $100^2 - 97^2$ b $50^2 - 48^2$ c $639^2 - 629^2$
d $98^2 - 45^2$ e $83^2 - 77^2$ f $1234^2 - 999^2$

Section 3: Apply your skills

Being able to factorise expressions and expand brackets is very helpful with problems involving shapes.

Using algebra, you can write expressions for lengths of sides or area, and use the skills of expanding and factorising to solve for a missing value.

EXERCISE 13E

1 Decide whether each statement is true or false.

	True or false?
a $(x + 11)(x - 5) = x^2 + 6x - 55$	
b $x^2 + 2x + 4$ is the square of a binomial.	
c $(a - 8)^2 = a^2 - 64$	
d $69^2 - 11^2$ can be solved by calculating $(69 + 11)(69 - 11)$	

2 Write an expression for the area of each shape.

a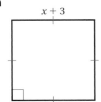
b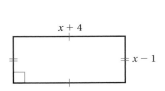
c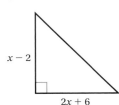

Tip

Remember:
Area of a square = side × side
Area of rectangle
       = length × breadth
Area of triangle
    = $\frac{1}{2}$ × base × height

3 The cost of rubber matting for a children's play area is £19.50 per square metre.

The rectangular play area is $(x + 4)$ metres wide and $(x + 7)$ metres long.

a Write an expression for the area to be covered by rubber matting.
b Write an expression for the cost of the rubber matting.
c Given that $x = 12$, find the cost of the rubber matting.

4 A carpet fitter has a square piece of carpet with sides of x metres.

He does the following sketches to work out how to carpet a rectangular area.

He plans to cut a 60 cm wide strip off the square carpet and place it along the adjacent side of the square as shown in the diagram.

The area marked with a cross will be cut off and thrown away.

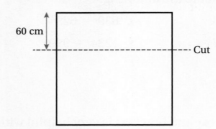
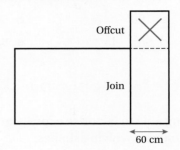

a Express the length and breadth of the rectangular carpet in terms of x.

b Write an expression for the area of the rectangular carpet.

c What type of expression is this?

d Form an expression and work out the difference in the areas of the original square carpet and the rectangular carpet.

5 The area of each rectangle and an expression for the length of one side are given.

Use this information to find an expression for the length of each missing side.

a Area = $8a + 12$

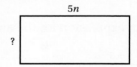

b Area = $10mn + 15$

c Area = $(2y)^2 - 49$

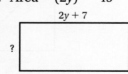

d Area = $x^2 + x - 30$

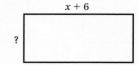

6 Fill in the blanks.

a $(x + 5)(x + 7) = x^2 + \Box x + \Box$

b $(x + \Box)(x + 6) = x^2 + 9x + \Box$

c $(x + 4)(x - \Box) = x^2 - 2x - \Box$

d $(2x + 3)(x + \Box) = 2x^2 + 7x + \Box$

e $(\Box x + \Box)(2x + 5) = 4x^2 + 12x + \Box$

7 Factorise these two quadratic expressions.

  a $x^2 - 11x + 24$

  b $x^2 - 25x + 24$

  c Write another two quadratic expressions with a first term of x^2 and a constant term of 24.

8 Use $a^2 - b^2 = (a + b)(a - b)$ to evaluate $1999^2 - 1998^2$

9 The area of a quadrilateral is expressed as $x^2 - 25$

  a Explain why this quadrilateral cannot be a square.

  b Another quadrilateral has an area of $x^2 + 10x + 25$

  Can it be a square? How do you know that?

Checklist of learning and understanding

Expanding binomials

- A binomial expression is one that contains two terms.
- Expand means to multiply out the terms in the brackets.
- After you expand binomials you simplify further by collecting any like terms.
- The square of a binomial is the binomial multiplied by itself.
  $(a + b)^2 \equiv (a + b)(a + b) \equiv a^2 + 2ab + b^2$
  $(a - b)^2 \equiv (a - b)(a - b) \equiv a^2 - 2ab + b^2$
- A binomial in the form of $(a - b)(a + b)$ gives a product that is a difference of two squares.
  $(a - b)(a + b) = a^2 - b^2$

Factorising

- Factorising is the inverse operation to multiplying out brackets.
- You can factorise by taking out a common factor.
  $6ab + 3ad \equiv 3a(2b + d)$
- You can factorise a quadratic expression by writing it as a product of its two binomial factors.
  $x^2 - x - 6 \equiv (x + 2)(x - 3)$
- The rules for multiplying positive and negative signs are important when you factorise quadratic expressions.

For additional questions on the topics in this chapter, visit GCSE Mathematics Online.

Tip

Expand and simplify each bracket first and then simplify by collecting like terms.

Chapter review

1 Expand and simplify by collecting like terms.

   a $(x - 2)^2 + (x - 4)^2$

   b $(x - 2)^2 + (x + 2)^2$

2 Fill in the blanks:

   a $(x + 3)(x - \square) = x^2 - 2x - \square$

   b $(x + 3)(x + \square) = x^2 + 10x + \square$

   c $(x + 2)(x - \square) = x^2 - x - \square$

   d $(x + 6)(\square + \square) = x^2 + 11x + 30$

   e $(x + 4)(\square + \square) = x^2 + 10x + 24$

3 **a** Expand and simplify $(x + 5)(x - 8)$ *(2 marks)*

   b Factorise $x^2 - 16$ *(1 mark)*

   ©*Pearson Education Ltd 2012*

4 A rectangular field has an area of $x^2 - 4x - 5$ metres.

   a Express the length and breadth of the field in terms of x.

   b Write an expression in simplest terms for the perimeter of the field.

5 Use the difference between two squares to simplify the expression $(x + 8)^2 - (x - 8)^2$

6 Write an expression in its simplest form for the area of the shaded part of this rectangle.

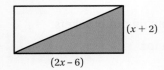

$(x + 2)$

$(2x - 6)$

7 The sides of a triangle are $(x + 8)$ cm, $(x + 6)$ cm and $(x - 1)$ cm.

If the square of the longest side is equal to the sum of the squares of the other two sides, the triangle is right-angled.

   a Which is the longest side in this triangle? How do you know that?

   b What is the square of the longest side?

   c Find the sum of the squares of the other two sides.

   d Show that, if $x = 9$, the triangle right-angled.

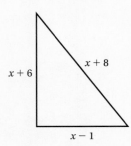

14 Equations

In this chapter you will learn how to ...
- solve linear equations and apply them in context.
- solve quadratic equations.
- set up and solve simultaneous equations.
- use graphs to find approximate solutions to equations.

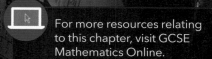

For more resources relating to this chapter, visit GCSE Mathematics Online.

Using mathematics: real-life applications

Accounting involves a great deal of mathematics. Accountants set up computer spreadsheets to calculate and analyse data. Programs such as Microsoft Excel® work by applying different equations to values in columns or cells, so you need to know what equations or formulae to use to get the results you need.

"Although the computer does the actual calculations, I have to insert different equations to tell it what operations to perform and in which order to perform them."

(Accountant)

Before you start ...

Ch 5	Check you can write an equation to represent a problem mathematically.	**1** Which of the equations below correctly represents this problem? "I think of a number, multiply it by 6 and add 1. The answer is 37. What is my number?" **A** $6x + 1 = 37$ **B** $y \times 6 = 37 + 1$ **C** $6a = 37$ **D** $6(x + 1) - 37 = 0$
Ch 1	You should be able to recognise and apply inverse operations.	**2** Complete the following statements. **a** $7 + \square = 0$ **b** $\square - 8 = 0$ **c** $-4a + \square = 0$ **d** $5 \times \square = 1$ **e** $\dfrac{1}{6} \times \square = 1$ **f** $\square \times 12x = x$
Ch 13	You need to know how to factorise quadratic expressions.	**3** Match each expression to its factors. **a** $x^2 - 5x + 6$ **A** $x(x + 3)$ **b** $x^2 + 3x$ **B** $(x + 5)(x - 5)$ **c** $x^2 - 25$ **C** $(x - 2)(x - 3)$ **d** $x^2 - 5$ **D** $(x + \sqrt{5})(x - \sqrt{5})$

Find answers at: cambridge.org/ukschools/gcsemaths-studentbookanswers

Assess your starting point using the Launchpad

STEP 1

1 Match each equation to its solution.

Equations
a $x + 7 = 19$ **b** $x - 6 = 11$ **c** $2x + 5 = 7$ **d** $8x = -24$ **e** $2 - 3x = 8$

Solutions
A $x = 1$ **B** $x = 17$ **C** $x = -2$ **D** $x = 12$ **E** $x = -3$

f How can you check whether a solution is correct?

2 Solve.
a $9a - 7 = 7a + 3$ **b** $3(x + 5) = 2(x + 6)$

3 When 16 is added to twice Jack's age, the answer is 44.
Write an equation and solve it to find Jack's age.

GO TO
Section 1:
Linear equations

STEP 2

4 If $x^2 - 2x - 3 = 0$, which pair of values is the solution?
A $x = 3$ or $x = -1$ **B** $x = -3$ or $x = 1$

5 What are the possible values of x given that $x^2 - 16 = 0$

GO TO
Section 2:
Quadratic equations

STEP 3

6 a How many positive whole number solutions can you find for $x + y = 6$?
b Which of those solutions are correct if $x + y = 6$ and $x - y = 4$?

GO TO
Section 3:
Simultaneous equations

GO TO
Step 4:
The Launchpad continues on the next page …

Launchpad continued ...

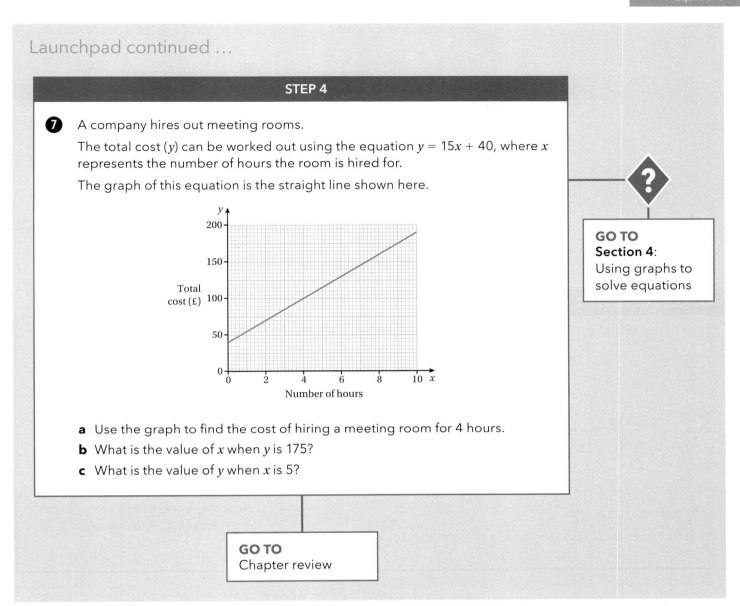

Section 1: Linear equations

An equation is a mathematical statement that contains an equal sign. For example:

$3 + 2 = 5$ $3 + x = 5$ $3 + 2 = x$ $2x + 3 = 6$

An **unknown** value can be represented by any letter but x and y are used most often.

The same letter can represent different values in different equations.

In the equation $x + 1 = 4$, the value of x is 3, but in the equation $x + 2 = 3$, the value of x is 1.

If the highest power of the unknown is 1 the equation is a **linear equation**.

Key vocabulary

unknown: part of an equation which is represented by a letter.

linear equation: an equation where the highest power of the unknown is 1, for example $x + 3 = 7$ (x means x^1, but you don't need to write the '1').

The = sign in an equation is like the pivot point of a set of balance scales.

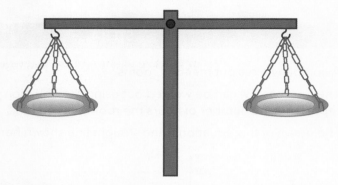

If you add or subtract, multiply or divide by the same amount to each pan, the scales will remain in balance.

WORKED EXAMPLE 1

Show that performing the same operations to both sides of the equation $x = 3$ gives an equivalent equation.

	$x = 3$
Add 5	$x + 5 = 3 + 5$
Now	$x + 5 = 8$
Subtract 1	$x + 5 - 1 = 8 - 1$
Now	$x + 4 = 7$
Multiply by 2	$2(x + 4) = 2(7)$
Now	$2x + 8 = 14$

To keep the equation balanced, do the same operation on both sides.

Is the solution to $2x + 8 = 14$ still $x = 3$?

$2(3) + 8 = 6 + 8 = 14$

The left-hand side (LHS) is equal to the right-hand side (RHS) so $x = 3$

Check this by substituting $x = 3$ into the left-hand side of the equation.

Solving an equation involves working out the value of the unknown letter.

In simple equations like $x + 3 = 7$ you can solve for x by inspection. You can see by looking that the only value that x can be is 4.

In more complex equations you can find the solution by carrying out inverse operations on both sides of the equation.

Tip

The symbol $\therefore$ is used a lot in mathematics to save time writing. It means 'therefore'.

WORKED EXAMPLE 2

Solve each equation for x.

a $2x + 8 = 14$ **b** $2(x + 1) = 10$

a $2x + 8 = 14$

$\therefore 2x + 8 - 8 = 14 - 8$ Subtract 8 from both sides.

$\therefore 2x = 6$

Continues on next page …

$$\therefore \frac{2x}{2} = \frac{6}{2}$$

2x means 2 × x, so divide each side by 2 to find x.

$$\therefore x = 3$$

Note: $2x + 8 = 14$ was the end result of Worked example 1, so you would expect the answer to be $x = 3$.

b $2(x + 1) = 10$

$$\therefore \frac{2(x+1)}{2} = \frac{10}{2}$$

Get rid of × 2 by dividing both sides by 2.

$$\therefore x + 1 = 5$$

$$\therefore x = 4$$

Subtract 1 from both sides.

Tip

The examples here show each step in detail to help you understand the process of solving an equation. In your own work you might find it more efficient to combine steps or leave out some of the working.

Tip

If the number multiplying the bracket is not a factor of the number on the right, you can expand the brackets and then proceed as in part **a**.

EXERCISE 14A

1 Work with another student.

Solve each equation by inspection.

Check your answers by substitution.

a $x + 11 = 8$ **b** $x - 6 = 11$ **c** $-2x = 16$

d $x + 7 = 29\frac{1}{2}$ **e** $3x = -24$ **f** $6x - 21 = 3$

2 Solve each equation by writing down at least two steps. Check your answers.

a $2a - 5 = 7$ **b** $3b + 4 = 19$ **c** $5d - 7 = 23$

d $3e - 2 = 16$ **e** $5h + 21 = 11$ **f** $6a + 17 = -1$

g $3a - 16 = -31$ **h** $4b + 12 = -16$ **i** $2t - 8 = 5$

3 Solve each equation. Try to do this without expanding the brackets.

a $2(x + 3) = 8$ **b** $3(x - 2) = 15$

c $4(b - 1) = 12$ **d** $5(p + 1) = 10$

e $3(x - 3) = 18$ **f** $2(y + 4) = 14$

g $2(3x + 4) = 0$ **h** $2(y - 3) = -8$

i Explain how not expanding the brackets in these equations means you can solve them in fewer steps.

4 Consider the equation $2(x + 1) - 3(x - 2) = 6$.

a Why does it make sense to expand the brackets before solving the equation in this example?

b Expand the brackets, collect like terms on the left of the equation and then solve for x.

Tip

When you are asked to solve an equation for x you cannot give an answer in the form of $-x = 3$ or $-x = -4$. If you end up with a negative unknown, divide both sides of the equation by negative 1 to get a positive value ($- \div - = +$). Or you can think of it as multiplying both sides by negative 1 ($- \times - = +$).

Equations with the unknown on both sides

Some equations have the unknown in expressions on both sides of the equation.

In these equations you take steps to get all the terms containing the unknown onto the same side of the equation.

The expressions may also contain brackets. If that is the case, you need to expand the brackets first.

> **Tip**
>
> Remember you can combine steps in your own working. These examples show all the steps in detail.

WORKED EXAMPLE 3

Solve for x.

a $5x - 5 = 3x + 1$ **b** $2y + 17 = 5 - 6y$ **c** $2(3x - 1) = 2(x + 1)$

a $5x - 5 = 3x + 1$
$5x - 5 - 3x = 3x + 1 - 3x$ — Subtract $3x$ from each side.
$2x - 5 = 1$ — Add like terms.
$2x - 5 + 5 = 1 + 5$ — Add 5 to each side.
$2x = 6$ — Simplify.
$x = 3$ — Divide both sides by 2.

b $2y + 17 = 5 - 6y$
$2y + 17 + 6y = 5 - 6y + 6y$ — Add $6y$ to both sides (this helps you get rid of negative signs).
$8y + 17 = 5$ — Add like terms.
$8y + 17 - 17 = 5 - 17$ — Subtract 17 from each side.
$8y = -12$ — Simplify.
$\dfrac{8y}{8} = \dfrac{-12}{8}$ — Divide both sides by 8.
$y = \dfrac{-3}{2}$ — Reduce the fraction to its simplest terms.

c $2(3x - 1) = 2(x + 1)$
$6x - 2 = 2x + 2$ — Expand the brackets paying attention to the signs.
$6x - 2 - 2x = 2x + 2 - 2x$ — Subtract $2x$ from each side.
$4x - 2 = 2$ — Add like terms.
$4x - 2 + 2 = 2 + 2$ — Add 2 to each side.
$4x = 4$ — Divide both sides by 4.
$x = 1$

EXERCISE 14B

1 Solve the following equations. Check each answer by substitution.
- **a** $13x + 1 = 11x + 9$
- **b** $5t - 4 = 3t + 6$
- **c** $4y + 3 = 2y + 8$
- **d** $5y + 1 = 3y + 13$
- **e** $3y + 10 = 5y + 3$
- **f** $12x + 1 = 7x + 11$
- **g** $5x - 2 = 3x + 6$
- **h** $5x + 12 = 20 - 11x$
- **i** $8 - 8a = 9 - 9a$
- **j** $5x + 3 = 2(x + 2)$

2 Solve these equations by expanding the brackets first.
- **a** $2(x + 1) + 4(x + 2) = 22$
- **b** $3(x + 1) = 2(x + 2)$
- **c** $5(t + 3) = 3(2t + 1)$
- **d** $7(x + 2) = 4(x + 5)$
- **e** $4(x - 2) + 2(x + 5) = 14$
- **f** $3(x + 1) = 2(x + 1) + 2x$
- **g** $-2(x + 2) = 4x + 9$
- **h** $4 + 2(2 - x) = 3 - 2(5 - x)$

3 a Try to solve these two equations.
  - **i** $2(x + 8) - 3x = x + 16$
  - **ii** $4(3 + x) + 4x = 4(2x + 3)$

b If an equation is true for any value of x, it is an identity.
   Which of these two equations is an identity? How do you know this?

c Do you think there is a solution for the other equation? Give a reason for your answer.

Forming and solving linear equations

When you set up an equation you must say what the letters stand for.

WORKED EXAMPLE 4

The sum of two numbers is 54. If one number is 14 more than the other number, find the two numbers.

Let one of the numbers be x.
 — Start by giving a letter to one of the unknown values.

So the other number is $(x + 14)$.
 — You are told that the other number is 14 more than this.

$x + (x + 14) = 54$
 — Use the information that the two numbers add up to 54 to form an equation.

Continues on next page ...

$2x + 14 = 54$
$2x = 54 - 14$
$2x = 40$
$x = 20$

Solve the equation to find one of the numbers.

$x = 20, \therefore x + 14 = 34$

Use the value of x to find the value of the other number.

The two numbers are 20 and 34.

Check that this works:
$20 + 34 = 54$
Yes, the solution is correct.

When the problem involves a shape it is useful to draw a sketch and label it to help you set up the equation.

WORKED EXAMPLE 5

In a triangle, the largest angle is four times the size of the smallest angle.

The third angle is 24° bigger than the smallest angle.

a Write an equation and solve it to find the size of each angle in this triangle.

b What type of triangle is this?

a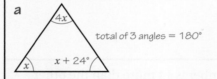

total of 3 angles = 180°

Draw a sketch and mark the angles using the information given.

Let the smallest angle be x.

The greatest angle is therefore $4x$.

The other angle is $(x + 24)$

$x + 4x + x + 24 = 180$
$6x = 180 - 24$
$6x = 156$
$x = 26$

You know that the sum of angles in a triangle is 180°.

So $4x = 4 \times 26 = 104$
And $x + 24 = 26 + 24 = 50$

Use $x = 26$ to find the size of the other angles.

The angles are 26°, 50° and 104°.

b The triangle is obtuse-angled and scalene.

WORKED EXAMPLE 6

The length of a rectangle is three times its width.

If the perimeter is 24 centimetres, find the area of the rectangle.

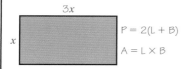

$P = 2(L + B)$
$A = L \times B$

Let the width be x.

∴ the length is $3 \times x = 3x$.

Draw a sketch and label it.

$2(x + 3x) = 24$
$x + 3x = 12$
$4x = 12$
$x = 3$

The width is 3 cm, the length is $3 \times 3 = 9$ cm.

The area is $3 \times 9 = 27$ cm²

Use the information on perimeter to solve for x.

Find the lengths of the sides so you can use them to work out the area.

Tip

Choose a letter to stand for the unknown value and say what that represents. We often choose x, but you can choose any letter.

Tip

Identify what value the letter is going to represent.

EXERCISE 14C

1 For each of the following, write an equation and solve it to find the unknown number.

  a Three times a certain number is 348, what is the number?

  b 7 less than a number is -2, what is the number?

  c 6 greater than a number is -4, what is the number?

  d Two less than four times a number is 66, what is the number?

  e Two consecutive numbers have a sum of 63, what are the numbers?

  f Three less than twice a number is -2. What is the number?

2 Form an equation and solve it to answer each question.

  a When 16 is added to twice Lucy's age, the answer is 44.

  How old is Lucy?

  b Stephen buys 8 pens and receives 80p change from £20.00.

  How much does a pen cost, assuming each pen costs the same amount?

  c Multiplying a number by 2 and then adding 5 gives the same answer as subtracting the number from 23.

  What is the number?

  d Nick is 20 years older than his daughter.

  His daughter has worked out that in five years' time, she will be half her father's age.

  How old is she now?

3 A square has sides of $3x$ cm. A parallelogram has sides of $2x$ cm and $(x + 9)$ cm.

  a Write an expression for the perimeter of the square.

  b Write an expression for the perimeter of the parallelogram.

  c Given that the two quadrilaterals have the same perimeter, form an equation and solve it to find the length of one side of the square.

Find answers at: cambridge.org/ukschools/gcsemaths-studentbookanswers

Key vocabulary

quadratic equation: an equation that contains a variable squared term, like x^2, but no variable term with a power greater than 2. $x^2 = 4$ and $x^2 + 2x - 6 = 0$ are quadratic equations. $x^3 - 1 = 0$ and $6x + 7 = 35$ are not quadratic equations.

Tip

Some equations have only one root. For example
$x^2 - 2x + 1 = 0$ and
$x^2 + 4x + 4 = 0$
$(x - 1)(x - 1) = 0$
$(x + 2)(x + 2) = 0$
$x = 1$
$x = -2$

The root is said to repeat. Here the graphs of the equations will just touch the x-axis at the root.

Others equations are said to have no real roots. This means that the graphs of these equations do not cut the y-axis.

In solving some problems using quadratic equations, only one of the roots will be valid.

Key vocabulary

roots: the individual values of x in a quadratic equation.

solution: both possible values of x in a quadratic equation.

Tip

You will deal with graphs of quadratic equations in more detail in Chapter 24.

4 The area of this rectangle is $10\,\text{cm}^2$.

Calculate the value of x and use it to find the length and width of the rectangle.

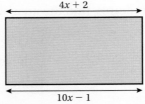

5 Jonty, Mary and Asit invest in a business. Jonty invests three times as much as Mary and Asit invests £80 more than Jonty. If the total investment is £290, how much do they each invest?

Section 2: Quadratic equations

To solve the simple quadratic equation $x^2 = 9$, you have to find the square root of 9. You know that $\sqrt{9} = 3$.

But, 3 is not a complete solution because $\sqrt{9}$ can also be -3. ($-3 \times -3 = 9$)

So, $\sqrt{9} = 3$ or -3, written as $\sqrt{9} = \pm 3$.

Because you are dealing with a squared variable, quadratic equations have two solutions, known as **roots**.

When you are asked to solve a quadratic equation you need to give a **solution** that contains both roots.

The solution to the equation $x^2 = 9$ is $x = \pm 3$

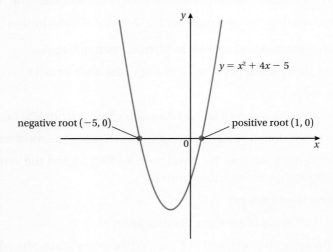

This diagram shows the quadratic function $y = x^2 + 4x - 5$. It cuts the x-axis in two places marked as the positive root and the negative root.

These are the points at which $y = 0$ so you can solve for x by writing the equation as $0 = x^2 + 4x - 5$.

Solving quadratic equations by factorising them

The general form of a quadratic equation is $x^2 + bx + c = 0$ (when the coefficient of x^2 is 1).

Before you learn how to solve quadratic equations by factorising them, you need to consider what it means if the product of two factors equals 0.

For example, if $7 \times a = 0$, then you know that $a = 0$.

Similarly, if $a \times b = 0$ then either $a = 0$ or $b = 0$ or both $a = 0$ and $b = 0$.

This means that if $(x - 1)(x - 2) = 0$ then either $(x - 1) = 0$ or $(x - 2) = 0$.

If $x - 1 = 0$, then $x = 1$.

If $x - 2 = 0$, then $x = 2$.

This zero factor principle allows you to factorise the left-hand side of a quadratic equation and use the fact that one of the factors must be zero to solve it and find the roots of the equation.

Tip

Make sure you remember these three methods of factorising quadratic expressions:

$x^2 - 6x \equiv x(x - 6)$ taking out a common factor
$x^2 + 5x + 4 \equiv (x + 1)(x + 4)$ writing a quadratic expression as a product of binomials
$x^2 - 100 \equiv (x + 10)(x - 10)$ applying the difference of two squares identity

Read through Chapter 13 again if you have forgotten anything.

WORKED EXAMPLE 7

Solve.

a $x^2 - 3x + 2 = 0$ **b** $x^2 - 4 = 0$ **c** $x^2 - 6x = 0$ **d** $x^2 - 8x = -12$

a $x^2 - 3x + 2 = 0$
$(x - 1)(x - 2) = 0$ ◁ Factorise the left-hand side.

Either $(x - 1) = 0$ or $(x - 2) = 0$ ◁ Apply the zero factor principle.

$x - 1 = 0 \qquad x - 2 = 0$ ◁ Solve both equations.
$\therefore x = 1 \qquad \therefore x = 2$

$x = 1$ or $x = 2$ ◁ State the solution.

b $x^2 - 4 = 0$
$(x + 2)(x - 2) = 0$ ◁ Factorise using the difference of squares identity.

Either $(x + 2) = 0$ or $(x - 2) = 0$ ◁ Apply the zero factor principle.

$x + 2 = 0 \qquad x - 2 = 0$ ◁ Solve the equations.
$\therefore x = -2 \qquad \therefore x = 2$

$x = -2$ or $x = 2$ ◁ State the solution.

Continues on next page …

c $x^2 - 6x = 0$
$x(x - 6) = 0$ — Factorise by taking out a common factor of x.

Either $x = 0$ or $(x - 6) = 0$ — Apply the zero factor principle.

$x - 6 = 0$
$\therefore x = 6$ — In this case you already have one value for x so you need only solve one equation.

$x = 0$ or $x = 6$ — State the solution.

d $x^2 - 8x = -12$
$x^2 - 8x + 12 = 0$ — Add 12 to each side to make the right-hand side = 0.

$(x - 2)(x - 6) = 0$ — Factorise the quadratic.

$(x - 2) = 0$ or $(x - 6) = 0$ — Apply the zero factor principle.

$x - 2 = 0 \qquad x - 6 = 0$
$\therefore x = 2 \qquad \therefore x = 6$ — Solve the equations.

$x = 2$ or $x = 6$ — State the solution.

After working through these examples you should be able to see a clear set of steps for solving a quadratic equation in the form of $x^2 + bx + c = 0$.

Step 1
If necessary, take all the terms to the left-hand side so the right-hand side is 0. Make sure the x^2 term is positive. You might need to change signs.

Step 2
Factorise the left-hand side:
- check for common factors
- check for difference of squares
- write quadratic expressions as a product of two terms.

Step 3
Write each factor equal to 0.
Solve the equations to find the roots.

EXERCISE 14D

1 Solve for x.
 a $x^2 - 5x = 0$
 b $x^2 - x = 0$
 c $4x^2 + 8x = 0$
 d $8x^2 - 2x = 0$
 e $5x^2 + 2x = 0$
 f $2x^2 + x = 0$

2 Solve for x.
 a $x^2 - 16 = 0$
 b $100 - x^2 = 0$
 c $x^2 - 1 = 0$
 d $9x^2 - 36 = 0$
 e $4x^2 - 1 = 0$
 f $9x^2 - 4 = 0$

3 Find the roots of each equation.

a $x^2 + 9x + 18 = 0$ b $x^2 - 10x + 9 = 0$ c $x^2 + 2x - 8 = 0$
d $x^2 - 9x + 20 = 0$ e $x^2 - 4x - 12 = 0$ f $x^2 + 4x - 21 = 0$
g $x^2 - 6x + 9 = 0$ h $x^2 + 10x + 25 = 0$ i $x^2 - 14x + 49 = 0$

4 Solve these equations.

a $x^2 - x = 12$ b $(x + 3)^2 = 25$ c $2x^2 + x = 0$
d $4x^2 = 3x$ e $x^2 + 4x = 5$ f $x^2 + 6x - 13 = 14$
g $18 - 3x = -x^2$ h $3x^2 - 21x + 36 = 0$

5 a Is it possible to factorise this quadratic equation applying one of the methods above?

$x^2 - 2x + 2 = 0$

Explain your answer.

b What can you say about the roots of this equation?

Forming and solving quadratic equations

As with linear equations, you can set up and solve quadratic equations.

Always define the letters you are using in your equation.

WORKED EXAMPLE 8

A number is added to its square and the result is 12. What could the number be?

Let the number be x.
∴ its square is x^2.

Define the unknown values.

$x + x^2 = 12$

Set up an equation using the information in the problem.

∴ $x^2 + x - 12 = 0$

Rearrange the equation so that it is in the general form $ax^2 + bx + c = 0$ and the right-hand side is 0.

∴ $(x + 4)(x - 3) = 0$

Factorise the quadratic.

Either $x + 4 = 0$ or $x - 3 = 0$
$x + 4 = 0$
∴ $x = -4$

Solve the equations.

$x - 3 = 0$
∴ $x = 3$

The number could be 3 or −4.

State the solution **in the context of the problem**.

When you use quadratic equations to model real-life situations you might find that one of the solutions is not possible.

For example, if x is the length of the side of a box in metres, and you get the roots $x = 2.5$ or $x = -1.75$ you can ignore the value of -1.75 as this cannot be the length of an object.

You use 2.5 as the length of the side.

Tip

If the problem involves square units then you can probably use a quadratic equation to solve it.

Find answers at: cambridge.org/ukschools/gcsemaths-studentbookanswers

WORKED EXAMPLE 9

A rectangle with an area of 28 cm² has one side 3 cm longer than the other.

How long are each of the shorter sides?

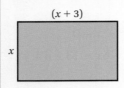

Draw a diagram and label the sides.

Let the shorter side be x.

∴ the longer side is $x + 3$.

$A = x(x + 3)$

Next form your equation.

The area of a rectangle is length × breadth.

$28 = x(x + 3)$

But you are told the area is 28.

$x(x + 3) = 28$
$x^2 + 3x = 28$
$x^2 + 3x - 28 = 0$

Rearrange the equation so that it is in the general form $ax^2 + bx + c = 0$ and the right-hand side is equal to 0.

$(x + 7)(x - 4) = 0$

Factorise.

Either $(x + 7) = 0$ or $(x - 4) = 0$

Apply the zero factor principle.

$x + 7 = 0 \quad x - 4 = 0$
∴ $x = -7 \quad$ ∴ $x = 4$

Solve the equations.

The shorter sides are 4 cm long.

State the answer.

In this problem you can ignore the $x = -7$ solution because -7 cm is not a possible length for the side of a rectangle.

EXERCISE 14E

1 Form an equation and solve it to find the unknown numbers.

  a The product of a certain whole number and four more than that number is 140.

  What could the number be?

  b The product of a certain whole number and three less than that number is 108.

  What could the number be?

c When three times a number is subtracted from the square of the number the answer is 10.

What are possible values of the number?

d The product of two consecutive positive even numbers is 48.

What are the numbers?

2 The base of a triangle is 4 cm longer than twice its height.

If the area of the triangle is 48 cm², calculate its height.

3 A rectangular hall has a floor area of 210 m².

The length of the hall is 1 m more than its width.

Form an equation and solve to find the dimensions of the hall.

4 A wall is 6 m longer than it is tall. If the area of the wall is 16 m², how high is the wall?

Tip

Remember to find the area of a triangle you use the formula

$A = \dfrac{1}{2}bh$

Section 3: Simultaneous equations

The equations in Sections 1 and 2 all had only one unknown (for example, x).

Some equations contain two unknowns. For example, $x + y = 4$.

You can't work out a single solution to this equation. You can find many values for x and y, for example, $1 + 3 = 4$ or $2 + 2 = 4$.

The equation $3x - y = 2$ also has many solutions.

But if you are told that $x + y = 4$ and $3x - y = 2$, you have a pair of equations in two unknowns that are true at the same time for a particular value of x and y.

These are called **simultaneous equations**.

You can use the fact that both these equations are true to solve them at the same time (simultaneously).

Key vocabulary

simultaneous equations: a pair of equations with two unknowns that can be solved at the same time.

WORKED EXAMPLE 10

There are 30 rose bushes in a nursery. Some are red and others are white.

There are twice as many red rose bushes as white.

How many rose bushes of each colour are there?

Let w = the number of white rose bushes, r = the number of red rose bushes.

$r + w = 30$
$w = 2r$

State the letters you will use for your variables.

To solve this you need to set up two equations.

There are 30 rose bushes altogether.

The number of red bushes is twice the number of white bushes.

Continues on next page …

$r + w = 30$

r	0	10	20	30
w	30	20	10	0

You can draw up a table for each equation to show some values of r and w.

$w = 2r$

r	0	5	10	15
w	0	10	20	30

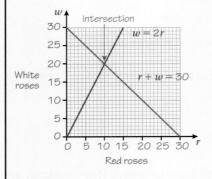

If you draw the graph of each set of values you get two lines that intersect in one place only.

Where the graphs intersect, the values on the axes must be the same.

The coordinates (r, w) of the point of intersection are $(10, 20)$, so $r = 10$ and $w = 20$.

This is the simultaneous solution of the two equations.

$r + w = 30$
$10 + 20 = 30$

Check by substitution in both equations

$w = 2r$
$20 = 2(10) = 20$

You can draw graphs to solve simultaneous solutions but this method takes a long time and it might not be very accurate, especially if the solution values are fractions.

There are two methods of solving simultaneous equations using algebra: substitution and elimination.

The method you choose depends on how the equations are written.

Solving simultaneous equations by substitution

 Tip

This method is suitable when one of the equations already has x or y on its own on one side or when you can easily rearrange one of the equations to have x or y on its own.

Problem-solving framework

Solve these simultaneous equations.

$x - 2y + 1 = 0$ and $y + 4 = 2x$

Steps for solving simultaneous equations by substitution	What you do in this example
Step 1: Number the equations (1) and (2).	$x - 2y + 1 = 0$ (1) $y + 4 = 2x$ (2)
Step 2: Rearrange one of the equations.	Equation (2) is simpler to rearrange. You get: $y = 2x - 4$ (2)
Step 3: Substitute the right-hand side of the rearranged equation into the other equation.	Substitute $(2x - 4)$ in place of y in (1). $x - 2(2x - 4) + 1 = 0$
Step 4: Solve the new equation (which now only has one unknown).	$x - 2(2x - 4) + 1 = 0$ $x - 4x + 8 + 1 = 0$ $-3x + 9 = 0$ $-3x = -9$ $x = 3$
Step 5: Substitute the solution into either of the original equations to find the other unknown.	Substitute $x = 3$ into (2) and solve. $y + 4 = 2(3)$ $y + 4 = 6$ $y = 6 - 4$ $y = 2$
Step 6: Check your solutions to both equations.	Check by substitituting the values $x = 3$ and $y = 2$ back into (1). $3 - 2(2) + 1 = 0$ ✓

EXERCISE 14F

1 Solve the following pairs of simultaneous equations by substitution.

Check that your solutions satisfy **both** equations.

- **a** $y = x - 2$
  $y = 3x + 4$
- **b** $y = 2x + 6$
  $y = 4 - 2x$
- **c** $y = x + 1$
  $x + y = 3$
- **d** $y = x - 2$
  $3x + y = 14$
- **e** $y = 2x + 1$
  $x + 2y = 12$
- **f** $y = 1 - 2x$
  $x + y = 2$

2 Solve these simultaneous equations.

- **a** $x + 2y = 11$
  $2x + y = 10$
- **b** $x - y = -1$
  $2x + y = 4$
- **c** $5x - 4y = -1$
  $2x + y = 10$
- **d** $3x - 2y = 29$
  $4x + y = 24$
- **e** $3x + y = 2$
  $9x + 2y = 10$
- **f** $3x - 2 = -2y$
  $2x - y = -8$

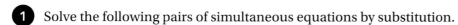
Find answers at: cambridge.org/ukschools/gcsemaths-studentbookanswers

Solving simultaneous equations by elimination

In this method you add or subtract the equations to get rid of one of the unknown variables.

You might need to multiply or divide one equation by a factor before you do this.

Problem-solving framework

Solve the simultaneous equations $2x + y = 8$ and $x - y = 1$.

Steps for solving simultaneous equations by elimination	What you do in this example
Step 1: Number the equations (1) and (2).	$2x + y = 8$ (1) $x - y = 1$ (2)
Step 2: Decide whether you can add or subtract to get rid of a variable.	(1) has a variable of y and (2) has a variable of $-y$. If you add the equations, these terms will cancel out. (1) + (2) $(2x + y) + (x - y) = 8 + 1$ $2x + y + x - y = 9$ $3x = 9$
Step 3: Solve the combined equation (which now only has one unknown.	$3x = 9$ $x = 3$
Step 4: Substitute the solution into either of the original equations to find the other unknown.	Substitute $x = 3$ into (2). $3 - y = 1$ $-y = 1 - 3$ $-y = -2$ $y = 2$
Step 5: Check your solutions to both equations.	Check by substituting the values $x = 3$ and $y = 2$ back into (1). $2(3) + 2 = 8$ ✓

In some examples, you will need to subtract the equations.

WORKED EXAMPLE 11

Solve the pair of simultaneous equations.

$x + 3y = 7$ (1) $x + y = 5$ (2)

$(x + 3y) - (x + y) = 7 - 5$ Addition will not eliminate one of the variables here so you need to subtract. Subtract (2) from (1).

$x + 3y - x - y = 2$ Remove the brackets.

$2y = 2$ Simplify.

$y = 1$ Solve for y.

$x + 1 = 5$ Substitute in equation (2).

$x = 4$ Solve for x.

The solutions are $x = 4$ and $y = 1$.   Check these values satisfy both equations.

> **Tip**
>
> Add if the signs are different and one variable can be eliminated.
>
> Subtract when the signs are the same, including when they are both negative.

The next examples show how you can use elimination when neither the x nor the y term has the same coefficient in the two equations.

WORKED EXAMPLE 12

Solve this pair of simultaneous equations by elimination:

$x + 2y = 7 \qquad 4x + 3y = 6$

$x + 2y = 7 \qquad (1)$
$4x - 3y = 6 \qquad (2)$

The coefficients for x and y are different in both equations, so you can't eliminate either from these equations as they are.

$4x + 8y = 28 \qquad (3)$

Multiply equation (1) by 4 to create a new version of equation (1) that has the same coefficient of x as equation (2). Call this equation (3).

$(4x + 8y) - (4x - 3y) = 28 - 6$
$4x + 8y - 4x + 3y = 22$
$11y = 22$
$y = 2$

Subtract (2) from (3) to eliminate the x term and solve for y.

$x + 2 \times 2 = 7$
$x + 4 = 7$
$x = 3$

Substitute $y = 2$ in one of the equations to find the value of y. Equation (1) is the simpler here. Solve for x.

The solutions are $x = 3$ and $y = 2$

Check these values satisfy both equations.

WORKED EXAMPLE 13

Solve this pair of simultaneous equations:

$3x - 2y = 5 \qquad (1)$

$4x + 3y = 18 \qquad (2)$

$9x - 6y = 15 \; (3)$
$8x + 6y = 36 \; (4)$

Multiply (1) by 3 and (2) by 2 to get equations (3) and (4).

This creates a pair of coefficients for y which you can eliminate.

$17x = 51$
$x = 3$

Add (3) and (4).

$4 \times 3 + 3y = 18$
$3y = 6$
$y = 2$

Substitute in (2).

So $x = 3$ and $y = 2$

Check these values satisfy both equations.

You need to develop a variety of strategies to solve simultaneous equations. The equations in Worked example 12 could also be solved by substitution, but this would be complicated for the equations in Worked example 13.

EXERCISE 14G

1 Solve the following pairs of simultaneous equations by elimination.
Check your solutions satisfy **both** equations.

- **a** $x - y = 2$
  $3x + y = 14$
- **b** $2x - 3y = 3$
  $x + 3y = 6$
- **c** $3x + y = 4$
  $2y - 3x = -10$
- **d** $x + 2y = 7$
  $3x - 2y = 13$
- **e** $2x - y = 14$
  $5x + y = 14$
- **f** $-x - y = 3$
  $x + 5y = -11$
- **g** $x + y = 5$
  $3x + y = 9$
- **h** $3x + 4y = 15$
  $x + 4y = 13$
- **i** $2x - y = 7$
  $4x - y = 15$

2 Solve each pair of simultaneous equations. Choose the most suitable method for doing this.

- **a** $2x + y = 7$
  $3x + 2y = 12$
- **b** $x + 2y = 7$
  $3x - 2y = 5$
- **c** $4x + 2y = 50$
  $x + 2y = 20$
- **d** $x + y = -7$
  $x - y = -3$
- **e** $y = 1 - 2x$
  $5x + 2y = 0$
- **f** $y = 2x - 5$
  $y = 3 - 2x$

Forming and solving simultaneous equations

Some problems can be described using a pair of simultaneous equations.

Once you have defined the variables and set up the equations you can use algebra to solve them.

> **Tip**
>
> If a problem asks for two different pieces of information then it means there are two unknowns and you will need two equations to solve it.

Problem-solving framework

A field contains a number of goats and chickens.
Altogether there are 60 heads and 200 legs.
How many are there of each type of animal?

Steps for solving problems	What you would do for this example
Step 1: Work out what you have to do. Start by reading the question carefully.	You have to find the number of goats (g) and the number of chickens (c).
Step 2: What information do you need? Have you got it all?	You need to know how many there are altogether. You're not told this, but you can work it out because each goat and each chicken must have one head. $g + c = 60$ Now you need another equation to link the goat and chickens. You've already used the number of heads, so it must be something to do with the legs. Goats have 4 legs each, so the number of goat legs is $4 \times g$. Chickens have 2 legs each, so the number of chicken legs is $2 \times c$. There are 200 legs in total so: $4g + 2c = 200$.

Continues on next page ...

Step 3: Decide what maths you can do.	There are two unknowns so you should use simultaneous equations.
Step 4: Set out your solution clearly. Check your working and that your answer is reasonable.	Let the number of goats = g and the number of chickens = c. $g + c = 60$ (1) $4g + 2c = 200$ (2) Make g the subject of equation (1). $g = 60 - c$ Substitute $(60 - c)$ for g in (2) and solve. $4(60 - c) + 2c = 200$ $240 - 4c + 2c = 200$ $240 - 2c = 200$ $240 - 200 = 2c$ $40 = 2c$ $20 = c$ Substitute $c = 20$ into equation (1) to find g. $g + 20 = 60$ so $g = 40$ Check your result by substituting c and g into original equation (2). $4(40) + 2(20) = 200$ $160 + 40 = 200$ Yes, this works.
Step 5: Check that you've answered the question.	There are 20 chickens and 40 goats.

EXERCISE 14H

1 A taxi company charges a flat fee plus an amount per mile.

A journey of 10 miles costs £7 and a journey of 15 miles costs £9.

What would you pay for a journey of 8 miles?

2 Sam got eighteen 5p and 10p coins totalling £1.65 as change when he went to the shop.

How many of each coin did he get?

3 Two children have a total of 264 stickers between them.

One child has 6 fewer stickers than 5 times the other child's stickers.

How many do they each have?

4 The sum of two numbers, a and b, is 120.

When b is subtracted from $3a$, the result is 160.

Find the value of a and b.

Find answers at: cambridge.org/ukschools/gcsemaths-studentbookanswers

Tip

You have to find two different values, so you need to write a pair of simultaneous equations. Multiply or divide to get equations you can solve by elimination.

Tip

You will work with graphs and equations again in Chapter 23 and Chapter 24.

5 Two numbers have a sum of 76 and a difference of 48.

What are the numbers?

6 Josh and Sanjita both spent £2.20 on sweets.

Josh bought five fizzers and four toffees.

Sanjita bought two fizzers and six toffees.

Work out the cost of each type of sweet.

Section 4: Using graphs to solve equations

If you have a graph you can use it to solve an equation or to answer questions based on the equations.

WORKED EXAMPLE 14

This is the graph of the equation $4x + y = 2$.

a Use the graph to estimate the value of y when:

  i $x = 0$

  ii $x = 1$

  iii $x = 2$

b What is the value of x when $y = -4$?

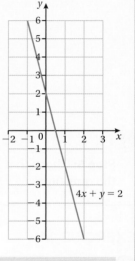

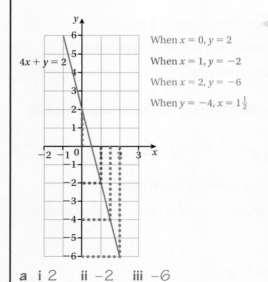

When $x = 0, y = 2$
When $x = 1, y = -2$
When $x = 2, y = -6$
When $y = -4, x = 1\frac{1}{2}$

Each point on the graph represents a value of x and y that work in this equation.

To find the solutions for different values of x or y, you need to use the value you have been given as one of the coordinates of a point. If you take a line from this point to the graph you can estimate the value of the other coordinate.

The solutions are shown on this graph in different colours.

a i 2   **ii** −2   **iii** −6
b 1.5

EXERCISE 14I

1 Use this graph of the equation $y = 3x - 2$ to estimate the value of y for the following values.

  a $x = 0$ **b** $x = 1$ **c** $x = 2$

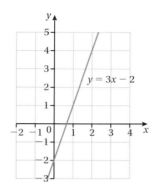

2 This graph shows the distance covered over time by a motorist in a car.

  a Estimate how long it took the driver to travel 70 km.

  b How far did the driver travel in the first 20 minutes?

  c By using two points on the graph, estimate the speed at which the driver was travelling.

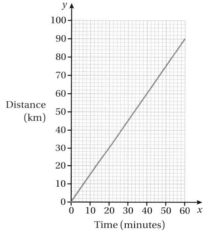

3 This graph represents the distance travelled by a cyclist over time.

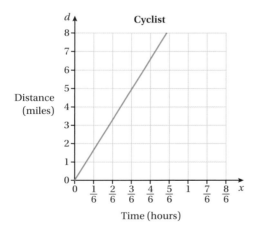

  a Use the graph to estimate how far the cyclist has travelled after 30 minutes.

  b How long did it take the cyclist to cover a distance of 8 miles?

  c The equation $s = \dfrac{d}{t}$ can be used to work out the speed (s) of the cyclist.

  Use values for d and t from the graph to estimate the speed at which this cyclist was travelling.

4 This graph shows how water drains from a tank at a constant rate.

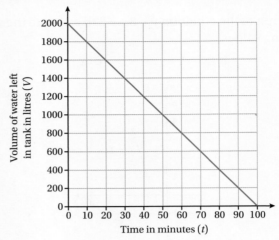

a How much water was in the tank to start with?

b How long did it take for the tank to empty?

c Zena says the equation for this graph is $y = 2000 - 20x$ and Leane says it is $y + 20x = 2000$.

Show using different points from the graph that they are both correct.

5 This graph shows the cost of producing goods and how much money is earned from sales (revenue).

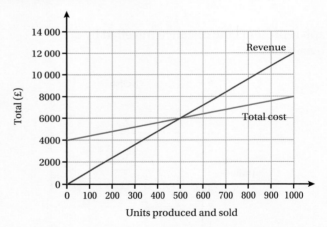

a Use the graph to estimate the point at which the costs and revenue are equal.

b The point at which costs and revenue are equal is called the break even point.

How does this information help a business owner?

6 This diagram shows the graphs of two linear equations
$y = 2x$ and $y = -2x + 8$.

Use the graph to find the simultaneous solution to the two equations.

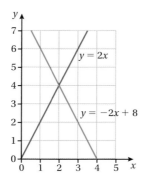

7 The graph on the right models a stunt rider's path in the air as he does a jump.

  a What do you think the axes represent in this case?

  b What values on the horizontal axis represent the rider taking off and landing again?

  c Why are these two values useful in terms of the equation?

  d Use the graph to estimate the coordinates of the maximum height reached during this jump.

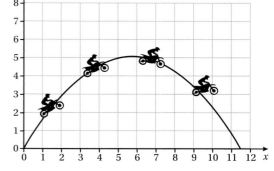

8 The graphs of two quadratic equations are given here.

Graph A   Graph B

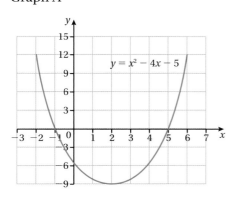
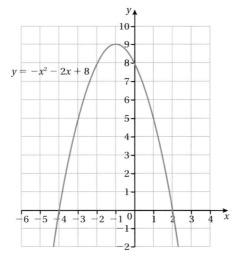

> **Tip**
>
> The solution of a quadratic equation is the value of x when $y = 0$. On a graph, these values are found where the graph intersects the x-axis.

Use the graphs to estimate the solution of the equations

a $y = x^2 - 4x - 5 = 0$

b $y = -x^2 - 2x + 8 = 0$

Check your solutions by substitution.

9 This is the graph of a quadratic equation but the equation is not given.

Explain how you can use the graph to find the roots of the equation even though you don't know what the equation is.

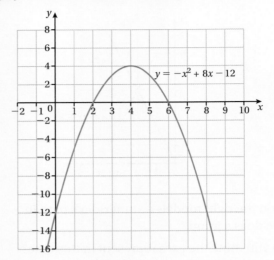

 Checklist of learning and understanding

Linear and quadratic equations

- A linear equation has one unknown and will have one unique value as a solution.
- A quadratic equation has a square as the highest power for the variable.
- Some quadratic equations in the form of $x^2 + bx + c = 0$, where b and c are given numbers, can be solved by factorising.
- Quadratic equations have two roots. Sometimes both values are the same and sometimes only one is sensible because the other value doesn't work in a context. Some quadratic equations have no real solutions.

Simultaneous equations

- Simultaneous equations are a pair of equations that have solutions that satisfy both equations.
- Simultaneous equations can be solved algebraically by substitution or by elimination.

Graphs and problems

- Equations are useful for setting up problems mathematically.
- You can find or estimate the solutions of equations using graphs.
- You need two graphs to solve simultaneous equations. The solution is the point of intersection of the graphs.

Chapter review

For additional questions on the topics in this chapter, visit GCSE Mathematics Online.

1 Solve for x.

  a $8x + 5 = 3(2x - 11)$
  b $x^2 = 8 - 2x$
  c $4(x + 2) = 3(x + 1)$
  d $-x^2 = 8x + 12$
  e $(x + 2)^2 = 36$
  f $(x + 4)(x + 3) = 6$

2 Study this graph.

  a What are the roots of the quadratic equation modelled by this graph?
  b What does the graph tell you about the selling price?

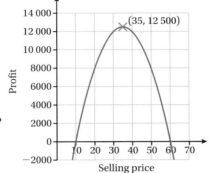

3 Is $y = x + 2$ equivalent to $5y = 5x + 10$?

How can you tell this without solving the equations?

4 **a** What happens when you try to solve this pair of equations?
  $x + y = 2$ $x + y = 4$

  b What would their linear graphs show?

5 If you drew the graph of $y = x^2 - x - 30$ where would the curve cut the x-axis?

6 The sum of two numbers is 19 and their difference is 5.

  a Write a set of equations in terms of x and y to show this.
  b Solve the equations simultaneously to find the two numbers.

7 Read each statement and decide whether it is true or false. If it is false, explain why.

  a $5(x - 2) = 15$ so $x = 2$.
  b The statement 'four is subtracted from a number x which has been multiplied by 5 and this is equal to the number x subtracted from 14' can be represented as $5x - 4 = 14 - x$.
  c The two solutions to the quadratic equation $x^2 - 2x - 15 = 0$ are $x = 3$ and $x = -5$.
  d $x = 2$ and $y = 1$ are the simultaneous solutions to the pair of equations $y = x + 4$ and $y = 2x + 3$.
  e The graph on the right represents a quadratic equation whose solutions are $x = 1$ and $x = -2$.

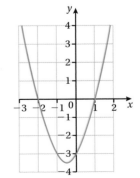

8 Solve.

  a $4a = 36$
  b $3x + 19 = 46$
  c $4(2x - 3) = 84$
  d $2x + 7 = x + 9$

Find answers at: cambridge.org/ukschools/gcsemaths-studentbookanswers

9. A rectangular floor has a width 8 m shorter than its length and a perimeter of 80 m.

   Find the length and width of the room by forming a linear equation and solving it.

10. The diagram (left) shows a parallelogram.

    The sizes of the angles, in degrees, are

    $2x \quad 3x - 15 \quad 2x \quad 2x + 24$

    Work out the value of x. *(3 marks)*

    ©*Pearson Education Ltd 2012*

11. Factorise and solve these quadratic equations:

    a $x^2 - 7x + 12 = 0$
    b $x^2 + 18x + 81 = 0$
    c $x^2 - 64 = 0$
    d $x^2 - 1 = 0$
    e $x^2 + 7x = 0$
    f $x^2 + 3x = 18$

12. Solve each pair of simultaneous equations by substitution.

    a $y = 3x - 5$
       $y = 6x - 11$
    b $y = 2x - 3$
       $y = 3x - 5$
    c $x = 2y - 4$
       $2x + y = 7$

13. Solve the following simultaneous equations by elimination.

    a $x + y = 2$
       $3x - y = 10$
    b $2x + y = 5$
       $x + y = 2$
    c $2x - 3y = 1$
       $3x + 3y = 9$

14. The total of two numbers is 5.

    The sum of three times one number and two times the other is 25. Find the two numbers.

15. A **line of best fit** has been drawn through this data to create a linear graph that plots BMI (Body Mass Index) against weight in pounds (lbs) for a specific group of people.

Key vocabulary

line of best fit: a line drawn through a group of points showing the general direction to represent the trend.

Tip

You can learn more about lines of best fit in Chapter 36.

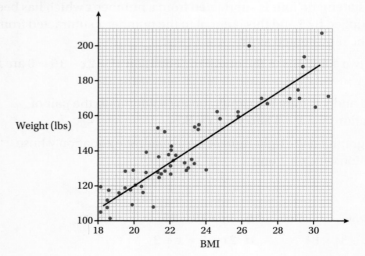

a Use the graph to estimate the weight of a person whose BMI is 24.5.

b Estimate the BMI of a person who weighs 165 lbs.

c Given that 1 kilogram is equivalent to 2.2 pounds, find the BMI of a person weighing 55 kg.

16 Naresh is going to study in the USA and he needs to buy a data package for his mobile so he can text home.

This graph shows the costs of three different options.

The number of text messages is shown on the x-axis and the related costs are shown on the y-axis in dollars.

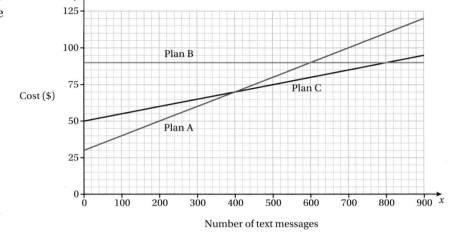

a Which plan represents a set monthly fee? How do you know?

b Use the graph to estimate the point at which Plan C and Plan A cost the same.

c What is the maximum number of texts you can send on Plan C before it begins to cost more than Plan B?

d Which plan is best for Naresh if he sends more than 800 texts per month? Why?

17 This diagram shows the graphs of two quadratic equations, $y = x^2 - 4x + 4$ and $y = -x^2 - 3x - 3$

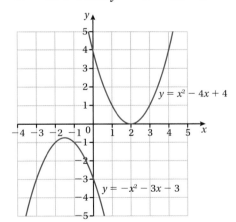

a What value does the graph give for the solution to $x^2 - 4x + 4 = 0$?

b Why is there only one root to this equation?

c Verify that both roots of this equation are the same by factorising the equation $x^2 - 4x + 4 = 0$

d Look at the graph of the equation $y = -x^2 - 3x - 3$

What do you think will happen if you try to find the roots of this equation? Why?

15 Functions and sequences

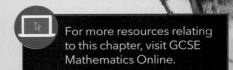

In this chapter you will learn how to …
- generate sequences and find unknown terms in a sequence.
- interpret expressions as functions with inputs and outputs.
- write rules or functions to find any term in a sequence.
- recognise and use a variety of special sequences.

For more resources relating to this chapter, visit GCSE Mathematics Online.

Using mathematics: real-life applications

Finding a pattern and working out how the parts of the pattern fit together is important in scientific discovery. Scientists use sequences to model and solve real-life problems, such as estimating how quickly diseases spread.

Tip

When you work with sequences you can draw diagrams, flow charts or tables to organise the patterns and make sense of them.

"When a new outbreak of a disease occurs I need to work out how quickly it is spreading. To do this I look at the sequence in which the numbers of victims are increasing. I use the sequence to predict how many people will become infected in a certain length of time." *(Medical researcher)*

Before you start …

Ch 4	You need to know your multiplication tables and recognise multiples of numbers.	**1**	a What are the first five multiples of 7? b Which of these are multiples of 6? 56, 66, 86, 18, 54, 36
Ch 4	You need to recognise square numbers and cube numbers.	**2**	a Which of these are square numbers? 1, 16, 66, 50, 25, 4, 6, 9, 49 b Which of these are **not** cube numbers? 9, 15, 27, 64, 1, 8, 125
KS3	You need to be able to spot and describe patterns.	**3**	a Describe this pattern in words. Shape 1 Shape 2 Shape 3 b How many matchsticks would you need to build the sixth shape in the pattern?

Assess your starting point using the Launchpad

STEP 1

1
a What are the next three numbers in the sequence 23, 35, 47, …?
b What is the rule for finding the next term in this sequence?

GO TO **Section 1**: Sequences and patterns

STEP 2

2
a What are the 10th, 20th and 100th terms in the sequence $3n - 1$?
b What is the expression for the nth term of a sequence that starts −1, 2, 5, 8, …?

GO TO **Section 2**: Finding the nth term

STEP 3

3 Draw an input-output flow diagram for the instruction 'multiply the input number by 2 and then subtract 4'.

4 The rule for generating a sequence is $y = x + 3$.
List the first five terms in the sequence.

GO TO **Section 3**: Functions

STEP 4

5 What is special about the sequence 2, 4, 7, 11, … ?

GO TO **Section 4**: Special sequences

GO TO Chapter review

Find answers at: cambridge.org/ukschools/gcsemaths-studentbookanswers

Section 1: Sequences and patterns

The term-to-term rule

> **Key vocabulary**
>
> **sequence**: a number pattern or list of numbers following a particular order.
>
> **term**: each number in a sequence is called a term.
>
> **consecutive terms**: terms that follow each other in a sequence.
>
> **first difference**: the result of subtracting a term from the next term.
>
> **term-to-term rule**: operations applied to any number in a sequence to generate the next number in the sequence.

A **sequence** is an ordered list or pattern.

Terms that follow each other in a sequence are called **consecutive terms**.

The first term in a sequence is called T(1), the second T(2) and so on.

You can find the next term in the sequence by working out what the difference is between each term. This is called the **first difference**.

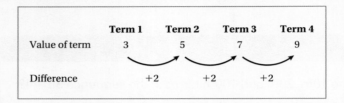

The **term-to-term rule** for this sequence is 'add two'.
9 + 2 = 11, so the next term in the sequence is 11.

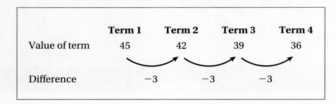

The term-to-term rule for this sequence is 'subtract three'.

3, 5, 7, 9, ... and 45, 42, 39, 36, ... are both **arithmetic sequences**.
In an arithmetic sequence the terms are generated by adding or subtracting a constant difference.

In a **geometric sequence**, the terms are generated by multiplying or dividing by a constant factor.
3, 6, 12, 24, ... and 1000, 500, 250, 125, ... are both geometric sequences.

> **Key vocabulary**
>
> **arithmetic sequence**: a sequence where the difference between each term is constant.
>
> **geometric sequence**: a sequence where the ratio between each term is constant.

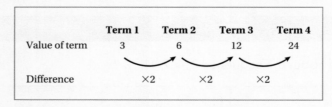

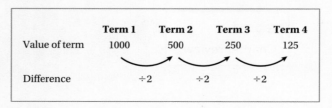

WORKED EXAMPLE 1

"I use term-to-term rules in my job. I know that each row of bricks will have three fewer bricks than the row below it, so I can work out how many bricks I need in each row."

(Bricklayer)

The first row of the wall has 57 bricks.

Use a term-to-term rule to find the number of bricks in the next three rows.

The term-to-term rule is 'subtract three'. ◁ The number of bricks decreases by 3 for each new row.

57 in the first row

57 − 3 = 54 in the second row ◁ Subtract 3 from 57 to get the next term.

54 − 3 = 51 in the third row ◁ Continue to do this for the next two terms.
51 − 3 = 48 in the fourth row

EXERCISE 15A

1 Find the next three terms in each of these sequences. Explain how you found them.

- **a** 4, 7, 10, 13, …
- **b** 38, 43, 48, 53, …
- **c** 27, 23, 19, …
- **d** 63, 57, 51, …
- **e** 1, 2, 4, 8, …
- **f** 64, 32, 16, …
- **g** 4, 12, 36, …
- **h** 729, 243, 81, …

2 Give the term-to-term rule for each of these sequences.

- **a** 7, 14, 21, 28, …
- **b** 19, 15, 11, 7, …
- **c** 2, 8, 32, 128, …
- **d** 84, 42, 21, …

3 Find the term-to-term rule for each of these sequences.
Use it to generate the next three terms in each sequence.

- **a** 3.5, 5.5, 7.5, …
- **b** 1.2, 2.4, 4.8, …
- **c** $1\frac{1}{2}$, 3, $4\frac{1}{2}$, …
- **d** 8, 5, 2, …
- **e** 72, 36, 18, …
- **f** −10, −7, −4, …

Tip

It can help to write out the sequence and label the difference between each term.

Tip

Look for the term-to-term rule to answer this question.

4 When a ball is dropped it bounces back to half its original height.

With each new bounce it bounces back to half the height of the previous bounce.

  a If a ball is dropped from 96 cm, how high will it bounce on its fourth bounce?

  b How many times will it bounce before it bounces to below 1 cm?

5 T(1) of a sequence is 4 and the term-to-term rule for the sequence is 'add x'.

Find a value for x so that:

  a every second term is an integer

  b every third term is a multiple of 4

  c T(2) is smaller than T(1).

Section 2: Finding the nth term

The position-to-term rule

Term-to-term rules are useful for generating the first few terms of a sequence and for finding the next term in a given sequence. They are less useful when you want to find the 50th or 100th term.

A **position-to-term** rule allows you to work out the value of any term in a sequence if you know its position in the sequence.

Key vocabulary

position-to-term rule: operations applied to the position number of a term in a sequence in order to generate that term.

The sequence 1, 3, 5, 7, ... can be generated using the position-to-term rule 'position number $\times 2$, subtract 1' by substituting the position number into the rule.

Position in sequence	Position-to-term rule 'position $\times 2, -1$'	Term
1	$(1 \times 2) - 1$	1
2	$(2 \times 2) - 1$	3
3	$(3 \times 2) - 1$	5
4	$(4 \times 2) - 1$	7
5	$(5 \times 2) - 1$	9
6	$(6 \times 2) - 1$	11
7	$(7 \times 2) - 1$	13
8	$(8 \times 2) - 1$	15
9	$(9 \times 2) - 1$	17
10	$(10 \times 2) - 1$	19

The nth term

The notation T(n) refers to 'any term' in the sequence, where n is the position of the term. T(n) is known as the 'nth term'.

Sometimes you will be given a sequence and you will need to find a rule to find any term, T(n), in that sequence. You can usually find a rule for a sequence by looking at the difference between consecutive terms.

Remember that this is called the **first difference**.

Once you have found the first difference you can compare it with number patterns you already know to find the rule for the sequence.

Problem-solving framework

Find an expression for the nth term of the sequence: 5, 8, 11, 14, 17, …

Steps for approaching a problem-solving question	What you would do for this example
Step 1: Identify what you have to do.	You are trying to find an expression to work out the value of any term in the sequence.
Step 2: If it is useful to have a table, draw one.	Draw a table showing the position and the term: | n	1	2	3	4	5 | |---|---|---|---|---|---| | $T(n)$ | 5 | 8 | 11 | 14 | 17 |
Step 3: Start working on the problem using what you know.	Label the table with the difference between each term: | n	1	2	3	4	5 | |---|---|---|---|---|---| | $T(n)$ | 5 | 8 | 11 | 14 | 17 | $+3 \quad +3 \quad +3 \quad +3$ The difference in this sequence is '+ 3'.
Step 4: Connect to other sequences and compare.	Another sequence that has the same term-to-term rule of 'add 3' is the multiples of 3. (If the difference was '+2' you would compare to $2n$; if it was '−4' you would compare it to $-4n$.) Add this sequence to your table: | n	1	2	3	4	5 | |---|---|---|---|---|---| | $T(n)$ | 5 | 8 | 11 | 14 | 17 | | **Multiples of $3n$** | 3 | 6 | 9 | 12 | 15 | $+2$ Compare the multiples of 3 to the terms of the original sequence. Each term is the corresponding multiple of 3 with 2 added. So the expression for the nth term of this sequence could be $3n + 2$.
Step 5: Check your working and that your answer is reasonable.	Test for $n = 5$: $(3 \times 5) + 2 = 15 + 2$ $15 + 2 = 17$ The fifth term is 17 so the expression is correct.
Step 6: Have you answered the question?	Yes. The expression for the nth term is $3n + 2$.

Tip

If you are struggling to find the rule for an arithmetic sequence, you can use the following formula: $(a + d(n - 1))$ where a = first term, and d = common difference. The **common difference** is the constant difference between terms.

WORK IT OUT 15.1

A sequence is defined by the rule $T(n) = 3n - 2$.

What are the first five terms of the sequence?

Only one answer below is correct. Explain why the other two are wrong.

Option A	Option B	Option C
−2, 1, 4, 7, 10	1, 4, 7, 10, 13	−1, 1, 3, 5, 7

EXERCISE 15B

1 A sequence is created using the position-to-term rule 'position ×3 subtract 1.'
  a What are the first six terms of the sequence?
  b What is the 20th term of the sequence?
  c Would the 40th term of the sequence be double the value of the 20th term? Explain your answer.

2 Find the value of the following terms for each position-to-term rule.
  i 1st term   **ii** 2nd term   **iii** 3rd term   **iv** 4th term
  v 10th term **vi** 20th term **vii** 100th term

  a $4n + 1$
  b $4n - 5$
  c $8n + 2$
  d $5n - \dfrac{1}{2}$
  e $\dfrac{n}{2} + 1$
  f $-2n + 1$

3 Find the expression for the nth term in the sequence that begins 5, 9, 13, 17,…

4 Find the expressions for the nth term of the following sequences.
  a 3, 5, 7, 9, …
  b 3, 7, 11, 15, …
  c −1, 4, 9, 14, …
  d 7, 12, 17, 22, …
  e −3, 0, 3, 6, …
  f −1, 6, 13, 20, …

5 Majid conducts an experiment in science and gets the following pattern of results:

67, 73, 79, 85

Write an expression for the nth term of Majid's results.

6 Sam has planted a sunflower and notices that it measures 4.5 cm at the end of the first week, 6.7 cm at the end of the second week and 8.9 cm at the end of the third week.
  a Find an expression for the nth term of this sequence, assuming the sunflower continues to grow at the same rate.
  b How big will the sunflower be at the end of the 100th week?
  c Explain why this is unlikely to be true.

7 Tammy has £100 saved.

She deposits £4 per week. At the end of the first week she has £104.
  a If she continues to save at this rate, how much will she have after 52 weeks?
  b How long will it take her to save £400?

8 A restaurant uses square tables that can seat four people.

Tables can be pushed together to create different seating arrangements.

  a How many people can fit around six tables pushed together in this way?

  b How many people can fit around 10 tables pushed together this way?

  c Is it possible to seat 31 people around tables pushed together like this with no empty seats?

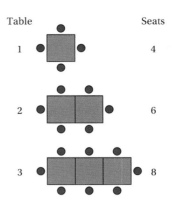

Section 3: Functions

A **function** is a rule for changing one number into another.

'Multiply by two', 'add three' and 'divide by 2 and then add 1' are examples of functions.

You can use algebra to write functions, for example 'multiply by 2' can be written as the expression '$2n$'.

Functions can be expressed in different ways:

$y = x + 3$ $x \rightarrow x + 3$

These mean the same thing: take any value of x and add 3 to it to get a result.

The steps you take to work out the value of a function can be shown as a simple flow diagram or function machine.

A function machine shows the input, operation and output for a given rule.

Input ⟶ | rule ⟩ ⟶ Output

In a function there is only **one** possible output for each input.

> **Key vocabulary**
>
> **function**: a set of instructions for changing one number (the input) into another number (the output).

> **Tip**
>
> You can think of a function as 'the answer you will get' if you apply this rule to a number.

Generating a sequence using a function

You can generate a sequence using a function.

The table shows the outputs when you input the values 1 to 10 into the function $y = 2n + 4$

$n \rightarrow$ | ×2 ⟩ ⟶ | +4 ⟩ ⟶ $2n + 4$

Input	Function	Output
1	× 2 + 4	6
2	× 2 + 4	8
3	× 2 + 4	10
4	× 2 + 4	12
5	× 2 + 4	14
6	× 2 + 4	16
7	× 2 + 4	18
8	× 2 + 4	20
9	× 2 + 4	22
10	× 2 + 4	24

So the function $y = 2n + 4$ generates the sequence
6, 8, 10, 12, 14, 16, 18, 20, 22, 24, …

If you know the values of a sequence, you can find the function that generates the sequence by finding the position-to-term rule.

WORKED EXAMPLE 2

A sequence has the following terms: 3, 8, 13, 18, 23 ...

Write down the function machine that generates this sequence.

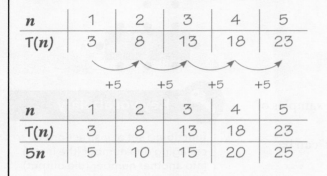

Draw a table showing the position and term. Work out the difference between consecutive terms. The terms increase by 5.

Compare the sequence with the sequence for $5n$. Each term in the sequence is 2 less than the term in the sequence $5n$.

$x \to \boxed{\times 5} \to \boxed{-2} \to 5x - 2$

$4 \to (4 \times 5) \to 20 - 2 \to 18$ ✓

The function could be 'multiply by 5, subtract 2'; test it using a term from the sequence.

EXERCISE 15C

1 The numbers 1 to 10 are the input for the function $x \to x + 3$. What sequence does this create?

2 Input the numbers 1 to 10 into each function to generate a sequence.
   a $x \to x - 5$ **b** $x \to 3x$ **c** $n \to n + 7$ **d** $n \to \dfrac{n}{2}$

3 Input the numbers 21 to 30 into each function to generate a sequence.
   a $y = 2x$ **b** $y = x - 8$ **c** $y = \dfrac{x}{3}$ **d** $y = x + \dfrac{1}{2}$

4 Look at this pattern.

   a Write the number of coins used to build each shape as a sequence.

   b How many coins would you need to build the next shape in the pattern?

   c Complete this function machine for calculating the number of coins used to make any shape in the pattern:

   input $(n) \to \boxed{} \to$ output

   d Use your function to work out the number of coins you would need to build the 6th and 10th shapes in the pattern.

5 Jess earns £6.25 per hour at her part-time job. Write a function for calculating how much she will earn if she works n hours.

6 Zena is a salesperson.
She earns a basic salary of £500 plus £3 for every item she sells.

a Write a function for calculating her salary if she sells *p* items.

b Apply your function to work out her earnings if she sells 10, 20, 30 or 40 items.

Section 4: Special sequences

Some patterns and sequences of numbers are well known.
You need to be able to recognise and use the following patterns.

Special sequence	Description
Simple arithmetic progression (or linear sequences)	The **difference** between each term is constant, for example, 3, 5, 7, … or 14, 11, 8, …
Geometric sequences	The **ratio** between each term is constant, for example, 3, 6, 12, 24, …
Triangular numbers	These are made by arranging dots to form equilateral triangles. 1 dot 3 dots 6 dots 10 dots 15 dots
Square numbers	A square number is the product of multiplying a whole number by itself. For example, $3^2 = 3 \times 3 = 9$ Square numbers form the sequence: 1, 4, 9, 16, 25, 36, …
Quadratic sequences	These sequences are linked to square numbers. A quadratic sequence has a position-to-term rule that involves squaring one of the variables. For example, the sequence formed by the rule $n^2 + 3$ is: 4, 7, 12, 19, 28, … The terms in a quadratic sequence do **not** increase or decrease by a constant amount. The first difference is **not** constant but the **second difference** is constant. The second difference is the difference between each term in the first difference.
Cube numbers	A cube number is the product of multiplying a whole number by itself and then by itself again. 64 is a cube number because $4^3 = (4 \times 4 \times 4) = 64$. Cube numbers form the sequence: 1, 8, 27, 64, 125, …
Fibonacci sequences	Leonardo Fibonacci was an Italian mathematician who developed the number pattern 1, 1, 2, 3, 5, 8, 13, 21, … while he was trying to work out how many offspring a pair of rabbits would produce over different generations. These numbers are now called Fibonacci numbers. If you start the sequence with the numbers 1 and 1, the term-to-term rule is 'add the previous two terms together'.

Did you know?

The Fibonacci pattern is found in many natural situations.

In the Fibonacci series, the sequence of numbers is created by adding the 1st and 2nd term together to make the 3rd term; adding the 2nd and 3rd terms together to make the 4th term and so on.

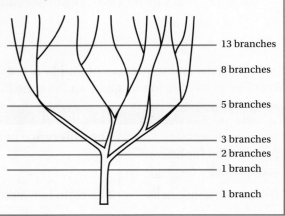

EXERCISE 15D

 a Write down the sequence of the first 10 square numbers.

   b Find the difference between consecutive terms in the sequence.

   c How could you use this new sequence to find the next two square numbers?

 a What type of number is shown on the left? Explain how you know this.

   b The picture represents the fifth term in the sequence.
   Write down the first 10 numbers in the same sequence.

   c Find the first and second differences between the terms of the sequence.

   d What type of sequence is this?

3 Honeybees live in colonies known as hives.

There is one queen bee, a female, who is the only one able to lay eggs.

All the other female bees are called workers. They are made when a male fertilises the queen's eggs. This means that a worker bee has a mother and a father.

The male bees are called drones. Drones are made when the queen's eggs hatch without being fertilised by a male. This means that a drone has a mother but not a father.

This is a family tree of a drone; a family tree shows each generation of bee in terms of their parents.

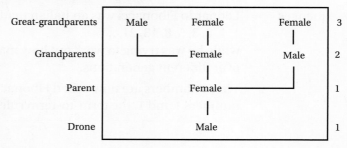

a How many great-great-grandparents does the male bee have?

b The family tree shows four generations of bee.
  i Copy and complete the tree so that it shows six generations.
  ii Use the family tree to write a sequence for the number of bees there are in each generation.
  iii Continue the sequence to find how many bees there are in the ninth generation.
  iv What is the relationship between the number of bees in each generation? Do you recognise the sequence?

4 a Find the first 10 terms of a sequence that follows the same rules as the Fibonacci sequence that starts with the numbers 3, 4, …

  b Start a Fibonacci sequence with the numbers −2 and 3. Write down the first 10 terms of the sequence.

  c Compare your sequence in part **b** with a Fibonacci sequence starting with 2, −3. What do you notice?

5 The sixth and seventh terms of a sequence that follows the same rules as the Fibonacci sequence are 31 and 50.
What are the first two terms?

> **Tip**
> You know the term-to-term rule of the Fibonacci sequence. How can you use this to help you?

6 Copy and complete the table below.

Position-to-term rule	1st term	2nd term	3rd term	5th term	10th term	20th term	50th term
$n^2 + 5$							
$n^2 - 3$							
$2n^2 + 1$							
$2n^2 - 7$							

7 Write down the first six terms of the sequence formed by the rule $n^3 + 1$.

8 Write down the first 10 terms of the sequence formed by using the rule $n^2 + n$.

9 Compare your answer to Question **8** with your answer from Question **2**.
  a What is the position-to-term rule for triangular numbers?
  b Use your rule to find the 10th and 25th triangular numbers.

Checklist of learning and understanding

Sequences
- Sequences can be formed using a term-to-term rule. Each term is generated by applying the same rule to the previous term.
- The position-to-term rule is used to find the value of any term, known as the nth term in a sequence using its position in the sequence.

 Find answers at: cambridge.org/ukschools/gcsemaths-studentbookanswers

Functions

- A function is an expression or rule for changing one number (the input) into another number (the output).
- A sequence can be generated by inputting an ordered set of numbers into a function.

Special sequences

- It is important to recognise familiar sequences such as square numbers, cube numbers, triangular numbers and Fibonacci numbers.
- In a quadratic sequence the first difference between the terms is not constant but the second difference is, and the pattern is linked to square numbers.

 For additional questions on the topics in this chapter, visit GCSE Mathematics Online.

Chapter review

1 For each sequence:

   a find the missing terms.

   b write an expression in terms of n to find any term in the sequence.

   c use your expression to find the 25th term in each sequence.

   i 1.5, 2, ☐, 3, 3.5, ☐, ...

   ii ☐, −8, −4, ☐, 4, 8, ☐, ...

   iii $\frac{1}{2}, \frac{1}{4},$ ☐, $\frac{1}{16}, \frac{1}{32},$ ☐, ...

   iv 5, ☐, 17, 23, 29, ☐, ...

2 Here are the first four terms of a number sequence.

   6   10   14   18

   a Write down the next term in this sequence. *(1 mark)*

   b Find the 10th term in this sequence. *(1 mark)*

   c The number **101** is not a term in this sequence. Explain why. *(1 mark)*

   d Write an expression, in terms of n, for the nth term of this sequence. *(2 marks)*

   ©Pearson Education Ltd 2012

3 The array of numbers on the left is called Pascal's triangle.

   Each number is the sum of the two numbers above it, except for the edges which are all 1.

   There are many different number patterns in the triangle.

   The first row (containing the number 1) is the 0th row.

   a Copy and complete Pascal's triangle to the 8th row.

   b Find the total for each row.

   Describe the sequence that is created by these totals.

   c What sequence is represented by the diagonal series of numbers that begins in the second row 1, 3, 6, 10, ...?

4. A virus is infecting the population of a village.

The table shows how the number of infections increased over 4 days.

Day	1	2	3	4
Number of infections	8	13	18	23

a If the virus continues to infect people at the same rate, how many people will be infected on day 5?

b Assume the infection rate is constant.

Find an expression to calculate how many people will be affected on any day.

c There are 126 people in the village.

How long will it be before everyone is infected?

5. Look at this pattern.

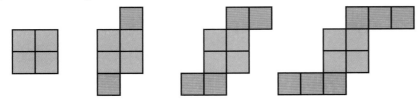

a Write the number of tiles used to build each shape as a sequence.

b How many tiles would you need to build the fifth shape in the pattern?

c Complete this function machine for calculating the number of tiles used to make any shape in the pattern:

input $(n) \to \boxed{} \to \boxed{} \to$ output

d Use your function to work out the number of tiles you would need to build the 20th, 25th and nth patterns.

6. Tamsyn is an ecologist.

She gets 28 days holiday every year plus $\frac{1}{2}$ a day holiday for each overnight bat survey she completes.

a Write a function for calculating how much holiday she will get if she does b overnight bat surveys in a year.

b Tamsyn expects to do 16 overnight bat surveys this year. How much holiday will she get?

16 Formulae

In this chapter you will learn how to …
- use formulae to express and solve problems.
- substitute numbers into formulae to find the value of the subject.
- change the subject of a formula.
- understand and use a range of formulae, including kinematics formulae.

For more resources relating to this chapter, visit GCSE Mathematics Online.

Using mathematics: real-life applications

Vets use formulae to make sure they are giving animals the correct dosage of medicines for their age and weight. A poodle weighing 6 kg needs a far smaller dose of medicine than a 35 kg retriever.

"I need to make sure I give my animal patients the correct amount of medicine. I do this by using formulae that take into account their age, weight and the ratio between any prescribed medicines."

(Veterinarian)

 Did you know?

Einstein's famous formula $E = mc^2$ looks deceptively simple. It describes the relationship between energy, mass and the speed of light (c). Find out more about this formula and why it is linked to the theory of relativity.

 Tip

Formulae often use letters for the variables that relate to the value they represent. For example, in formulae for area, A is often used to represent the value for **a**rea and h to represent the value for **h**eight.

Before you start …

Chs 1 and 5	You need to be able to substitute values into expressions.	**1** Evaluate $\dfrac{x + 2y}{z}$ when: **a** $x = 7$, $y = 4$ and $z = 2$ **b** $x = 7$, $y = -2$ and $z = 2$ **c** $x = -7$, $y = 4$ and $z = 4$ **d** $x = -7$, $y = -2$ and $z = 2$
Ch 14	You should be able to solve simple equations.	**2** Solve. **a** $6x = x + 35$ **b** $5x = 64 - 3x$ **c** $2(2x + 3) = x + 7$ **d** $5x - 8 = 3x + 12$
Ch 2, 5, 10, 13 and 14	You should be familiar with some formulae already, and be able to identify the subject, variable(s) and any constants.	**3** Look at these two formulae for finding the area of shapes. $A = \dfrac{1}{2}bh \quad A = \pi r^2$ **a** What do the variables represent? **b** What is the constant in each formula? **c** What is the subject of each formula? **d** What tells you that the second formula applies to circles?

16 Formulae

Assess your starting point using the Launchpad

STEP 1

1 In an isosceles triangle the base angles labelled *b* are equal.

If the angle at the apex is *a*, write a formula that could be used to calculate the value of one base angle.

2 A plumber charges £20 an hour plus a £50 call-out fee.

Write a formula for the total cost of calling a plumber out.

3 A chicken requires 40 minutes cooking time per kg, plus an extra 30 minutes.

Write a formula to work out the total cooking time, *T* minutes, for a chicken with mass *m* kg.

GO TO Section 1: Writing formulae

STEP 2

4 a Use the formula $C = 2\pi r$ to calculate the circumference (*C*) of a circle with $r = 2.5$ cm:

    i in terms of π **ii** correct to 2 decimal places.

b Is π a constant or a variable?

5 The volume *V* of a cuboid of length *l*, width *w* and height *h* can be calculated by the formula $V = lwh$

If $l = 2$ cm, $w = 3$ cm and $h = 10$ cm, find the volume of the cuboid.

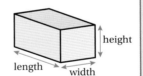

GO TO Section 2: Substituting values into formulae

STEP 3

6 The formula for calculating average speed is $s = \dfrac{d}{t}$

a Rearrange this formula to make *d* the subject.

b Rearrange the formula to make *t* the subject.

GO TO Chapter review

GO TO Section 3: Changing the subject of a formula

Section 4: Working with formulae

Find answers at: cambridge.org/ukschools/gcsemaths-studentbookanswers

Section 1: Writing formulae

Key vocabulary

formula: a general rule or equation showing the relationship between unknown quantities; the plural is formulae.

A *formula* is a rule that shows you how to work out one quantity by combining other quantities.

For example, the rule for working out cooking time for a roast chicken might be 'cook for 40 minutes per kilogram plus an extra 20 minutes'.

The rule for finding the area of a triangle is to multiply half the length of the base by the height.

In mathematical language, the formula for the area of a triangle is $A = \frac{1}{2}bh$.

To make sense of a formula you need to know what the variables represent. In this example:

- A is the area in square units.
- b is the length of the base of the triangle.
- h is the perpendicular height of the triangle.

A, b and h are variables. The letters can be replaced by many different values. $\frac{1}{2}$ is a constant. No matter what value you use for b, you have to multiply it by $\frac{1}{2}$ in the formula.

Tip

For a reminder about variables, see Chapter 5.

A single variable on one side of the formula is called the **subject** of a formula.

The subject is the variable (or quantity) that is being **expressed in terms of** the other variables in the formula.

In $A = \frac{1}{2}bh$, A is the subject of the formula.

In $y = 3x$, y is the subject of the formula.

Writing formulae to represent real-life contexts

Key vocabulary

subject: the variable which is expressed in terms of other variables; it is the variable on its own on one side of the equals sign. In the formula $s = \frac{d}{t}$, s is the subject.

In Chapter 14 you formed equations to represent information given in problems and then solved them to get the answer to a question. You use the same procedures to set up formulae.

WORKED EXAMPLE 1

Mary and Peter are sharing half a circular pizza.

Peter cuts himself a slice that makes an angle of a with the straight edge of the pizza half.

He tells Mary that he will cut another slice the same size, and then she can have what is left.

Write a formula for the size of Mary's share of the pizza.

Subject = Size of Mary's share
Variable = Peter's share
Constant = Size of starting slice

> List the quantities involved.
> Work out whether they represent the subject, a variable, a constant or a coefficient.

Continues on next page ...

Peter's share = $2a$

Angles on a straight line equal 180°
180° − $2a$ = size of Mary's share

Let the size of Mary's share = M
$M = 180° − 2a$

> Establish the relationship between each quantity. What is the subject being expressed in terms of?
>
> Peter takes two slices, each with an angle of a, so you can write an expression for the total size of Peter's share.
>
> You know that angles on a straight line add up to 180°.
>
> The size of Mary's share is expressed in terms of the size of the pizza half and the size of Peter's slice.
>
> Write the formula as concisely as possible using the language of algebra.

Problem-solving framework

A group of sixth form students are planning to run a day conference.

The local university offers conference rooms for hire at a daily rate.

The students think they will have a maximum of 60 delegates and want to offer refreshments costing £4 per delegate.

The largest room they can hire costs £160 for the day.

They want to charge each delegate enough to cover the costs of running the conference.

Write a formula for calculating how much they should charge each delegate.

Steps for approaching a problem-solving question	What you would do for this example
Step 1: If it is useful, draw a diagram.	A diagram is not particularly useful for this question.
Step 2: Identify what you have to do.	Calculate how much to charge each delegate at the conference by writing a formula using the information given.
Step 3: Test the problem with what you already know.	You know a formula is a general rule showing the relationship between quantities.
Step 4: What maths can you do?	**1** List the quantities involved, and establish if they are the subject, a variable, a constant or a coefficient: Number of delegates (d) = 60 (variable, as this can change). Cost of refreshments = £4 per delegate (coefficient, as this value is multiplied by how many delegates there are). Cost of the room = £160 (constant, this is fixed by the university). Cost to charge each delegate = C (subject, this is what we want to find out).

Continues on next page ...

Step 4 (ctd): What maths can you do?	**2** Establish what the relationship is between each quantity.
	The cost to charge each delegate is the same as the cost of having each delegate at the conference, which is the total cost divided by the number of delegates.
	The total cost is equal to the cost of refreshments for each delegate and the cost of the room hire.
	Now put this together using algebra:
	Cost of conference = $4d + 160$
	Cost of one delegate is this value divided by the number of delegates (d).
	3 Write the formula as concisely as possible using the language of algebra.
	$$C = \frac{4d + 160}{d}$$

EXERCISE 16A

1 Construct a formula for:

a S in terms of C and P, where £S is selling price, £C is cost price and £P is profit.

b D in terms of n, where D is the number of degrees in n right angles.

c m in terms of h, where m is the number of minutes in h hours.

2 When y is the subject and x is an unknown value, write a formula to determine y, when y is:

a three more than x

b six less than x

c ten times x

d the sum of -8 and x

e the sum of x and the square of x

f twice x more than x plus 1.

3 To cook a chicken you need to allow 20 minutes per $\frac{1}{2}$ kg and another 20 minutes.

A chicken weighs x kg.

Write a formula to show the number of minutes m required to cook a chicken.

4 An electrician charges a call-out fee, £f, and then charges £r per hour (h) for the job. Write a formula for the total cost of a job.

Section 2: Substituting values into formulae

To find the value of the subject (or any variable) in a formula you need to know the value of all the other variables. The values are substituted into the formula to work out the missing value.

Substitute means replace the letters with the numbers you have been given.

Evaluate means work out the numerical value of a given calculation.

Key vocabulary

substitute: to replace letters with numbers.
evaluate: to calculate the numerical value of something.

WORKED EXAMPLE 2

The perimeter of a square can be found using the formula $P = 4s$, where s is the length of a side.

What is the perimeter of a square with sides of:

a 10 cm **b** 2.5 mm?

a $P = 4 \times 10$
  $P = 40$ cm — Substitute (replace) s with 10

b $P = 4 \times 2.5$
  $P = 10$ mm — Substitute (replace) s with 2.5

WORKED EXAMPLE 3

The volume of an object can be found using the formula $V = \frac{1}{3}Ah$

Find the volume of an object when $A = 30\,\text{cm}^2$ and $h = 6\,\text{cm}$.

$V = \frac{1}{3} \times 30 \times 6$ — Substitute $A = 30$, and $h = 6$.

$= \frac{1}{3} \times 180$ — Remember volume is given in cubic units.

$= 60\,\text{cm}^3$

Tip

It is good practice to show you can correctly substitute values into a formula before calculating the answer. Write down what you need to work out and then calculate the value.

Kinematics equations

Kinematics is the study of how objects move.

You have already learned that the average speed of an object can be found using the formula

$$\text{speed} = \frac{\text{distance travelled}}{\text{time taken}}$$

Now you are going to use some different formulae to work out how fast and how far objects are moving.

In kinematics there are three important terms.

- **Velocity** – a measure of how fast an object is moving in a direction (in everyday terms this is speed).

Find answers at: cambridge.org/ukschools/gcsemaths-studentbookanswers

> **Tip**
>
> Velocity, acceleration and displacement are vector quantities because they have both size and direction.
>
> You will learn more about vectors in Chapter 22.

- **Acceleration** – the rate at which the velocity of an object changes (in everyday terms this is increasing speed).
- **Displacement** – a change in the position of an object (in everyday terms this is distance travelled).

You will work with these three kinematics formulae:

$$v = u + at \qquad s = ut + \frac{1}{2}at^2 \qquad v^2 = u^2 + 2as$$

where,

v = final velocity	velocity is measured in units of distance and time (m/s)
u = initial velocity	
a = acceleration	acceleration in units of distance and time (m/s/s or m/s^2)
s = displacement (distance travelled)	displacement in units of length (m)
t = time taken	time in seconds (s)

> **Tip**
>
> These kinematics equations apply only to an object moving in a straight line at a constant acceleration.

WORK IT OUT 16.1

Two students studying physics were given the equation $v = u + at$ and asked to calculate the value of v when u = 12 metres per second, a = 10 metres per second^2, t = 3 seconds.

Which student has calculated v correctly, and written the correct units for their answer?

Student A	Student B
$v = 12 + 10 \times 3$	$v = 12 + 10 \times 3$
$v = 66$ m	$v = 42$ m/s

EXERCISE 16B

1 Evaluate these expressions.

   a $2a(a - 3b)$ when:

      i $a = 2$ and $b = -5$ **ii** $a = -3$ and $b = -2$ **iii** $a = \frac{1}{3}$ and $b = \frac{1}{2}$

   b $x^2 - 2y$ when:

      i $x = -7$ and $y = 2$ **ii** $x = -\frac{1}{3}$ and $y = \frac{5}{6}$

2 Calculate the value of the subject in each formula.

   Where units are given, include the correct units in your answer.

   a $A = lw$, where $l = 6$ m, $w = 9$ m

   b $s = \frac{d}{t}$, where $d = 72$ km, $t = 3$ hours

   c $A = \frac{1}{2}(a + b)h$, where $a = 4$ cm, $b = 7$ cm, $h = 9$ cm

   d $x = \sqrt{ab}$ (answer to 3 decimal places), where $a = 12$, $b = 13$

   e $V = \pi r^2 h$ (answer to 3 decimal places), where $r = 3.5$ cm, $h = 17$ cm

   f $t = a + (n - 1)d$, where $a = 10$, $n = 6$, $d = 2$

> **Tip**
>
> You will probably recognise some of these formulae. Knowing what the value you calculate represents makes it easier to see if the answer makes sense. Section 4 lists some common formulae.

3. The formula for converting temperature from degrees Fahrenheit (F) to degrees Celsius (C) is

$$C = \frac{5}{9}(F - 32°)$$

What is the value of C when $F = 76°$?

Tip

In questions working with formulae and associated units, remember to write the correct units in your answer.

4. The stopping distance of a car, in metres, once the brakes are applied is given by the formula

$d = 0.2v + 0.005v^2$,

where v km/h is the speed of the car when the brakes are applied.

In how many metres will a car stop if the brakes are applied when $v = 130$ km/h?

5. The Greek mathematician Hero showed that the area of a triangle with sides a, b and c is given by the formula

$A = \sqrt{s(s-a)(s-b)(s-c)}$,

where $s = \frac{1}{2}(a + b + c)$.

Use Hero's formula to find the area of this triangle.

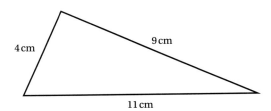

6. The simple interest payable when £P is invested at a rate of R% per year for T years is given by $I = \frac{PRT}{100}$.

Calculate the simple interest payable when £2000 is invested at 1.5% per year for 6 years.

Section 3: Changing the subject of a formula

There are times when you might want to find the value of a variable that is not the subject of a formula.

To do this, rearrange the formula to make the letter you are trying to find the subject of the formula.

Since a formula is a type of equation, you can use inverse operations to 'solve' the formula for the letter you want.

WORKED EXAMPLE 4

Make r the subject of the formula $C = 2\pi r$.

$\frac{C}{2\pi} = r$ Divide both sides by 2π.

$r = \frac{C}{2\pi}$ Swap the whole formula round so that the subject is on the left.

Tip

When u has a positive value, the square root is positive. However, when u can be positive or negative, it is usually written as $u = \pm\sqrt{v^2 - 2as}$.

WORKED EXAMPLE 5

Given the kinematic formula $v^2 = u^2 + 2as$:

a make s the subject of the formula

b make u the subject of the formula.

a $u^2 + 2as = v^2$
$2as = v^2 - u^2$
$s = \dfrac{v^2 - u^2}{2a}$

◁ Swap the sides to get $2as$ on the left.
Subtract u^2 from each side.
Divide both sides by $2a$.

b $u^2 + 2as = v^2$
$u^2 = v^2 - 2as$
$\sqrt{u^2} = \sqrt{v^2 - 2as}$
$u = \sqrt{v^2 - 2as}$

◁ Swap the sides to get u^2 on the left.
Subtract $2as$ from each side.
Take the square root of each side to get u on its own.

You can calculate the value of **any** variable in a formula provided all the other variables are known.

If the variable you are looking for is not the subject of the formula, you can change the subject of the formula before you start.

You can also substitute numbers into the formula **before** rearranging it.

WORKED EXAMPLE 6

Given the formula $v^2 = u^2 + 2as$, find the value of s when $u = 8$, $v = 10$ and $a = 3$.

$v^2 = u^2 + 2as$
$v^2 - u^2 = 2as$
$\dfrac{v^2 - u^2}{2a} = s$

◁ Rearrange the formula to make s the subject:
- subtract u^2 from both sides
- divide by $2a$.

$s = \dfrac{v^2 - u^2}{2a}$

◁ Writing the subject on the left is the usual convention.

$u = 8, v = 10$ and $a = 3$

$s = \dfrac{10^2 - 8^2}{2 \times 3}$

$s = \dfrac{100 - 64}{6} = \dfrac{36}{6}$

$s = 6$

◁ Substitute in the values you know.

Continues on next page ...

$$10^2 = 8^2 + 2 \times 3s$$
$$100 = 64 + 6s$$
$$100 - 64 = 6s$$
$$36 = 6s$$
$$36 \div 6 = s$$
$$6 = s$$
$$s = 6$$

You could, alternatively, substitute the numbers first.

EXERCISE 16C

1 These formulae have been rearranged to make the given variable the subject.

In each case, choose the correct answer.

a $A = \frac{1}{2}bh$ $b = ?$ **A** $\frac{A}{2h}$ **B** $\frac{2A}{h}$ **C** $2Ah$

b $V = IR$ $I = ?$ **A** VR **B** $\frac{V}{R}$ **C** $\frac{R}{V}$

c $V = \pi r^2 h$ $h = ?$ **A** $V\pi r^2$ **B** $\frac{V}{\pi r^2}$ **C** $\frac{Vr^2}{\pi}$

d $A = \frac{1}{2}(a + b)h$ $a = ?$ **A** $\frac{2A - b}{h}$ **B** $\frac{2Ah}{b}$ **C** $\frac{2A}{h} - b$

> **Tip**
>
> You might recognise some of these formulae. See the list in Section 4.

2 Rearrange the formulae to make the variable in brackets the subject.

a $y = mx + c$ (c) b $V = \frac{1}{3}\pi r^2 h$ (h)

c $v = u + at$ (t) d $A = 2\pi r^2 + 2\pi rh$ (h)

e $s = ut + \frac{1}{2}at^2$ (a) f $c^2 = a^2 + b^2$ (b)

3 The formula for the perimeter, P, of a rectangle, l by w, is $P = 2(l + w)$.

If $P = 20$ cm and $l = 7$ cm, what is the length w?

4 The area, A cm², enclosed by an ellipse is given by $A = \pi ab$

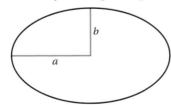

Calculate to 2 decimal places the length a cm, if $b = 3.2$ and $A = 25$.

5 When an object is thrown into the air at a speed of u metres per second, its height h metres above the ground and time t seconds of flight are related by

$h = ut - 4.9t^2$ (ignoring air resistance).

Find the speed at which a javelin was thrown if it reached a height of 15 metres after 5 seconds.

6 The kinetic energy E joules of a moving object is given by
$$E = \frac{1}{2}mv^2$$
where m kg is the mass of the object and v m/s is its speed.

   a Rearrange the formula to make m the subject.

   b Use the new formula to calculate the mass of the object when $E = 300$ joules and $v = 10$ m/s.

7 The formula for the sum, S, of the interior angles in an n-sided polygon is $S = 180(n - 2)$.

Rearrange the formula to make n the subject and use this to find the number of sides of the polygon if the sum of the interior angles is 2160°.

Section 4: Working with formulae

Common formulae

You should memorise these formulae. Use them in the following exercise.

Perimeter of a rectangle	$P = 2l + 2w$ or $P = 2(l + w)$ l = length, w = width
Area of a rectangle	$A = lw$ l = length, w = width
Area of a parallelogram	$A = bh$ b = length of base, h = perpendicular height
Circumference of a circle	$C = \pi d$ or $2\pi r$ d = diameter, r = radius
Volume of a cuboid	$V = lwh$ l = length, w = width, h = height
Surface area of a cuboid	$SA = 2lw + 2wh + 2lh$ l = length, w = width, h = height
Area of a triangle	$A = \frac{1}{2}bh$ b = length of base, h = perpendicular height
Area of a trapezium	$A = \frac{1}{2}(a + b)h$ $a + b$ = sum of lengths of parallel sides, h = height
Area of a circle	$A = \pi r^2$ r = radius
Volume of a cylinder	$V = \pi r^2 h$ r = radius, h = height

Volume of a cone	$V = \frac{1}{3}\pi r^2 h$ r = radius, h = height
Converting degrees Celsius into degrees Fahrenheit	$F° = \frac{9}{5}C + 32$
Converting degrees Fahrenheit into degrees Celsius	$C° = \frac{5}{9}(F - 32)$

EXERCISE 16D

1 Decide whether each statement is true or false. Correct any false statements to make them true.

  a If $A = \frac{1}{2}bh$, when $b = 4.5$ cm and $h = 8$ cm, then $A = 18\,\text{m}^2$.

  b When the formula to calculate speed $S = \frac{D}{T}$ is rearranged to make T the subject, it becomes $T = \frac{D}{S}$.

  c A formula to calculate the number n half way between two numbers x and y can be written as $n = \frac{x + y}{2}$.

  d In the formula to calculate the area of a circle $A = \pi r^2$, π, A and r are the variables.

2 Use the appropriate formula from the table to complete these problems.

  a Calculate the perimeter P of a rectangle if the length $l = 6.5$ m and width $w = 8$ m.

  b What is the height of a parallelogram that has an area $A = 45\,\text{cm}^2$ and a base $b = 2.5$ cm?

  c What is the surface area of rectangular solid (cuboid) if the length $l = 20$ cm, the width $w = 3.5$ cm and the height $h = 7.2$ cm?

  d What is the area A of a circle with a radius $r = 12.3$ cm?

  e What is the height h of a cone that has a volume $V = 45\,\text{cm}^3$ and a radius $r = 5$ cm?

3 **a** Find the Fahrenheit temperature $F°$ when $C = 8$

  b Find the Celsius temperature $C°$ when $F = 72$

  c Freezing point is 0 °C. What is the equivalent temperature in Fahrenheit?

4 Given $s = ut + \frac{1}{2}at^2$
Find the value of s when $u = 10$, $t = 5$ and $a = 0.27$

5. In an electrical circuit the voltage V is the product of the current I and the resistance R.

   $V = IR$

   a Calculate the voltage V when $I = 25$ milliamps and $R = 330$ ohms.

   b Calculate the current I when $V = 9$ volts and $R = 330$ ohms.

   c Calculate the resistance R when $V = 9$ volts and $I = 18$ milliamps.

6. Use the formula $A = 4\pi r^2$ to find A correct to 1 decimal place when:

   a $r = 3$ b $r = 25.6$

7. In general, temperature decreases with height above sea level. This formula shows how temperature and height above sea level are related.
   $T = \dfrac{h}{200}$, where T is the temperature decrease in °C and h is the height increase in metres.

   a If the temperature at a height of 500 m is 23 °C, what will it be when you climb to 1300 m?

   b What increase in height would result in a 5 °C decrease in temperature?

8. Rearrange each of these formulae to make the letter in brackets the subject.

   a $A = 4\pi r^2$ (r)

   b $W = \sqrt{\dfrac{3V}{\pi h}}$ (V)

9. A large wooden crate has to be moved by a car. The crate must be kept upright. Its base is 150 cm long by 50 cm wide.

   Which of the following cars could be used? Which would be the best car to use?

Car	A	B	C
Area of boot (cm²)	11 250	8642	9480
Width of car boot (cm)	72	58	62.5

10. Two friends, Shona and Chelsea, are planning to go on holiday together. Shona prefers resorts where the average daytime temperatures are below 27 °C, and Chelsea likes the temperature to be at least 20 °C. Convert the following average daytime temperatures for three possible holiday resorts from Fahrenheit to Celsius. Which of the resorts would be acceptable to both friends?

Holiday resort	Average daytime temperature
Sunshine Beach	84.2 °F
Palm Tree Bay	77 °F
Silver Sands	80.6 °F

11 Driver A covers a distance of 87 miles in 1.4 hours travelling on a motorway and Driver B covers a distance of 174 miles in 2.4 hours on the same motorway. Work out the average speed of each driver. Which driver is likely to have broken the speed limit of 70 mph for a motorway?

Checklist of learning and understanding

Writing formulae
- A formula is a general rule showing the relationship between quantities.
- You can use formulae to represent real-life problems as long as you define the variables you are using.

Substituting values into formulae
- To evaluate a formula, you need to know the value of all but one of its variables.
- Substitute given values to find the unknown variable.

Changing the subject of a formula
- In any formula you can 'change the subject' by rearranging the formula in the same way as you rearrange equations.

Chapter review

1 Substitute the given values into these formulae and evaluate.
  a Find v when $v = u + at$ and $u = 50$, $a = 10$ and $t = 2$.
  b Find s when $s = ut + \frac{1}{2}at^2$ and $u = 0$, $a = 6$ and $t = 10$.
  c Find v when $v^2 = u^2 + 2as$ and $u = 5$, $a = 6$ and $s = 12$.

2 Rearrange the formula for the area of a circle, to calculate the radius of a circle with area 25 cm² (leave π in the answer).
The area of a circle is $A = \pi r^2$

3 $A = 4bc$
$A = 100$
$b = 2$
  a Work out the value of c. *(2 marks)*
$m = \dfrac{k+1}{4}$
  b Make k the subject of the formula. *(3 marks)*

©Pearson Education Ltd 2013

4 The cost £C of hiring a reception room for a function is given by the formula $C = 12n + 250$, where n is the number of people attending the function.

 a Rearrange the formula to make n the subject.

 b How many people attended the function if the cost of hiring the reception room was:

  i £730   **ii** £1090   **iii** £1210   **iv** £1690?

5 Rearrange the following formulae to make the letter in the brackets the subject.

 a $A = \pi ab$ (a) **b** $P = 2a + 2b$ (a)

 c $\sqrt{ab} = c$ (b) **d** $a\sqrt{b} = c$ (b)

 e $\sqrt{(b+c)} = c$ (b) **f** $\sqrt{(x-b)} = c$ (b)

 g $\sin \theta = \dfrac{o}{h}$ (o) **h** $\dfrac{x}{\sqrt{y}} = c$ (y)

17 Volume and surface area

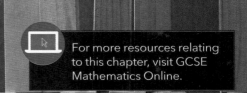

In this chapter you will learn how to ...
- calculate the volume and surface area of cuboids and prisms.
- calculate the volume and surface area of cylinders.
- calculate the volume and surface area of cones and spheres.
- calculate the volume and surface area of pyramids.
- solve volume and surface area problems involving composite shapes.

For more resources relating to this chapter, visit GCSE Mathematics Online.

Using mathematics: real-life applications

Freight costs are dependent upon the volume of material being transported. Freight rates are calculated using the container volume measured against the length of the container. The longer the container the higher the freight cost.

"To transport a container full of apples from Felixstowe, England, to Le Havre in France, I have to let the freight operator know the volume of apples I have to transport as well as the dimensions of the crates. I am then quoted a transport cost."
(Apple producer)

Before you start ...

Chs 2 and 3	You need to be able to recognise and identify solid objects.	**1** Name each object as accurately as possible from the description. **a** A 3D object with six identical square surfaces. **b** A 3D solid with two parallel circular faces. **c** An object with a square base and triangular side faces that meet at an apex. **d** A 3D object with a circular base and one vertex. **e** A 3D object with many flat surfaces that are polygons. **f** A polyhedron with two triangular and three rectangular faces.
Ch 12	You must be able to calculate the area of plane shapes.	**2** What is the formula for the area of a circle? **3** What is the area of a right-angled triangle with sides of 3 cm, 4 cm and 5 cm?
Chs 2 and 3	You should understand and use the properties of solids.	**4** A shape has 6 faces, 8 vertices and 12 edges. **a** What could it be? **b** What additional information do you need to name the shape more accurately?

Find answers at: cambridge.org/ukschools/gcsemaths-studentbookanswers

Assess your starting point using the Launchpad

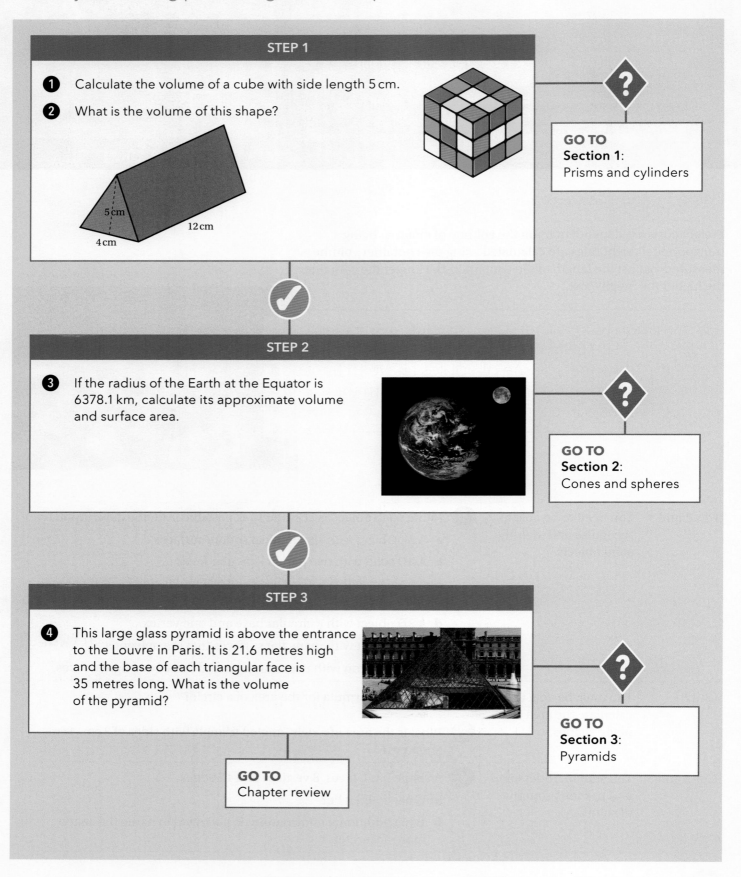

STEP 1

1. Calculate the volume of a cube with side length 5 cm.
2. What is the volume of this shape?

GO TO **Section 1**: Prisms and cylinders

STEP 2

3. If the radius of the Earth at the Equator is 6378.1 km, calculate its approximate volume and surface area.

GO TO **Section 2**: Cones and spheres

STEP 3

4. This large glass pyramid is above the entrance to the Louvre in Paris. It is 21.6 metres high and the base of each triangular face is 35 metres long. What is the volume of the pyramid?

GO TO **Section 3**: Pyramids

GO TO Chapter review

Section 1: Prisms and cylinders

A prism is a 3D object with a uniform cross-section along its length, where the cross-section is a polygon.

This shape sorter on the right is really a prism sorter. You should be able to name the prisms sticking out of it.

The diagram below shows examples of **right prisms**. One of the end faces is known as the base of the object. The sides are rectangles perpendicular to the base.

A cube (square prism)   A rectangular prism   A triangular prism

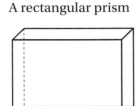 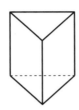

Key vocabulary

right prism: a prism with sides perpendicular to the end faces (base).

Volume

Tip

A rectangular prism is also called a cuboid.

The volume of an object is the three-dimensional space that it takes up. Volume is given in cubic units, such as mm^3, cm^3 and m^3 (for solids). For liquids, the volume of a container is referred to as capacity. Capacity is measured in litres.

You can find the volume of any right prism by finding the area of one end face (base) and multiplying this by its length or height.

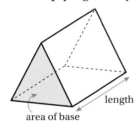 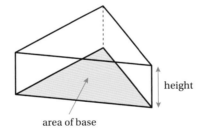

Surface area

Surface area is the total area of the faces of a three-dimensional object.

Sketching a rough net of the object can help you to see what faces to include when you calculate the surface area.

Cubes and cuboids

Tip

You learnt about nets of solids in Chapter 3. Revise that section if you need to.

The net of a cuboid shows that the surface area is the total area of the six faces.

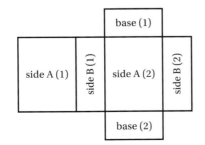

The opposite faces match, so:

Surface area = 2(area of side A) + 2(area of side B) + 2(area of base)

A cube is a special case of a cuboid; it has six identical square faces. You can use the following formulae for volume and surface area:

Volume of a cube = x^3

Surface area of a cube = $6x^2$

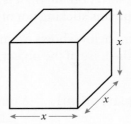

WORKED EXAMPLE 1

Calculate the volume and surface area of this cuboid.

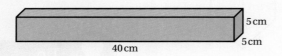

Tip

Remember to check that all measurements are in the same units, for example all cm or all m.

Volume = **lbh** = 40 × 5 × 5 = 1000 cm³

For a cuboid, any of the surfaces could be chosen as the 'base'. The volume of a cuboid is usually calculated using the formula $V = lbh$, where l, b and h are the three different dimensions of the shape.

Surface area = 4 × (40 × 5) + 2 × (5 × 5)
= 800 + 50
= 850 cm²

Look for faces that are identical. This can cut down on the working you need to do.

Prisms with bases of other shapes

WORKED EXAMPLE 2

Find the volume and surface area of this triangular prism.

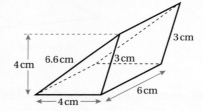

Area of the base = $\frac{1}{2}bh = \frac{1}{2} \times 4 \times 4 = 8$ cm²

The base is a triangle.

Volume = 8 × 6 = 48 cm³

Volume of prism = area of base × length

Surface area of the two base triangles = 8 + 8 = 16 cm²

Surface area of side face 1 = 3 × 6 = 18 cm²
Surface area of side face 2 = 4 × 6 = 24 cm²
Surface area of side face 3 = 6.6 × 6 = 39.6 cm²

The other faces are all rectangles. Work systematically around the shape.

Total surface area = 97.6 cm²

Surface area of triangular prism
= 2(area of triangular base) + area of three side faces

Prisms with bases that are trapeziums are quite common. Rubbish skips, wheelbarrows, planters and many other containers take this shape.

Area of the trapezium base = $\frac{1}{2}(a + b) \times h$.

Volume of prism = area of the trapezium × length

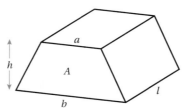

WORK IT OUT 17.1

What is the volume of soil that can be contained in this skip?

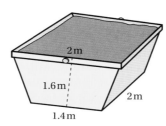

Which of the following is the correct calculation?

Calculation A	Calculation B	Calculation C
Area of the trapezium Area = $\frac{1}{2}(a + b) \times h$ Area = $\frac{1}{2} \times 3.4 \times 1.6 = 2.72 \text{ m}^2$ Volume = area of the trapezium × length Volume = $2.72 \times 2 = 5.44 \text{ m}^2$	Area of the trapezium Area = $\frac{1}{2}(a - b) \times h$ Area = $\frac{1}{2} \times 0.6 \times 1.6 = 0.48 \text{ m}^2$ Volume = area of the trapezium × length Volume = $0.48 \times 2 = 0.96 \text{ m}^3$	Area of the trapezium: Area = $\frac{1}{2} \times b \times h$ Area = $\frac{1}{2} \times 1.4 \times 1.6 = 1.12 \text{ m}^2$ Volume = area of the trapezium × length Volume = $1.12 \times 2 = 2.24 \text{ m}^3$

Rearranging the formulae

You can change the subject of the formula to find the length of a prism if you know the volume and area of the base.

WORKED EXAMPLE 3

The volume of a triangular prism is 100 cm³ and the area of the end is 25 cm². How long is the prism?

V = area of the triangle × length   Rearrange the formula.
so *V* ÷ *A* = *L*

L = 100 ÷ 25 = 4 cm

Problem-solving framework

You are painting a room and it needs two coats of paint. The room is 10 m long, 7 m wide and 3 m high. There is a door which is 2 m high and 1.5 m wide and a window that is 2.6 m high and 2.2 m wide. You want to colour it blue and it should take 5 days to paint it.

Paint covers 5 m^2 per litre and you can buy it in 5-litre pots. Each pot costs £14.99. How many 5-litre paint pots will you need to buy?

You should add 10% into your calculations for special circumstances.

Steps for solving problems	What you would do for this example
Step 1: What have you got to do?	Paint a room with two coats of paint. Work out how many 5-litre paint pots are needed. A useful estimate could be calculated at first. The room is roughly 100 m^2 and the door and window are roughly 10 m^2. So coverage is 2 × 90 m^2 = 180 m^2. Plus 10% takes is roughly to 200 m^2. This means 200 ÷ 5 = 40 litres. Eight 5-litre pots are needed.
Step 2: What information do you need?	Room dimensions are needed for surface area calculations. Door and window dimensions need to be taken away from the coverage area.
Step 3: What information don't you need?	The colour of paint, the length of time taken and the cost of the paint are not needed.
Step 4: What maths can you do?	Calculate the surface area of the room: Walls 1 & 3: 10 × 3 = 30 m^2 Walls 2 & 4: 7 × 3 = 21 m^2 Total surface area = 30 + 30 + 21 + 21 = 102 m^2 Door area = 2 × 1.5 = 3 m^2 Window area = 2.6 × 2.2 = 5.72 m^2 Total surface area for painting 102 − 3 − 5.72 = 93.28 m^2. Two coats required = 93.28 × 2 = 186.56 m^2 Add 10% = 18.656 m^2 Total paint coverage required 205.216 m^2 1 litre = 5 m^2 coverage of paint 205.216 ÷ 5 = 41.0432 litres of paint are required. 5-litre pots of paint can be bought 41.0432 ÷ 5 = 8.208 64 You will need to buy nine pots.
Step 5: Have you done it all?	Yes: room size less the door and window sizes two coats of paint + 10% total divided by the coverage of one paint pot
Step 6: Is it correct?	Yes: checked against estimate

Cylinders

Cylinders are not prisms, but you can find their volume and surface area in the same way as you do with prisms.

The net of a cylinder shows that the curved surface forms a rectangle when it is flattened out. The length of the rectangle is equivalent to the circumference of the circular base.

The surface area of a cylinder is calculated using the formula:

$S = 2\pi rh + 2\pi r^2$

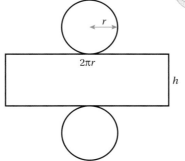

> **Tip**
>
> A cylinder is not a prism because its cross-section is a circle, not a polygon.

WORKED EXAMPLE 4

A road roller has a roller on the front which is filled with water to make it heavy. The tank of water in the roller has a radius of 0.95 m and a length of 2.4 m. Calculate the volume (or capacity) of the tank.

Volume of cylinder = area of the circle × length
Volume = $\pi r^2 \times l$
Volume = 3.14 × 0.95 × 0.95 × 2.4 = 6.801 24 m³
 = 6.8 m³ (1 decimal place)

> **Tip**
>
> As with prisms, you can find the radius or diameter of the base of a cylinder when the other dimensions are known by changing the subject of the formula.

EXERCISE 17A

1 Calculate the volume and surface area of each object. (Each object is a closed object.)

a 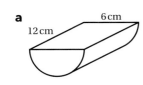   **b** 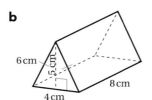   **c**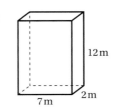

d 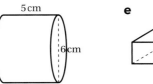   **e** 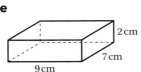   **f**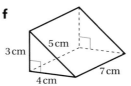

2 What is the capacity, in litres, of the aquarium shown on the left, when filled to the top? (1 litre = 1000 cm³)

3 The volume of a cube is 144 m³. What is the length of each side?

4 The dimensions of an Olympic-sized swimming pool are 50 m long, 25 m wide, and the water is 2 m deep. What is the volume of water in an Olympic-sized swimming pool?

5 What is the volume of this triangular prism to 2 decimal places?

6 What is the surface area of one side of this solar-panelled roof?

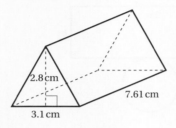

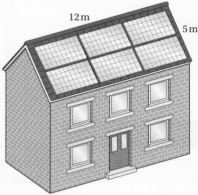

7 Calculate the volume of the object below.

8 Calculate the volume of this solid piece of wood with a cylindrical hole drilled through the middle.

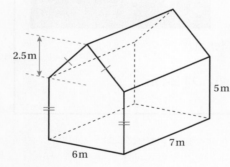

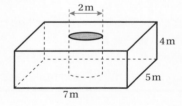

Section 2: Cones and spheres

Cones

The formula for the volume of a cone is $\frac{1}{3} \times$ area of circular base $\times h$, where h is the perpendicular height from the base to the apex of the cone.

The area of the base can be found using the formula for the area of a circle, πr^2.

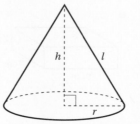

Tip

In some problems involving cones you might need to use Pythagoras' theorem to find the perpendicular height using the radius and the slant height. You will meet this in Chapter 30.

WORKED EXAMPLE 5

Find the volume of a cone of radius 12 cm with a perpendicular height of 14 cm.

Volume $= \frac{1}{3}(\pi r^2)h = \frac{1}{3}(3.14 \times 12 \times 12) \times 14$
$= 2110 \text{ cm}^3$ (to 3 significant figures)

The area of the curved surface of a cone is πrl, where r is the radius of the base, and l is the **slant height** of the cone.

Therefore, the total surface area (S) of a cone is:

S = area of curved surface + area of base
$= \pi rl + \pi r^2$

Spheres

A sphere is any perfectly round object.

The volume of a sphere is equal to $\frac{4}{3}\pi r^3$, where r is the radius of the sphere.

Many objects include spheres or parts of spheres in their structure.

The surface area of a sphere is equal to $4\pi r^2$, where r is the radius of the sphere.

WORKED EXAMPLE 6

Find the surface area and volume of a sphere with radius 3 cm. Use 3.14 as an approximate value of π. Give your answers to 2 decimal places.

Surface area $= 4 \times 3.14 \times 3 \times 3$
$= 113.04 \text{ cm}^2$

Volume $= \frac{4}{3} \times 3.14 \times 3 \times 3 \times 3$
$= 113.04 \text{ cm}^3$

Notice that this is numerically the same answer as for the area, so it is important that you include the correct units.

WORKED EXAMPLE 7

The radius of the Earth is approximately 6378.1 km.

a Find the approximate volume and surface area of the Earth. Use the calculator value of π. Give your answers to 4 significant figures.

b 70% of the surface of Earth is covered with water. What is the surface area of land?

a $V = \frac{4}{3} \times \pi \times 6378.1^3 \approx 1\,087\,000\,000\,000 \text{ km}^3$

Surface area $= 4 \times \pi \times 6378.1^2 \approx 511\,200\,000 \text{ km}^2$

b 30% of the Earth's surface area is land.
$0.3 \times 511\,200\,000 \approx 153\,400\,000 \text{ km}^2$

Tip

In calculations with such large values you would normally give your answers in Standard form. You will deal with this in Chapter 27.

Find answers at: cambridge.org/ukschools/gcsemaths-studentbookanswers

EXERCISE 17B

1 Calculate the volume and surface area of each object. (The objects are all closed.)

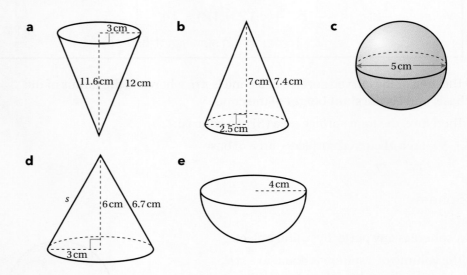

2 Earth's moon has a mean radius of 1738 km. Use the calculator value of π to find its approximate volume.

3 The table below gives some standard diameters of spherical balls used in different sports. Calculate the surface area of each ball. Assume they are round and ignore any dimples on the surface.

Give your answers to 3 significant figures.

	Sport	Standard diameter (cm)
a	snooker	5.3
b	tennis	6.5
c	football	21.6
d	golf	4.3
e	bowling	12.5
f	basketball	23.8
g	hockey	7.4
h	baseball	7.3
i	cricket	7.2

4 A factory needs to calculate the volume and outer sloped surface area of open plastic cones which have the dimensions (in cm) given in the table. Calculate each volume and surface area.

Give your answers correct to 3 significant figures.

	Radius (r)	Slant height (l)	Perpendicular (h)
a	5	10	$\sqrt{75}$
b	18	34	$\sqrt{832}$
c	7	21	$\sqrt{392}$
d	16	22	$\sqrt{228}$
e	60	64	$\sqrt{496}$
f	9	26	$\sqrt{595}$
g	30	52	$\sqrt{1804}$

5 A cone and a sphere both have a diameter of 8 cm. If they have the same volume, how tall is the cone?

Composite solids

Objects in real life are very rarely composed of just one kind of geometric object. Most buildings involve a combination of solid shapes, and modern buildings often incorporate unusual shapes into their designs.

This is the winning design for the air traffic control tower at Newcastle airport. The design incorporates cut-off conical shapes around a cuboid-shaped cement tower.

This is the design of the North Gate Bus Station in Northampton. You can see that many different solids have been used in the design.

When you worked with area in Chapter 12 you split composite shapes into known shapes and found the area of each shape separately. You can use the same technique to find the volume of composite solids.

To find the total surface area of a composite solid you need to find the area of each section separately. However, you cannot just automatically add the areas because the area of some faces will overlap and not form part of the 'outside' area of the solid.

> **Tip**
>
> It is useful to develop a system for checking that you have included all the surfaces when you are finding the surface area of a composite shape.

Find answers at: cambridge.org/ukschools/gcsemaths-studentbookanswers

WORKED EXAMPLE 8

Calculate the total volume and surface area of the object shown on the right.

Use the calculator value of π in your calculations and give final answers correct to 2 decimal places.

The object consists of a cone and a cylinder.

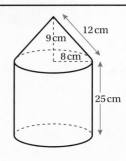

Volume

Volume of cone = $\frac{1}{3}(\pi r^2) \times h$

$= \frac{1}{3} \times \pi \times (8)^2 \times 9$

$= 603.19 \, cm^3$

Volume of cylinder = $\pi r^2 h$

$= \pi \times (8)^2 \times 25$

$= 5026.55 \, cm^3$

Total volume = $5629.74 \, cm^3$

Surface area

Conical top is a cone without a base.
Curved surface area = $\pi r l$

$= \pi \times 8 \times 12$

$= 301.59 \, cm^2$

Cylinder with one base only.

S = area of base + area of curved side

$= \pi \times (8)^2 + 2 \times \pi \times 8 \times 25$

$= 1457.70 \, cm^2$

Total surface area = $1759.29 \, cm^2$

EXERCISE 17C

1. Find the surface area of each solid. Give your answers correct to the nearest m² or mm² (the upper part of shape **b** is a semi-circle).

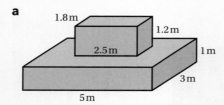
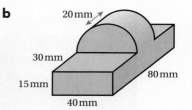

2. Calculate the volume of each solid to the nearest unit (the upper part of shape **b** and the lower parts of shape **c** are semi-circles).

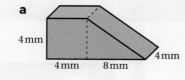
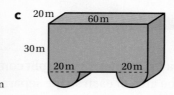
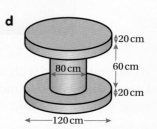

3. A sphere of diameter 1.6 m is cut through the centre to form two hemispheres. What is the total surface area of the two hemispheres?

Section 3: Pyramids

Pyramids are named according to the shape of their base.

The volume of a pyramid is $\frac{1}{3}$ of the volume of a prism with the same base area and height.

Volume of a pyramid = $\frac{1}{3}$ area of base × perpendicular height.

The surface area of a pyramid is the total area of the base plus the area of each triangular side.

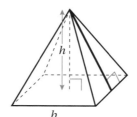

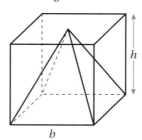

WORKED EXAMPLE 9

Calculate the volume and the surface area of this square-based pyramid.

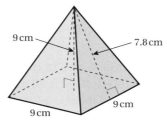

Volume = $\frac{1}{3}$ × area of base × h

= $\frac{1}{3}$ × 9 × 9 × 9 = 243 cm³

Surface area = area of square base + 4 × area of triangular sides

Surface area = b^2 + 4 × ($\frac{1}{2}$ × slant height × b)

= (9 × 9) + 4 × ($\frac{1}{2}$ × 7.8 × 9)

= 81 + 140.4 = 221.4 cm²

The perpendicular height of each triangular face can be calculated using Pythagoras' theorem.

Tip

The slant height of a pyramid is the perpendicular height from the middle of the base of a lateral or sloping side to the top of that side

EXERCISE 17D

1 These three pyramids have square bases.

Calculate the surface area and the volume of each one.

a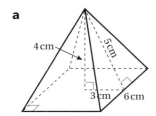

b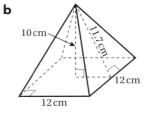

c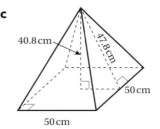

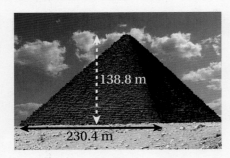

2 What is the volume of the Great Pyramid in the photograph (left)? It has a square base.

3 What is the difference in the volumes of a pyramid with a square base of side 6 m and a pyramid with an equilateral triangle with side 6 m as a base, if both have a perpendicular height of 8 m? Take the perpendicular height of the triangular base to be 5.2 m.

4 The solid wooden sculpture (right) is a triangular-based pyramid. The base is an equilateral triangle with sides of 1 m and a perpendicular height of 86.6 cm. The height of the sculpture is 2 m. Calculate the volume of wood that makes up the sculpture.

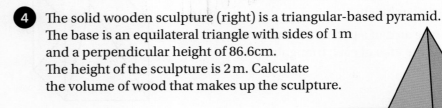

5 Ancient Egyptians used objects called obelisks in their architecture. They consisted of a square-based column with a pyramidal structure on the top.

Calculate the volume and surface area of the Obelisk of Queen Hapshetsut in the photograph (left). It is 30 m high, the square base has an area of 5 m² and the pyramid itself is 1.5 m high. Take the slant height of the pyramid to be 1.87 m.

Checklist of learning and understanding

Volume

- Volume is the amount of space a 3D object occupies.
- Volume is calculated in cubic units.
- The volume of a prism and a cylinder is the area of base × length.
- Volume of a cone = $\frac{1}{3}$ × area of base × height.
- Volume of a sphere = $\frac{4}{3}\pi r^2$.
- Volume of a pyramid = $\frac{1}{3}$ × area of base × height.
- The volumes of standard solids can be calculated by remembering only three formulae:
  - sphere - which is a 'one off'
  - prism - which is base area × height
  - pyramid - which is $\frac{1}{3}$ × base area × height.
- As long as you know how to calculate the area of the relevant base, all the volume formulae fall easily into place.

Surface area

- The surface area of a solid is the total area of all the external faces.

17 Volume and surface area

Chapter review

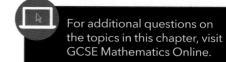

For additional questions on the topics in this chapter, visit GCSE Mathematics Online.

1 How much canvas is in this tent? (Assume the shape is a triangular prism and that there is no base sheet.)

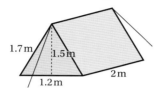

2 These solids have been built from cubes with side length 2 cm. Find the total surface area of each solid.

a   b 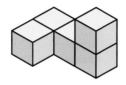 c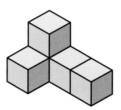

3 A solid prism is shown on the right.

Work out the volume of the prism. *(3 marks)*

©*Pearson Education Ltd 2013*

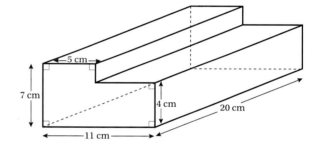

4 What is the volume of this model house?

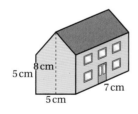

5 What is the volume of this piece of art sculpture (right)? It is made of a cube with a cylinder cut out through the middle of it.

6 The dimensions of a cube are whole numbers. If the volume of this cube is 64 cm³, which of the following whole numbers could be a side length?

A 4   B 10   C 8   D 16   E 5

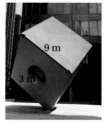

7 Calculate:
 a the volume of a tin of dog food.
 b the surface area of the printed label.

Find answers at: cambridge.org/ukschools/gcsemaths-studentbookanswers

18 Percentages

In this chapter you will learn how to ...

- work interchangeably with fractions, decimals and percentages.
- calculate a percentage of an amount.
- express a quantity as a percentage of another.
- increase and decrease amounts by a given percentage.
- solve problems involving percentage change.

For more resources relating to this chapter, visit GCSE Mathematics Online.

Using mathematics: real-life applications

Percentages are often used in daily life to express fractions. For example, you might see adverts claiming that 76% of pets prefer a particular brand of food or that 90% of dentists recommend a particular type of toothpaste. Sale price-reductions, discounts and interest rates are usually given as percentages.

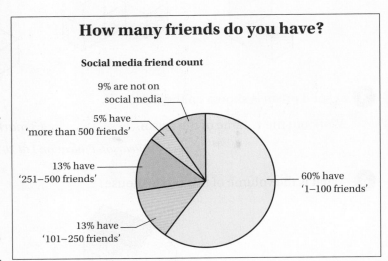

"Statistics in the media are often reported as percentages. This makes it easier to understand, but percentages can also be misleading – 60% sounds like a lot, but it could just mean 3 out of 5 people interviewed." *(Statistician)*

Before you start ...

Ch 7	You need to be able to multiply and divide by 100.	① Where should the decimal point go in each answer? **a** $210 \div 100 = 21$ **b** $21 \div 100 = 21$ **c** $0.24 \times 100 = 24$ **d** $0.024 \times 100 = 24$
Ch 6	You need to be able to cancel to express fractions in simplest terms.	② Match each fraction in box A to its equivalent from box B. **Box A**: $\frac{16}{36}$, $\frac{15}{35}$, $\frac{30}{36}$, $\frac{9}{36}$, $\frac{39}{52}$, $\frac{13}{39}$ **Box B**: $\frac{1}{4}$, $\frac{3}{4}$, $\frac{1}{3}$, $\frac{5}{6}$, $\frac{3}{7}$, $\frac{4}{9}$
Ch 7	You should be able to express any percentage as a decimal.	③ Are the following statements true or false? **a** $20\% = 0.02$ **b** $25\% = 1.4$ **c** $3\% = 0.3$ **d** $12.5\% = 0.125$ **e** $1.25\% = 0.125$

Assess your starting point using the Launchpad

STEP 1

1 Write each percentage as a fraction.
  a 34%   **b** 115%

2 Write each set of numbers in order from smallest to biggest.
  a 12%, 0.125, $\frac{7}{50}$, $\frac{5}{12}$, 19%
  b $2\frac{3}{4}$, 200%, 2.5%, 12.5%, 1.08, 1.25

GO TO Section 1: Review of percentages

STEP 2

3 What is 19 out of 25 marks as a percentage?

4 What is 50% of 128?

5 In a population of 12 500 000 people of working age, 3 400 000 are unemployed.
What is the unemployment rate as a percentage?

6 Express 25p as a percentage of £7.50

GO TO Section 2: Percentage calculations

STEP 3

7 Increase £20 by 9.5%.

8 Pete wants to buy a second-hand car marked at £2800.
The dealer offers him a 7.5% discount if he pays cash. What will the cash price be?

9 Mandy bought a book in a 25% off sale for £2.55. What was the original price of the book?

GO TO Section 3: Percentage change

GO TO Chapter review

Find answers at: cambridge.org/ukschools/gcsemaths-studentbookanswers

Section 1: Review of percentages

Percentages, fractions and decimals

Percentages, fractions and decimals are different ways of showing a part of a whole.

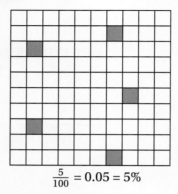

$\frac{5}{100} = 0.05 = 5\%$

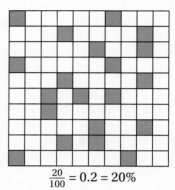

$\frac{20}{100} = 0.2 = 20\%$

During 2014, a mobile phone maker did a survey of 872 people.

92% of people said they would feel stressed if their phone battery ran out and 81% of people said that running out of power on their phones had led to them having a bad experience.

Since 92% means 92 out of every 100 and 81% means 81 out of every 100, percentages can be changed to fractions and to decimals.

	Write the percentage as a fraction with a denominator of 100	For a fraction, simplify if possible.	For a decimal, put the decimal point in the correct position.
92%	$= \frac{92}{100}$	$= \frac{23}{25}$	$= 0.92$
81%	$= \frac{81}{100}$	Cannot be simplified further	$= 0.81$
125%	$= \frac{125}{100}$	$= 1\frac{25}{100}$	$= 1.25$

The mobile phone survey also showed that nearly $\frac{1}{2}$ of the people surveyed could remember no more than three phone numbers.

Another survey showed that $\frac{2}{3}$ of people aged 11 to 17 take an internet-connected device to bed with them every night.

Fractions can be written as percentages.

$\frac{1}{2} = \frac{50}{100} = 50\%$	If the denominator is a factor of 100, it can be written as an equivalent fraction with a denominator of 100. Do not simplify, write the fraction of 100 as a percentage
$\frac{2}{3}$ [2] [÷] [3] [×] [100] = 66.6666666667	Using a calculator. Your display will show 66.666666667 This is the percentage. Write it as 66.67% (correct to 2 decimal places).

Tip

When you use a calculator to convert a fraction to a percentage you are actually first changing $\frac{2}{3}$ to a decimal (2 ÷ 3 = 0.6666666667) and then converting the decimal to a percentage. You do not enter the percentage sign in the calculation because the values you are entering are not percentages. The percentage is the answer you get.

To change a decimal to a percentage write it as a fraction with a denominator of 100, or use your calculator to multiply it by 100.

$0.3 = \frac{3}{10} = \frac{30}{100} = 30\%$ $0.3 \times 100 = 30\%$

$0.025 = \frac{25}{1000} = \frac{2.5}{100} = 2.5\%$ $0.025 \times 100 = 2.5\%$

$3.75 = \frac{37.5}{10} = \frac{375}{100} = 375\%$ $3.75 \times 100 = 375\%$

> **Tip**
>
> Simple rules:
> - to change any fraction or decimal into a percentage you just multiply by 100
> - to change any percentage into a fraction or decimal you just divide by 100.

Comparing percentages, fractions and decimals

Some common conversions:

Fraction	Decimal	Percentage
$\frac{1}{2}$	0.5	50%
$\frac{1}{4}$	0.25	25%
$\frac{3}{4}$	0.75	75%
$\frac{1}{3}$	$0.3\dot{3}$	$33\frac{1}{3}\%$
$\frac{2}{3}$	$0.6\dot{6}$	$66\frac{2}{3}\%$

To compare a mixed set of percentages, fractions and decimals, change them all to percentages.

> **Tip**
>
> You can also change all the values to decimals or equivalent fractions to compare them if the numbers are easier.

WORKED EXAMPLE 1

Write the following in ascending order.

$35\%, \frac{1}{3}, 0.38, \frac{2}{5}, \frac{2}{7}$

35%	$\frac{1}{3}$	0.38	$\frac{2}{5}$	$\frac{2}{7}$
35%	$\frac{1}{3} \times 100$ = 33.33%	0.38×100 = 38%	$\frac{2}{5} \times 100$ = 40%	$\frac{2}{7} \times 100$ = 28.57%

The order is: $\frac{2}{7}, \frac{1}{3}, 35\%, 0.38, \frac{2}{5}$

Convert all the fractions and decimals to percentages. Remember to round the values that are not exact to a suitable level of accuracy.

Remember to use the original fractions when you write the answer, not the percentages you have changed them to.

EXERCISE 18A

1 Express the following as percentages. Use fractional ($32\frac{1}{2}\%$) or decimal (2.5%) percentages where you need to.

a $\frac{5}{100}$ b $\frac{27}{50}$ c $\frac{11}{25}$ d $\frac{17}{20}$

e $\frac{1}{2}$ f $\frac{2}{3}$ g $\frac{5}{8}$ h $\frac{92}{50}$

i 0.3 j 0.04 k 0.47 l 1.12

m 2.07 n 2.25 o 0.035 p 0.007

2 Write each of the following percentages as a common fraction in its simplest terms.

a 25% b 80% c 90% d 12.5%

e 50% f 98% g 60% h 22%

3 Write the decimal equivalent of each percentage.

a 82% b 97% c 45% d 28.6%

e 0.05% f 0.08% g 0.006% h 0.0007%

i 125% j 300% k 7.28% l 9.007%

4 State whether the following are true or false.

a $\frac{3}{5} > 70\%$ b $\frac{7}{9} < 83\%$ c $\frac{1}{3} = 30\%$ d $67\% > 0.666$

5 a If 93.5% of the students in a school have WiFi at home, what percentage do not?

b If $\frac{2}{3}$ of all the SIM cards sold in a mobile phone shop are pre-paid, what percentage are not pre-paid?

c 0.325 of computer users back up their work every day. What percentage do not do this?

6 Zack spends 24.7% of a day playing computer games, 0.138 of the day doing homework and $\frac{3}{8}$ of the day playing sport.

What percentage of the day is spent doing other things?

7 What percentage of each pie chart is shaded?

a b 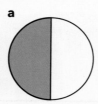 c  d

8 Write each set of values in ascending order.

a $\frac{1}{20}$, 30%, 0.1, $\frac{3}{5}$, 0.8%

b 0.75, 57%, 0.88, $\frac{1}{4}$, 0.15

c $\frac{2}{3}$, 0.75, 60%, $\frac{9}{10}$, 0.25

d $\frac{3}{7}$, 0.43, 45%, 0.395, $\frac{4}{9}$

e $\frac{5}{6}$, 80%, $\frac{19}{25}$, 55%, 49.3%

9 A media company states that 83.5% of its customers read the news online every day.

What fraction of the customers is this?

10 Anna pays 0.06 of her salary into her credit card account.

What percentage of her salary is this?

11 During one shift at work, Sandy spent $\frac{9}{20}$ of her time texting on her phone. What percentage of the shift was she not texting?

12 Angie gets the following marks for three maths assignments: $\frac{31}{40}$, $\frac{27}{30}$ and $\frac{13}{15}$.

a Which of these marks is the highest?

b What is her mean result for the three assignments as a percentage?

Section 2: Percentage calculations

WORK IT OUT 18.1

9% of 400 is 36.

Which of the following methods will give you the correct answer?

Explain why the other methods won't work.

Method A	Method B	Method C	Method D	Method E
$\frac{9}{100} \times 400$	$\frac{400}{9} \times 36$	0.009×400	9×400	$\frac{9}{400} \times 100$

> **Tip**
> The word 'of' means multiply.

To find a percentage of an amount you have to multiply by the percentage. Unless you use a calculator, you have to write the percentage as a fraction (with a denominator of 100) or a decimal.

WORKED EXAMPLE 2

What is 12% of 700?

Using fractions	Using decimals	Using a calculator
$\frac{12}{100} \times 700$ $= 84$	0.12×700 $= 0.12 \times 100 \times 7$ $= 12 \times 7$ $= 84$	7 0 0 × 1 2 % 84

Calculator tip

Make sure you know how to use the button on your calculator.

You might need to enter 12% × 700 or 700 × 12% (some calculators will work both ways).

On some calculators you need to press the = but on others you might not have to. Check how your calculator works by finding 12% of 350. The answer should be 42.

You do enter the percentage sign in these calculations because one of the values you are working with is a percentage.

EXERCISE 18B

1 Calculate.

- **a** 5% of 250
- **b** 9% of 400
- **c** 20% of 120
- **d** 65% of 4500
- **e** 12% of 75
- **f** 75% of 360
- **g** 32% of 50
- **h** 110% of 60
- **i** 150% of 90

Tip

Remember your answer will have a unit not a percentage sign. You are not working out a percentage here; you are working out what a given percentage of a quantity is.

2 Calculate, giving answers as mixed numbers or decimals as necessary.

- **a** 19% of £50
- **b** 60% of 70 kg
- **c** 45% of 35 cm
- **d** 90% of 29 kg
- **e** $3\frac{1}{2}$% of £400
- **f** 2.6% of 80 minutes
- **g** 7.4% of £1000
- **h** 3.8% of 180 m
- **i** $9\frac{2}{3}$% of 600 litres

3 Annie got 85% for a test that was out of 80 marks. What was her mark out of 80?

4 A salesperson at a mobile phone shop estimates that about 3% of phones come back for some sort of repair in the first week.

If the shop sells 180 phones, how many can they expect to come back for repairs in the first week?

5 46% of residents in an area throw out the local free newspaper without even looking at it; the rest read some or all of it.

If there are 2450 residents, how many people:

- **a** don't look at the paper
- **b** read some or all of it?

6 Of 240 trains arriving at King's Cross, 2.5% arrived early and 13.75% arrived late.

How many trains arrived exactly on time?

7 A tablet computer is advertised for sale for £899 excluding VAT.

Nisha wanted to buy it when VAT was 17.5% but she didn't get round to it and VAT was increased to 20% before she bought it.

How much would she have saved if she had bought it when VAT was 17.5%?

8 7.5% of a 620 m² market garden is set aside for growing tulips and the rest is used to grow vegetables.

How many square metres of land is used to grow:

a tulips **b** vegetables?

9 The population of a town in Cornwall increases by about 24.8% each summer.

If the population of the town is normally 12 760, how many people move in during the summer?

10 Pure gold contains 24 parts (called carats) of gold in every 24 parts.
$\frac{24}{24} = 100\%$ gold.

18 carat gold contains 18 parts pure gold per 24 parts and 9 carat gold contains 9 parts pure gold per 24 parts.

a Work out the percentage of pure gold in 9 carat and 18 carat gold.

b If Naz buys an 18 carat gold ring that weighs 7.3 grams, how much pure gold does it contain?

c If Vishnu buys a 9 carat gold pendant that has a mass of 16.3 grams, how much pure gold does it contain?

d Do you think it is accurate to label 9 carat gold as gold? Explain your answer.

Expressing one quantity as a percentage of another

You can write one quantity as a percentage of another quantity by writing the first quantity as a fraction of the other and then multiplying by 100 to get a percentage. The two quantities must be in the same units before you write them as a fraction.

Tip

You expressed one quantity as a fraction of another in Chapter 6. Read through that work again if you cannot remember how to do this.

WORK IT OUT 18.2

Brian has run 1500 m of a 5 km race when he gets a cramp in his foot.

What percentage of the race has he completed at this stage?

Which of these two students has got the correct answer? Why is the other one wrong?

Student A	Student B
$\frac{1500}{5} \times 100$	$\frac{1500}{5000} \times 100$
$= 300 \times 100$	$= \frac{3}{10} \times 100$
$= 300\%$	$= 30\%$

Tip

When you convert quantities to get them to the same unit you can avoid decimal values by choosing the smaller units (for example, making both units metres in this example rather than making them both kilometres).

EXERCISE 18C

1 Express the first amount as a percentage of the second.

Give your answer correct to no more than 2 decimal places.

- **a** 400 m of 5 km
- **b** 45 m of 3 km
- **c** 150 m of 1 km
- **d** 8 cm of 2 m
- **e** 14 mm of 4 cm
- **f** 19 cm of 3 m
- **g** 25p of £4
- **h** 66p of £3.50
- **i** 20 seconds of a minute
- **j** 25 seconds of 1.5 minutes
- **k** 750 g of 23 kg
- **l** 800 g of 1.5 kg
- **m** 4 days of a week
- **n** 3 days of 6 weeks
- **o** 800 kg of 3 tonnes
- **p** 8.4 tonnes of 50 000 kg
- **q** 500 mm of 2 m
- **r** 90 mm of 14 cm
- **s** 350 ml of 2 litres
- **t** 5 ml of 0.5 litres

2 Sandra got 19 out of 24 for an assignment and Nina got 23 out of 30.

Which girl got the higher percentage mark?

3 In a local election there were 5400 registered voters. Of these, 3240 voted.

What was the percentage voter turnout?

4 Mel improved his running time for the 400 m race by 3 seconds.

If his previous running time was 50 seconds, what is his percentage improvement?

5 Kenny had a box of 40 chocolates. He ate 32 of them.

What percentage of the chocolates remain?

6 Sylvia keeps a record of how many sets she wins when she plays tennis against her sister.

In the past month she won 19 out of 27 sets. What percentage of the sets did she lose?

7 The longest kiss lasted 58 hours, 35 minutes and 58 seconds and was achieved by Ekkachai Tiranarat and Laksana Tiranarat at an event organised by Ripley's Believe It or Not!, in Pattaya, Thailand, on 12–14 February 2013.

What percentage of the three-day event was this?

8 Read the label (left) and answer the questions.

- **a** Calculate the combined percentage of fat and sugar in a serving.
- **b** What percentage of a serving is sodium?

Nutritional values
(Per 30 g serving)

Carbohydrates	19 g
(of which sugars)	6.2 g
Fat	3.8 g
Sodium	93 mg

Section 3: Percentage change

You will often see changes (increases or decreases) in amounts expressed as percentages. For example, you might read that the price of petrol is to increase by 5.5% or that the cost of mobile broadband has decreased by 15% over the past year.

Increasing or decreasing an amount by a percentage

WORK IT OUT 18.3

A school population of 650 students increases by 12%.

At the same time, the registration fee of £120 decreases by 15%. Work out:

a the new student population

b the new registration fee.

Look at these examples to see how two different students solved these problems.

Which method seems easier to you?

Could you use your calculator to do these calculations? How?

	Student A	Student B
a	650 increased by 12% 12% of 650 = $\frac{12}{100} \times 650$ $= 78$ $650 + 78 = 728$ There are now 728 students.	650 increased by 12% Old population = 100% New population = old + increase $= 100 + 12\% = 112\%$ $112\% = \frac{112}{100} = 1.12$ $1.12 \times 650 = 728$ The new student population is 728.
b	120 decreased by 15% 15% of 120 = $\frac{15}{100} \times 120$ $= 18$ £120 − £18 = £102 The new registration fee is £102.	£120 decreased by 15% Old fee = 100% New fee = 100% − 15% = 85% $85\% = \frac{85}{100} = 0.85$ $0.85 \times 120 = 102$ The new registration fee is £102.

Tip

You can express any % increase or decrease as a multiplier.

To increase a number by $x\%$, multiply it by $1 + \frac{x}{100}$. So, to increase by 12%, multiply by 1.12.

To decrease a number by $x\%$, multiply it by $1 - \frac{x}{100}$. So, to decrease by 15%, multiply by 0.85.

EXERCISE 18D

1 Increase each amount by the given percentage.

    a £48 increased by 14%   **b** £700 increased by 35%

    c £30 increased by 7.6%   **d** £40 000 increased by 0.59%

    e £90 increased by 9.5%   **f** £80 increased by 24.6%

 Find answers at: cambridge.org/ukschools/gcsemaths-studentbookanswers

2 Decrease each amount by the given percentage.

   a £68 decreased by 14%
   b £800 decreased by 35%
   c £90 decreased by 7.6%
   d £20 000 decreased by 0.59%
   e £85 decreased by 9.5%
   f £60 decreased by 24.6%

3 A building which cost £125 000 to build, increased in value by $3\frac{1}{2}\%$, what is it worth now?

4 Josh currently earns £3125 per month.

If he receives an increase of 3.8%, what will his new monthly earnings be, correct to the nearest pound?

5 Sally earns £25 per shift. Her boss says she can either have £7 more per shift or a 20% increase.

Which is the better offer?

6 The membership of a sports club increased by 26% one year.

If they had 284 members the previous year, how many will they have now?

7 Sammy bought £2500 worth of shares.

At the end of the first month their value had decreased by 4.25%.

At the end of the second month Sammy checked the value again and found it had gone up 1.5% from the previous month.

Work out the value of the shares at the end of each month.

8 Amira earns £25 000 per year plus 12% commission on any sales she generates.

Calculate her annual earnings if she sold £145 250 worth of goods.

Finding original values

If you know the percentage by which an amount has increased or decreased, you can use it to find the original amount. Problems involving original values are often called reverse or inverse percentages. When you work with these problems you need to remember that you are dealing with percentages of the original values.

WORKED EXAMPLE 3

A shop is offering a 10% discount on all sale goods.

Jessie bought a bike in the sale and paid £108.

What was the original price of the bike?

90% of x = £108

> If the cost is reduced by 10% then you are actually paying 90%.
>
> If you let the original amount be x, you can write an equation and solve it to find x.

Continues on next page ...

$$\therefore \frac{90}{100}x = 108$$
$$\therefore 90x = 100 \times 108 = 10\,800$$
$$\therefore x = \frac{10\,800}{90} = 120$$

The original price was £120.

You could write $0.9x = 108$

0.9 is a multiplying factor.

Then, $x = \dfrac{108}{0.9}$

 Tip

Undoing a 10% decrease is not the same as just increasing the sale price by 10%. If you add 10% to the sale price of £108 you will get £118.80 which is **NOT** the right answer.

WORKED EXAMPLE 4

Sameen sells her shares and receives £3450. This gives her a profit of 15%. What did she pay for the shares originally?

Let the cost price be x.

$1.15x = 3450$
$x = \dfrac{3450}{1.15}$
$x = 3000$

15% profit means an increase of 15%, so the selling price = 115% of the cost.

The multiplying factor is 1.15.

She paid £3000 for the shares.

Check this by increasing 3000 by 15%.

$3000 \times 1.15 = 3450$.

EXERCISE 18E

1 Find the original value if:

  a 25% is £30
  b 8% is 120 g
  c 120% is 800 kg
  d 115% is £2000

2 VAT of 20% is added to most goods before they are sold.

Some tourists to the UK can claim back the VAT when they leave the country.

Work out the price of each of these items without VAT to see what a tourist would pay for them.

The prices given here include VAT.

  a necklace £1200
  b camera £145.50
  c painting £865
  d boots £54.99

3 Misha paid £40 for a DVD box set in a 20% off sale.

What was the original price of the DVD set?

4 In a large school 240 pupils are in Year 10. This is 20% of the school population.

  a How many pupils are there in total in the school?
  b How many pupils are in the other years at this school?

Find answers at: cambridge.org/ukschools/gcsemaths-studentbookanswers

5 Susie was told that her pay had increased by 15%. Her new pay is £172.50.

What was her pay before the increase?

6 9 carat gold is 37.5% pure gold.

A piece of 9 carat gold jewellery is tested and found to contain 97.5 grams of pure gold.

What did the piece of jewellery weigh?

7 Julia is training for a marathon and she reduces her weight by 5% over a three-month period.

If she weighs 58 kg at the end of the period, what did she weigh at the start?

8 In a particularly hard ultramarathon, only 310 runners completed the course within the cut-off time.

If this represents 62% of the runners, how many runners started the race?

Checklist of learning and understanding

Review of percentages
- 'Per cent' means 'parts per hundred'.
- To convert a percentage to a fraction write the percentage with a denominator of 100 and simplify.
- To convert percentages to decimals divide by 100.
- To order a mixture of fractions, decimals and percentages change them all to percentages or decimals.

Percentage calculations
- To find a percentage of an amount, express the percentage as a fraction over 100 and then multiply the fraction by the amount.
- To express one quantity (A) as a percentage of another quantity (B), make sure the units are the same and then calculate $\frac{\text{quantity A}}{\text{quantity B}} \times 100$.

Percentage change
- To increase or decrease an amount by a percentage, find the percentage amount and add or subtract it from the original amount.
- Or, use a multiplier:
  - to increase an amount by $x\%$, the multiplier is $(1 + \frac{x}{100})$
  - to decrease an amount by $x\%$, the multiplier is $(1 - \frac{x}{100})$.
- To find an original value when you know the percentage increase or decrease and the new amount, make an equation and use reverse percentages to solve for x.

Chapter review

1 Write each percentage as a fraction.
   a 25%   **b** 30%   **c** 3.5%

2 Express each of these as a percentage.
   a $\frac{1}{20}$ **b** $\frac{1}{8}$ **c** $\frac{8}{15}$
   d 0.5   **e** 1.25   **f** 0.005

3 The value of an investment increased from £120 000 to £124 800.
What percentage increase is this?

4 The population of New Orleans was 484 674 before Hurricane Katrina.
Afterwards, the population had decreased by 53.9%.
What was the population afterwards?

5 Shaz works 30 hours per week. She wants to increase this by 12%.
How many hours will she then work per week?

6 Express as a percentage
   a 3 hours of one day   **b** 750 metres of 2 km.

7 Petra booked a family holiday.
The total cost of the holiday was £3500 **plus** VAT at 20%.
Petra paid £900 of the total cost when she booked the holiday.
She paid the rest of the total cost in 6 equal monthly payments.
Work out the amount of each monthly payment.   *(5 marks)*
©*Pearson Education Ltd 2013*

8 The price of a plane ticket was reduced by 8% to £423.20.
What was the original price of the ticket?

9 Nick sold his shares for £1147.50 and made a 35% profit.
What did he pay for the shares?

19 Ratio

In this chapter you will learn how to …
- work with equivalent ratios.
- divide quantities in a given ratio.
- identify and work with fractions in ratio problems.
- apply ratio to real-life contexts and problems, such as those involving conversion, comparison, scaling, mixing and concentrations.

For more resources relating to this chapter, visit GCSE Mathematics Online.

Using mathematics: real-life applications

Converting between different currencies, working out which packet of crisps is the best value for money, mixing large quantities of cement and scaling up a recipe to cater for more people all involve reasoning using ratios.

"Every day customers bring me paints to match. I have to understand how changing the ratio of base colours affects the colour of the paint and how to scale the quantities up and down for larger or smaller amounts of paint. If I get it wrong, customers will have patches of different colours and their walls would look quite strange." *(Paint technician)*

Before you start …

Ch 6	You need to be able to identify and simplify fractions.	**1**	**a** In a class of 35 pupils 21 are boys. What fraction of the class are girls? **b** What fraction of this shape is shaded? Write your answer in its simplest form.
Ch 6	You need to be able to find a fraction of a quantity.	**2**	Find $\frac{2}{3}$ of 42
Ch 6	You need to be able to find an original amount given a fraction.	**3**	There are 51 parents of students in the audience at a school play. These parents make up $\frac{3}{4}$ of the audience. How many people are in the audience?

19 Ratio

Assess your starting point using the Launchpad

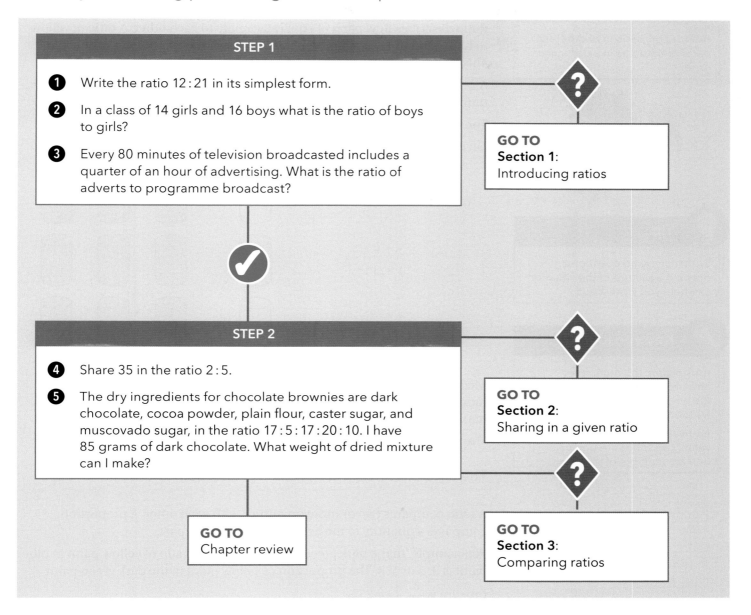

STEP 1

1. Write the ratio 12 : 21 in its simplest form.
2. In a class of 14 girls and 16 boys what is the ratio of boys to girls?
3. Every 80 minutes of television broadcasted includes a quarter of an hour of advertising. What is the ratio of adverts to programme broadcast?

GO TO Section 1: Introducing ratios

STEP 2

4. Share 35 in the ratio 2 : 5.
5. The dry ingredients for chocolate brownies are dark chocolate, cocoa powder, plain flour, caster sugar, and muscovado sugar, in the ratio 17 : 5 : 17 : 20 : 10. I have 85 grams of dark chocolate. What weight of dried mixture can I make?

GO TO Section 2: Sharing in a given ratio

GO TO Chapter review

GO TO Section 3: Comparing ratios

Section 1: Introducing ratios

Most colours of paint can be mixed from the four base colours: blue, yellow, red and white.

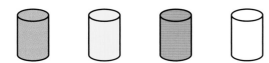

The amount of each base colour is important for getting the same shade of, say, green at different times.

Find answers at: cambridge.org/ukschools/gcsemaths-studentbookanswers

Key vocabulary

ratio: a comparison of different parts or amounts in a particular order.

Tip

With many ratio questions, drawing a picture of the situation can help you work it out.

Key vocabulary

proportion: the number or amount of a group compared to the whole, often expressed as a fraction, percentage or ratio.

equivalent: having the same value, two ratios are equivalent if one is a multiple of the other.

Paint technicians can mix the same shade of green over and over by mixing yellow and blue paints in a particular **ratio**.

Ratio describes how parts of equal size relate to each other. A ratio of yellow to blue paint of $1:3$ means one unit of yellow for every three units of blue. This would give a very dark green.

A ratio of yellow to blue paint of $5:1$ means five units of yellow for every one unit of blue. This would give a much lighter green.

The diagram shows a ratio of yellow to blue of $3:9$.

Dividing both parts by three simplifies the ratio to give $1:3$

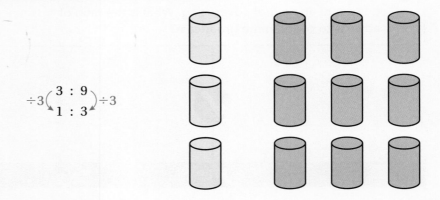

Mixing paint in the ratio $3:9$ would give the same colour as mixing it in the ratio $1:3$ because the colours are mixed in the same ratio. The yellow paint makes up the same **proportion** of the mix in both cases.

The ratios $3:9$ and $1:3$ are **equivalent** ratios.

The difference between ratio and proportion

A ratio compares two or more quantities with each other. A proportion compares a quantity to the 'whole' of which it is a part.

For example, in the dark green paint mixture, the ratio of yellow paint to blue paint is $3:9$ or $1:3$. The proportion of yellow paint in the dark green paint mixture is $\frac{3}{12}$, $\frac{1}{4}$ or 25%.

EXERCISE 19A

1 36 girls 45 boys and 9 teachers went on a school trip.

   a What is the ratio of boys to girls?

   b What is the ratio of pupils to teachers?

   c What is the ratio of pupils to people on the trip?

   d The school policy is that each teacher can be responsible for no more than 10 pupils. Does this trip meet this requirement?

2. In each diagram, what is the ratio of shaded squares to unshaded squares? Write the answers in simplest form.

a    b 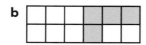   c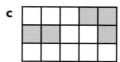

3. In each diagram, what is the ratio of shaded squares to total squares? Write the answers in simplest form.

a 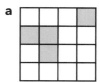   b 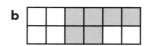   c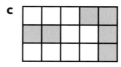

4. The ratio of shaded to unshaded squares in this diagram is 1:3. How many more squares need to be shaded to make the ratio 2:3?

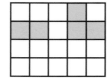

5. The distance between the post office and the bank on the local high street is represented as 5 cm on a map. In real life this distance is 20 m. What is the scale of the map (as a ratio)?

6. On a scale drawing of a cruise ship a cabin is 8 cm from the restaurant. On the actual ship the distance is 76 m. Express the distances as a ratio.

Tip

Ratios do not include units. To compare measured amounts you need to make sure they are written in the same units.

7. A natural history programme lasts 90 minutes. The crew recorded 60 hours of footage. What is the ratio of used footage to recorded footage?

8. a Use the diagram to find the ratio of:
   i side AB to side AC   ii side EB to side DC   iii side AE to side AD

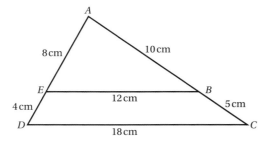

   b What does this tell you about triangles ABE and ACD?
   c What does it tell you about lines EB and DC?
   d What is the ratio of angle AEB to angle EDC?

9. A jam recipe uses 55 g of fruit for every 100 g of jam. The rest is sugar. What is the ratio of fruit to sugar?

10 An adult ticket for the cinema is one and a half times the price for a child's ticket. What is the ratio of the price of an adult ticket to the price of a child's ticket?

11 According to recent statistics $\frac{3}{5}$ of 16 year olds have a mobile phone. What is the ratio of 16 year olds with mobiles to those without?

Section 2: Sharing in a given ratio

WORKED EXAMPLE 1

A group of three office workers form a lottery syndicate. Together they buy eight lottery tickets a week. Simon pays £1 a week, Oliver £3 and Lucy £4. They win £32 000.

Should they each get an equal share of the winnings? If not, how should they share their winnings? What is the fairest way?

The winnings should be distributed between Simon, Oliver and Lucy in the ratio 1 : 3 : 4.

> This is the fairest way, as it is in the same ratio as they bought tickets.

> Every box has to have the same quantity in it. In total there are eight boxes, in which you have to share £32 000.
>
> Each box gets £32 000 ÷ 8 = £4000
>
> So:
>
> Simon receives £4000.
>
> Oliver receives 3 × £4000 = £12 000
>
> Lucy receives 4 × £4000 = £16 000

£4000 + £12 000 + £16 000 = £32 000

> Check that the shared quantities sum to the original amount.

This also shows that Simon gets $\frac{1}{8}$ of the winnings, Oliver $\frac{3}{8}$ and Lucy $\frac{1}{2}$.

Tip

The box method shown in the example is useful for working out shares in a given ratio problem.

EXERCISE 19B

1 Share 144 in each of the given ratios.

a 1 : 3   b 4 : 5   c 11 : 1
d 2 : 3 : 1   e 1 : 2 : 5   f 2 : 7 : 5 : 4

2 To make mortar you mix sand and cement in the ratio of 4 : 1.
  a How much sand is needed to make 25 kilograms of mortar?
  b What fraction of the mix is cement?

3 The first two-colour £2 coin was issued in 1998. The inner circle is made of cupronickel. This is copper and nickel in the ratio 3 : 1. The inner circle weighs 6 grams. How much copper is used to make the centres of ten £2 coins?

4 Flaky pastry is made by mixing flour, margarine and lard in the ratio 8 : 3 : 3 and then adding a drizzle of cold water.
  a How much of each ingredient is needed to make 350 g of pastry?
  b What fraction of the pastry is made up of fats (margarine and lard)?

5 The sides of a rectangle are in the ratio of 2 : 5. Its perimeter is 112 cm.
  a What are the dimensions of the rectangle?
  b Use these dimensions to calculate its area.

6 Orange squash is made by mixing one part cordial to five parts of water. How much squash can you make with 750 ml of cordial?

7 Two-stroke fuel is used to power small engines. It is produced by mixing oil and petrol in the ratio of 1 : 20. How much oil needs to be mixed with 10 litres of petrol to make two-stroke fuel?

8 Tiffin is a sweet made by crushing biscuits and mixing them with dried fruit, butter and cocoa powder. The ratio of biscuit to dried fruit to butter to cocoa powder is 5 : 6 : 2 : 2. How much of each ingredient is needed to make 600 g of tiffin?

9 In a music college the ratio of flute to oboe to string to percussion players is 7 : 2 : 15 : 1. If the college has 175 students, how many play an oboe?

10 The ratio of red to green to blue to black to white pairs of socks in a drawer is 2 : 3 : 7 : 1 : 4. If there are eight pairs of white socks, how many pairs are there altogether?

11 Potting compost is made by mixing loam, peat and sand in the ratio of 7 : 3 : 2. If a gardener has 4.5 kg of peat and plenty of loam and sand, how much potting compost can she make?

Section 3: Comparing ratios

It is often useful to write ratios in the form $1 : n$ (where n represents a number) so that they are in the same form and you can compare them by size.

Tip

The scale of maps is given as a ratio in the form of $1 : n$. For example $1 : 25\,000$.

WORKED EXAMPLE 2

Red and white paint can be mixed to make pink paint.

Which of the mixes below will give the lightest shade of pink?

A 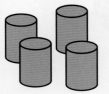   **B** 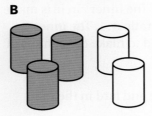   **C**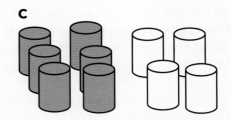

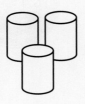

The ratios of red to white paint are:

A 4:3 B 3:2 C 6:4

A $\frac{4}{4} : \frac{3}{4} = 1:0.75$ B $\frac{3}{3} : \frac{2}{3} = 1:0.67$ C $\frac{6}{6} : \frac{4}{6} = 1:0.67$

> Change these to form $1:n$ by dividing both parts of the ratio by the first part.
>
> Give the answers as decimals to make the comparison simpler. (Round to 2 decimal places.)

Paint A has the greatest amount of white paint per unit of red paint, 0.75 tins of white for 1 tin of red, so this will make the lightest shade of pink.

Ratios in the form of $1:n$ are also useful for converting from one unit to another.

For example, the ratio of inches to centimetres is $1:2.54$.

This means that 1 inch is equivalent to 2.54 cm.

So, 2 inches = 2×2.54 cm and 12 inches = 12×2.54 cm.

This is a linear relationship and it can be shown as a straight-line graph.

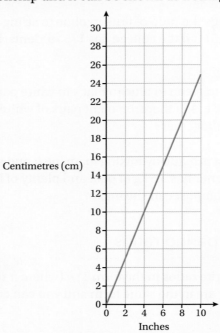

EXERCISE 19C

1 Different types of coffee are made by mixing espresso shots, hot water and milk in specified ratios.

Espresso	1 : 0 : 0
Double espresso	2 : 0 : 0
Flat white	1 : 2 : 1
Cappuccino	1 : 0 : 2
Latte	1 : 0 : 4

Put the drinks in order of strength, weakest first.

2 When Jon was going on holiday he used this graph to convert between pounds and euros.

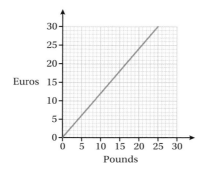

a What is the ratio of pounds to euros? Express this in the form $1 : n$.

b What is the ratio of euros to pounds? Express this in the form $1 : n$.

3 This graph shows the relationship between ounces and grams.

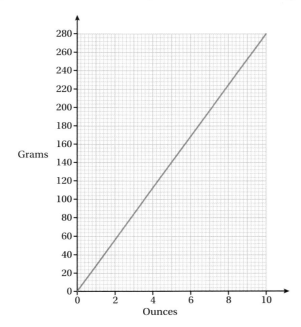

a What is the ratio of ounces to grams? Express this in the form $1 : n$.

b What is the ratio of grams to ounces? Express this in the form $1 : n$.

Sausage casserole (serves 6)

12 sausages

3 tins of tomatoes

450 g potatoes

9 tsp mixed herbs

600 ml vegetable stock

4. Three siblings; Daisy, aged 5, Patrick, 8, and Imogen, 12, share sweets in the same ratio as their ages. Imogen gets 21 more sweets than Daisy.
   a How many sweets were there to begin with?
   b What fraction of the sweets did Patrick get?

5. The ratio of kilometres to miles is approximately 8:5. A car travels at 60 miles per hour for 30 minutes. How many kilometres does it travel?

6. This recipe for sausage casserole (left) serves 6.

   The ratio of people to sausages is 6:12, this can be simplified to 1:2. Hence if the recipe needs adapting for 4 people, the people to sausages ratio of 1:2 means that you would need 8 sausages.

   Using this as an example, find the quantities of ingredients needed to serve 4 people. Show all the steps in your answer.

7. Gill and her sister Bell share a box of sweets. Bell gets $\frac{1}{3}$ of the box. Gill shares her sweets with her best friend Katy in the ratio 4:3. Katy gets 12 sweets. How many sweets were there in the box?

8. A box of chocolates contains white, milk and dark chocolates. A quarter of the box are white chocolate. The ratio of dark to milk chocolates is 2:5.

   If there are seven white chocolates, how many more milk chocolates than dark chocolates are there?

Golden ratio

The golden ratio has been studied and used for centuries. Artists, including Leonardo da Vinci and Salvador Dali often produced work using this ratio. The ratio can also be seen in buildings, such as the Acropolis in Athens. The golden ratio is said to be the most aesthetically pleasing way to space out facial features.

The diagram shows how the golden ratio can be worked out using the dimensions of a 'golden' rectangle. The large rectangle ACDF is similar to BCDE. Hence the ratio of $a : a + b$ is equivalent to $b : a$.

An approximate numerical value for this ratio can be found by measuring.

Tip

You will do more work on similar shapes in Chapter 28.

Tip

You worked with the special sequence of Fibonacci numbers in Chapter 15. This sequence follows the golden ratio. If you calculate the ratio of consecutive numbers in the Fibonacci sequence you will find that the ratio gets closer and closer to the actual golden ratio $\frac{1+\sqrt{5}}{2}$ (this is equal to 1.618 to 3 decimal places) as you get further along the sequence.

EXERCISE 19D

How golden are your hands?

Measure the distances A, B and C.

Now calculate these ratios and write them in the form $1:n$.

  Distance B : Distance C

  Distance A : Distance B

  Length of your hand : Distance from your wrist to your elbow

Can you see anything special about these ratios?

The closer your results are to 1.618 the more golden your hand!

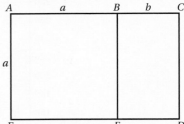

EXERCISE 19E

1. What is the ratio of the side length of a square to its perimeter in the form $1:n$?

2. What is the ratio of the diameter of a circle to its circumference in the form $1:n$?

3. The three angles of a triangle are in the ratio $3:3:4$. What information can you give about the triangle?

4. The ratio of the five angles in a pentagon are $1:1:1:1:1$. What information does this tell you about the pentagon? How do you know this?

5. The ratio of the angles in a triangle is $1:2:1$. What information can you give about the triangle?

6. Gareth and John share a box of chocolates. Gareth gets $\frac{3}{5}$ of the box.
   The ratio of white to milk to dark chocolates in John's share is $1:2:1$, he gets four white chocolates. The ratio of milk to dark to white chocolates in Gareth's share is $2:1:5$. How many types of each type of chocolate were in the box?

> **Tip**
>
> To answer the questions about triangles, pentagons and rectangles you may need to look again at Chapter 2. For the question on circles, you could refer to Chapter 11.

Checklist of learning and understanding

Notation
- The order in which a ratio is written is important. A ratio of $2:5$ means 2 parts to 5 parts. Each part is equal in size.

Simplifying ratios
- Two ratios are equivalent if one is a multiple of the other.
- Ratios can be simplified, by dividing both parts of the ratio by a common factor.
- Expressing ratios in the form $1:n$ makes it easy to compare ratios.

Sharing in a given ratio
- The box method can be used to tackle problems which involve sharing a quantity in a given ratio. To share quantity Q in the ratio $a:b:c$, divide the quantity evenly into $a+b+c$ boxes.

Chapter review

1. What is the ratio of vowels to consonants in the English alphabet?

2. What is the ratio of prime numbers to square numbers between (and including) 1 and 20 in its simplest form?

3. Share 360 in the ratio $3:5:1$.

For additional questions on the topics in this chapter, visit GCSE Mathematics Online.

4 Talil is going to make some concrete mix.

He needs to mix cement, sand and gravel in the ratio 1 : 3 : 5 by weight.

Talil wants to make 180 kg of concrete mix.

Talil has

15 kg of cement

85 kg of sand

100 kg of gravel

Does Talil have enough cement, sand and gravel to make the concrete mix?

(4 marks)

©*Pearson Education Ltd 2012*

5 Using the graph, express the ratio of miles to kilometres in the form $1 : n$.

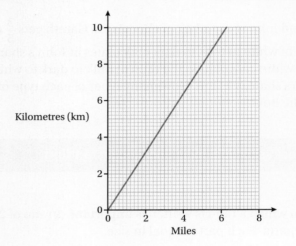

Tip

Don't forget about the silver cars!

6 In a car park, three quarters of the cars are not silver, but are blue, red, black or yellow. The proportion of blue to red to black to yellow cars is 6 : 2 : 3 : 1. There are six more black cars than yellow cars. How many cars of each colour are in the car park?

20 Probability basics

In this chapter you will learn how to ...

- use the language of probability and the 0 to 1 probability scale.
- calculate the probability of events happening or not happening.
- carry out experiments, record outcomes and use results to predict future probabilities.

For more resources relating to this chapter, visit GCSE Mathematics Online.

Using mathematics: real-life applications

Data is collected by many professionals and used to find the probability of particular things happening. For example, in a fertility clinic data collected over a period of many years can be used to draw a graph that shows the probability that a woman of a certain age will be successful at falling pregnant.

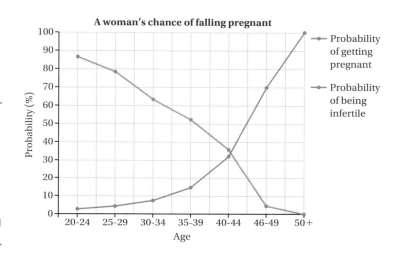

"A 20-year-old woman has an 86% chance of falling pregnant and only a 3% change of being infertile. This means there is a high probability that she will get pregnant if she is trying to have a baby."   *(Fertility doctor)*

Before you start ...

Chs 6, 7, 18	You need to be able to calculate with fractions, decimals and percentages	**1** Choose the correct answer without doing the calculations. a $\frac{1}{2} + \frac{1}{4}$ **A** $\frac{2}{3}$ **B** $\frac{2}{6}$ **C** $\frac{3}{4}$ b 0.2×0.3 **A** 0.5 **B** 0.06 **C** 0.6 c 20% of 40 **A** 0.8 **B** 8% **C** 8
Chs 6, 7, 18	You should be able to find equivalent fractions, decimals and percentages.	**2** Choose the correct sign: <, = or > a $\frac{12}{25} \square 8\%$ b $0.8 \square 8\%$ c $\frac{24}{50} \square 0.5$

 Tip

In probability calculations you will need to work with fractions, decimals and percentages, often interchangeably. Remember:

- a fraction is a number in the form of $\frac{a}{b}$.
- you can change fractions to decimals and decimals to fractions (see Chapter 7).

Find answers at: cambridge.org/ukschools/gcsemaths-studentbookanswers

Assess your starting point using the Launchpad

STEP 1

1 Match the events in the box with each of these probabilities.
  a A probability of 0.
  b A 100% chance of happening.
  c A probability of about 0.5.
  d A probability of about 80%.

> **A** Getting heads when you toss a coin.
> **B** May following June this year.
> **C** Choosing a letter of the alphabet and not getting a vowel.
> **D** Getting a number from 1 to 6 when you roll an ordinary dice.

GO TO
Section 1: The probability scale

STEP 2

2 Josh has six green sweets, two red sweets and three yellow sweets in a packet. If he puts his hand into the packet and chooses the first sweet he touches, what is the probability that it will be green?

GO TO
Section 2: Calculating probability

STEP 3

3 A doctor keeps records of how many patients who have the flu prevention injection actually get the flu. Over a three-year period he found that of 2500 people who had the injection, three still got the flu.
  a Estimate the probability that people who have the injection will get the flu.
  b If 7300 people have the flu injection, how many would you expect to get the flu?

GO TO
Section 3: Experimental probability

GO TO
Section 4: Mixed probability problems

GO TO
Chapter review

Section 1: The probability scale

You may hear people saying things like:

- There is a good chance it will rain tomorrow.
- It is impossible for this football team to win the league this year.
- There's a fifty-fifty chance that she will make the team.

These statements describe the chance, or **probability**, of something happening (or not happening).

Impossible means there is no chance that something will happen, while **certain** means it will definitely happen. A **fifty-fifty chance** means that it is just as likely to happen as it is to not happen (also called an **even** chance).

If you flip a coin, there are two possible outcomes:

Outcome 1: heads   Outcome 2: tails.

If the event you want is getting a head, then heads is a favourable outcome.

The chance of an event happening is called the probability of the event. Mathematicians use the short form P(E) to represent 'the probability of the event'.

Expressing probability in numbers

Probability can be described in numbers using a scale from 0 to 1.

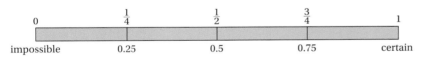

A probability of:	An example:
0 means that the event cannot happen; it is **impossible**.	The probability that the day immediately after a Friday will be a Tuesday is 0.
1 means the event will definitely happen; it is **certain**.	The probability that the day immediately after a Sunday will be a Monday is 1.
0.5 or $\frac{1}{2}$ means that the event has the same chance of happening or not happening (this is sometimes called an **even** chance).	The probability of getting heads when you flip a coin is 0.5

The smaller the fraction, the less likely it is that the event will happen. An event with a probability of 0.5 (or $\frac{1}{2}$) is more likely to happen than an event with a probability of 0.25 (or $\frac{1}{4}$) or 0.1 (or $\frac{1}{10}$).

Tip

Fifty: fifty, or 50 : 50, really means 50% : 50%. Because 50% equals $\frac{1}{2}$, this is called 'evens'.

Key vocabulary

event: the thing to which you are trying to give a probability.

outcome: a single result of an experiment or situation.

Tip

Remember that E in this notation means the event, so if the event is getting heads when you toss a coin, then you would write P(Heads) or P(H) as the short form of 'the probability of getting heads'.

Tip

You cannot get a negative number or an answer greater than 1 when you calculate probability. If you do, then you know you have made an error.

Probability as a percentage

A probability can be given as a decimal, fraction or percentage.

Some probabilities are clearer when given in a particular form.

The probability of winning the jackpot on a three-barrel slot machine could be given as:

0.000 38%   and also $\dfrac{1}{266\,144}$

Both mean that in about a quarter of a million goes on the slot machine, you're only likely to win once, but here the presentation of the probability as a fraction allows you to see that the chance of winning is very small.

In reality, someone could win from just playing once and another time nobody may win even after playing a million times because the frequency of events is not the same as the calculated probability (see Section 3: Experimental probability).

EXERCISE 20A

1. Jane has three red scarves, two yellow scarves and a black scarf.

   It is dark when Jane gets dressed, so she grabs a scarf from her drawer hoping for the black one.

   a What is the event in this example?

   b What are the possible outcomes?

   c Does Jane have a high or low probability of a favourable outcome? Why?

   d Which outcome has the highest probability? Why?

2. Estimate and then write the probability of each of these events as a fraction, a decimal and a percentage.

   a You'll get heads when you toss a coin.

   b You'll get a 7 when you roll a normal dice.

   c You'll win the lotto when you haven't bought a ticket.

   d You'll win the lotto when you have bought a ticket.

   e A newborn baby will be a girl.

   f You'll get an odd number when you roll a dice.

   g You will see a famous person this week.

   h It will rain tomorrow.

> **Tip**
>
> If any probability question involves dice, assume that it refers to a normal, six-sided dice with the sides numbered 1 to 6, unless the question tells you otherwise.

3 Work with a partner.

Discuss whether the following statements are true or false and give reasons for your answers.

- **a** It can either rain or be sunny, so there is a 50% chance of rain tomorrow.
- **b** The lotto grand prize has rolled over for three weeks now, so someone is certain to win the grand prize this week.
- **c** You have to be very lucky to get doubles when you roll two dice.
- **d** If a woman has already given birth to three boys, she has an excellent chance of having a girl next time round.
- **e** It is unlikely that I will ever be able to have a credit card because I have a bad credit history.
- **f** Two football teams playing against each other have an equal chance of winning.

Section 2: Calculating probability

Equally likely, random outcomes

You have the same chance of getting heads as you have of getting tails, when you flip a fair coin, so you say the outcomes are **equally likely**.

This does not mean that if you flip a coin six times in a row you will get three heads and three tails. You could get six heads in a row, or four heads in a row followed by one tail and then one head.

Although the outcomes are equally likely, they are also **random**.

However, the more often you flip the coin, the closer you will get to an equal number of heads and tails.

Theoretical probability

When the outcomes are equally likely, you can calculate the probability of each event using the formula:

probability of an event = $\dfrac{\text{number of favourable outcomes}}{\text{total number of outcomes}}$

Tip

A fair, or unbiased, coin, spinner or dice is one that is symmetrical, not damaged or unbalanced in ways that make it fall more often on one side than the other. Each outcome is equally likely to occur.

Key vocabulary

equally likely: having the same probability of happening.

random: not predetermined.

Find answers at: cambridge.org/ukschools/gcsemaths-studentbookanswers

WORK IT OUT 20.1

A teacher puts the names of six students (Anna, Basil, Candy, David, Eliza and Fatimah) into a container and chooses one at random to decide which student is going to give the answers to a homework task.

Calculate the probability that the teacher draws the following:

a Basil   **b** a girl's name   **c** a name other than Basil.

Which of these solutions is correct? Why are the others incorrect?

Solution A	Solution B	Solution C
a P(Basil) = $\frac{1}{5}$ = 0.2	**a** P(Basil) = $\frac{1}{6}$ = 0.17	**a** P(Basil) = $\frac{1}{6}$ = 1.7%
b P(Girl) = $\frac{3}{6}$ = $\frac{1}{2}$ = 0.5	**b** P(Girl) = $\frac{4}{6}$ = $\frac{2}{3}$ = 0.67	**b** P(Girl) = $\frac{4}{6}$ = 6.7%
c P(Not Basil) = 1 − 5 = −4	**c** P(Not Basil) = 1 − $\frac{1}{6}$ = $\frac{5}{6}$ = 0.83	**c** P(Not Basil) = $\frac{5}{6}$ = 8.3%

The probability of an event not happening

The sum of all the possible outcomes is 1.

Consider a bag containing two blue and three red beads.

The probability of picking each bead is $\frac{1}{5}$.

There are five beads, so there are five outcomes, $\frac{1}{5} + \frac{1}{5} + \frac{1}{5} + \frac{1}{5} + \frac{1}{5} = 1$

The probability of picking a blue, P(B), is $\frac{2}{5}$.

The probability of **not** picking a blue is the same as the probability of picking a red, P(R), $\frac{3}{5}$.

Since $\frac{2}{5} + \frac{3}{5} = 1$, the probability of **not** picking a blue = 1− the probability of picking a blue.

If the probability of a given outcome is P(E), then the probability of that outcome **not** happening is 1 − P(E).

Mutually exclusive events

Events that cannot happen at the same time, are known as **mutually exclusive**.

When you roll a normal dice, you cannot roll a six **and** a four at the same time.

A single card, drawn from a standard pack of playing cards, cannot be both a king **and** a queen.

So, you cannot roll a six **and** a four **but** you could roll a six **or** a four.

To find the probability of a favourable outcome from mutually exclusive events, you add the probabilities of each of the possible favourable outcomes.

P(rolling six or four) = $\frac{1}{6} + \frac{1}{6} = \frac{2}{6} = \frac{1}{3}$

Tip

Remember, the probability scale goes from 0 to 1.

Key vocabulary

mutually exclusive: events that cannot happen at the same time.

Tip

Drawing a card that is red and the number two is possible so is **not** mutually exclusive.

EXERCISE 20B

1 For each of the following:

  i identify all the possible outcomes.

  ii state whether the outcomes are equally likely or not and explain why.

  a A fair dice is rolled.

  b 100 raffle tickets numbered from 1 to 100 are placed in a barrel and one is drawn at random.

  c A drawing pin is dropped to see whether it lands point up or point down.

  d A coin is drawn from a bag containing twelve £1 coins and twenty-five 50p coins.

  e A student from your class is chosen to speak at a school assembly.

  f Two coins are tossed at the same time.

2 A couple plan to have three children.

Given that they have an equal chance of having a boy or a girl, what is the probability that their first child will be a girl?

3 A wallet contains three £1 coins and four £2 coins.

Assuming that all the coins have an equal chance of being chosen, what is the probability that the first coin taken at random from the wallet is:

  a £1   **b** not £1   **c** 50p.

4 In a game at a school fete, any of the numbers from 1 to 20 is equally likely to be chosen. What is the probability of choosing:

  a an even number   **b** a number < 13   **c** a multiple of 6

  d 26   **e** a number that is not 19?

5 Scrabble tiles with one of each of the 26 letters of the alphabet are placed in a bag and a tile is drawn at random. What is the probability of getting:

  a a vowel

  b a consonant

  c a letter from the word 'square'

  d a letter from the name John or a letter from the name Ali

  e a letter from the name Nicky or a letter from the name Sue

  f a letter from the name Gary and from the name Ben?

6 A single card is drawn from a pack of standard playing cards. What is the probability that it is a club or a red card?

Section 3: Experimental probability

Predicting outcomes

The probability of getting heads when you toss a coin is $\frac{1}{2}$. But what happens if you toss a coin twice? It doesn't always land once on heads and once on tails.

If you roll a dice it can land on any of the numbers from 1 to 6, but if you rolled the dice six times, you probably wouldn't get 1, 2, 3, 4, 5 and 6.

Experiments with coins, counters, dice, spinners or playing cards involve a number of trials. For each trial, the outcome is recorded to see how often the favourable outcome occurred.

EXERCISE 20C

Work in pairs.

You will need two identical coins and a small container. Shake the coins in the container and then drop them onto your desk. You will do this 40 times each (80 times in total).

This table shows the three possible outcomes in this event.

Possible outcomes	Predicted frequency	Tally	Actual frequency
Heads, Heads (HH)			
Heads, Tails (HT)			
Tails, Tails (TT)			

a Predict how many times you think each outcome will occur in 80 throws. Write your prediction in the table.

b Do the experiment. Take turns to drop the coins and use tallies to record the outcomes.

c Total the tallies and write the actual frequency of each outcome (in other words, how many times each outcome happened).

d How do your results compare with your predictions?

e How many times would you expect to get two heads if you dropped these two coins 10 000 times? Why?

f Will your results be the same if you do the experiment again (or will they be the same as another pair)? Why?

In this experiment you found the actual frequency of different outcomes.

Relative frequency is the number of times you get a favourable outcome (e.g. 'two heads') out of all the outcomes. It is useful for estimating probability.

$$\text{Relative frequency} = \frac{\text{number of favourable outcomes}}{\text{total number of outcomes}}$$

If the favourable outcome is 'two heads' and two heads came up 32 times in the experiment, the relative frequency of heads is:

Relative frequency (HH) = $\frac{32}{80}$ = 0.4

Tip

Increasing the number of trials gives a relative frequency that is closer to the theoretical probability of the event.

Predictions based on empirical evidence

People collect data (empirical or experimental evidence) about all sorts of things to make predictions about what might happen in the future.

An insurance company might collect data about the age of drivers who have car accidents and find that drivers aged from 17 to 23 years have a higher relative frequency of being involved in accidents.

The insurance company estimates the probability of these drivers getting into an accident and uses this to calculate their premiums.

Probabilities based on empirical evidence are used:

- in weather forecasting.
- in market research to predict what people might buy.
- by medical professionals to work out the risk of different people contracting a disease.
- by sports teams and betting agencies to work out the chance of a team or player winning their match.
- by insurance companies to work out life expectancy and health-related problems.
- by industry to work out the probability that a manufactured part might be faulty.

WORK IT OUT 20.2

A laboratory tested 500 batches of tablets and found four to be contaminated. What is the probability that a batch of tablets produced in this laboratory would be:

a contaminated **b** not contaminated?

Which of these answers is the correct answer to the following question? Why is the other one wrong?

Option A	Option B
a P(Contaminated) = $\frac{4}{500}$ = 0.8%	**a** P(Contaminated) = $\frac{4}{500}$ = 0.008
b P(Not contaminated) = $\frac{496}{500}$ = 92%	**b** P(Not contaminated) = 1 − 0.008 = 0.992

EXERCISE 20D

1 What is the experimental probability of getting heads with a coin if you have done an experiment with the coin and it has landed heads up on 35 out of 60 tosses?

2 Nick rolled up balls of paper and threw them from his desk into the bin. The paper balls landed in the bin 175 out of 200 throws.

What is the experimental probability that the paper will land in the bin the next time Nick throws?

3 Paul has six red T-shirts, a green T-shirt and a yellow T-shirt. He says the probability of picking a red T-shirt at random is $\frac{6}{3}$ because there are three possible colours and six red T-shirts to choose from. Paul's reasoning is incorrect. How would you explain this to him?

Find answers at: cambridge.org/ukschools/gcsemaths-studentbookanswers

4. Mrs Noonan drives a taxi on the same route every morning. Over a period of 290 days in a year, she has been stopped at a level crossing by a train crossing the road 58 times.

   Calculate the experimental probability that she will be stopped by the train crossing on her morning route.

5. A pharmacist kept a record of which painkiller brand was bought by 80 customers in a month. These are her results:

Brand	Frequency
'Stopthepayne'	27
'Make-it-go-away'	22
'Painless'	20
Shop's own brand	11

   a Based on this data, what is the experimental probability of:
      i a customer buying an unbranded (own brand) painkiller
      ii a customer not buying a 'Stopthepayne' painkiller?
   b Do you think this data is sufficient to predict what brand of painkiller would be chosen by most customers in Britain? Give a reason for your answer.

6. Another pharmacist only stocks the 'Make-it-go-away' brand and the shop's own brand. He estimates that three times as many customers choose 'Make-it-go-away' over the shop's own brand.

   a If 381 customers chose the 'Make-it-go-away' brand, estimate how many chose the own brand.
   b Copy and complete this table.

Brand	Number sold	Relative frequency
'Make-it-go-away '	381	
Shop's own brand		
total		

   c Use the data in your table to estimate the probability that the next person who buys a painkiller will choose:
      i the 'Make-it-go-away' brand   ii the own brand.

7. Amira kept a record of the weather forecast for ten days and compared it with the actual weather on each day. These are her results:

Day	Forecast	Actual weather	Was the forecast correct?
1	Rain	Rain	Yes
2	Some showers	Sunny and warm	No
3	Cloudy	Cloudy	Yes
4	Sunny and warm	Sunny and warm	Yes
5	Some showers	Sunny and warm	No
6	Some showers	Some showers	Yes
7	Cloudy and windy	Cloudy and windy	Yes
8	Rain	Rain	Yes
9	Sunny and warm	Some showers	No
10	Sunny and warm	Sunny and warm	Yes

a Calculate the experimental probability that the weather forecast is correct.

b What is the chance that the weather forecast is wrong?

c You plan to go for a hike the next day. The weather forecast for that day is rain. Would you take rain gear with you? Explain why or why not.

8 Mica is a film-maker based in Edinburgh. She needs to know what the weather is going to be like in advance because bad weather can cause delays in her schedule and this is very expensive. She uses predictions from a specialist weather website because she finds its forecasts to be accurate 99% of the time. This is the meteogram she downloaded on 24 March.

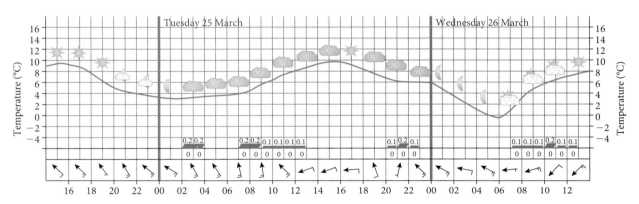

a What information is shown on the graph?

b Mica can only film when it is not raining. Should she schedule filming for Tuesday?

c Is Mica right to trust this website? Access a site to download the meteogram or long-term forecast for your area. Decide how you will test the reliability of the data and work out how accurate it is for your area.

Tip

You can do this activity using any weather forecasting site, and you could compare two or more to see which one has the highest probability of being correct for your area.

Organising outcomes – tables and frequency trees

There are many ways of recording and organising information to show how many favourable events there are, how many events there are in total and to find the figures you need to make decisions and calculate probabilities. Tables and simple diagrams called frequency trees are used to do this.

A doctor was interested in whether his patients knew the difference between having a cold and having the flu. One winter he kept track of 42 patients who came to see him because they thought they had a cold or the flu. Of these, 11 said they had a cold and 31 said they had the flu. Only 19 of those who said they had the flu actually had flu, and 4 of those who said they had a cold actually had the flu.

Here are two ways of organising the information from this experiment:

Two-way table   **Frequency tree**

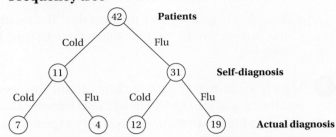

A frequency tree shows the actual frequency of different events.

The branches of the tree show the paths or decisions and the 'leaves' show the actual number of data for each path. Both the table and the frequency tree show the same information, but the frequency tree is clearer because it shows how many patients thought they had a cold or flu without you having to add the data in the table.

EXERCISE 20E

Tip

Frequency trees are organisational tools and they are often used in computer programming (they are sometimes called binary trees). They are not the same as probability tree diagrams which you will deal with in Chapter 26.

1 A hotel chain keeps track of which customers make use of its in-house spa facilities. Here are its results.

Gender	Spa use	
	Use the spa	Don't use the spa
Female	780	232
Male	348	640

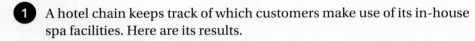

a Complete the frequency tree to show this data.

b Are male or female guests more likely to use the spa?

2 Of 60 patients visiting a doctor's rooms, 42 are convinced they will need prescription medication, the others think they probably won't need a prescription. Of those who think they will need a prescription 13 do not. Altogether 36 patients do need a prescription.

Complete the frequency tree to show the actual numbers.

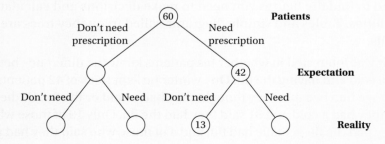

3 Eighty volunteers take an HIV-test to help the medical researchers work out how accurate the test is. Of the volunteers 17 people are HIV-positive, the others are not. The results show that one of the HIV-positive people gets a negative result on the test and two of the HIV-negative people get a positive result.

Draw a frequency tree to show the actual results.

Section 4: Mixed probability problems

In real life, people tend to use probability quite informally to explain things and to predict what will happen in the future. You might have heard people say things like:

- It is never sunny here in February.
- Most people prefer to wear sandals in summer.
- We are only selling 10 000 tickets so you have an excellent chance of winning the car.
- Young people who haven't had a driving licence for very long have more accidents than older drivers.

Understanding probability allows you to think more critically about statements like the ones above and to work out more accurately what the chance is of different things happening.

WORKED EXAMPLE 1

Zunaid read on a travel website that September was a good time to holiday in Italy because there was little chance of rain and the chance of sunny weather was highly likely.

He went on the Internet and found the following average weather data for the first 28 days of September.

Sunny days	Cloudy days	Rainy days
11	9	8

Was the travel website claim correct?

The relative frequency of rainy days was $\frac{8}{28} = \frac{4}{14}$ or 28.6%

This is not such a low chance of rain.

The relative frequency of sunny days was $\frac{11}{28} = 39.3\%$

This is less than an even 50–50 chance of sun.

In reality, it rained more than 1 out of every 4 days and there were far more cloudy and rainy days than sunny days.

He might find it better to go in a different month.

Sometimes the way a problem is worded can be confusing, but the actual calculations in probability are generally fairly simple additions or multiplications.

When you have to solve a word problem involving basic probability you can generally do this by organising your work and following the steps as shown in the example below.

Problem-solving framework

Nick is throwing a ball randomly at a wall on the side of a building. The side of the building is 2 m high and 10 m wide. There are three windows on the side of the building, each window is 2 m wide and 1 m high. Assuming that Nick hits the wall each time he throws, what is the probability that Nick will hit a window when he throws the ball at the wall? Express your answer as a percentage.

Steps for solving problems	What you would do for this example
Step 1: What are you trying to work out?	The probability of hitting any of the windows.
Step 2: What do you need to work out before you can find this?	The area of the wall and the area of the windows. Area of wall = 10 m × 2 m = 20 m² Area of windows = 3 × (2 m × 1 m) = 3 × 2 m² = 6 m²
Step 3: Apply the formula and calculate the probability. Convert the answer to a percentage.	P(Hits window) = $\frac{6}{20} = \frac{3}{10}$ $\frac{3}{10} \times 100 = 30\%$ There is a 30% probability that Nick will hit a window.

EXERCISE 20F

1 Calculate the theoretical probability of each outcome.

  a Tossing a coin and getting tails.

  b Rolling a dice labelled 1 to 6 and getting 2.

  c Randomly picking a red counter from a bag that contains three green, one red and five blue counters.

  d Rolling two dice and getting seven as the total score.

2 Zara has 10 black, five white, six red and three green sweets in a packet. She offers the packet to her friend Anna who takes a sweet without looking.

  a What colour is Anna most likely to pick?

  b Which colour has the lowest chance of being picked?

  c What is the probability that she picks a red sweet?

  d What is the probability that the sweet is white or green?

3 Nina and Maria made up a game with an eight-sided dice. The sides of the dice are labelled 6, 24, 9, 29, 15, 7, 18 and 12.

  They take turns to roll the dice. Nina wins the roll if the dice shows a multiple of 2. Maria wins the roll if the dice shows a multiple of 3.

  a Is this a fair game? Give a reason for your answer.

  b What is the theoretical probability that the dice will show a multiple of 3?

4 During a netball competition, the same coin was tossed 20 times. Busi claimed the coin was unfair because it landed on tails only five out of the 20 times.

She says the probability of getting tails when you toss a coin is 0.5, so if you toss the coin 20 times you should get 20 × 0.5 = 10 tails.

Was she correct? Explain your answer.

5 Grey College has a sports tournament against St George's College every year. The weather on the day of the tournament can be described as sunny and dry, cloudy and humid, or rainy. Grey College keeps a record of the weather on the day and whether it won or drew the tournament. Here are its results for the past 30 years.

Weather	Wins	Draws	Tournaments played
Sunny and dry	4	1	7
Cloudy and humid	3	2	10
Rainy	3	3	13
Total	10	6	30

  a What is the relative frequency of rain on tournament days?

  b A student from Grey College says they have a better chance of winning if it is sunny. Is that a correct statement? Support your answer.

  c A student from St George's says they have an almost even chance of winning the tournament, no matter what the weather. Is that a correct statement? Support your answer.

  d Calculate the experimental probability that Grey College will draw a tournament. Express this as a probability in words.

6 The chart below is a 10-day weather forecast in April for Cardiff.

Today	Sun 8	Mon 9	Tue 10	Wed 11	Thu 12	Fri 13	Sat 14	Sun 15	Mon 16
Rain	Showers	Sunny	Sunny	Sunny	Sunny	Partly cloudy	Mostly sunny	Cloudy	Scattered showers
Chance of rain:									
100%	80%	10%	0%	0%	0%	0%	0%	10%	30%

  a What is the probability that it will rain on:

   i Sunday 8 April   **ii** Sunday 15 April?

  b What does a 100% chance of rain mean?

  c Rhys wants to go hiking on Monday 9 April. Should he pack rain clothes? Give a reason for your answer.

  d The 10-day forecast for Plymouth for this period shows a 0% chance of rain every day. Does this mean it definitely won't rain in Plymouth in this period? Explain why or why not.

Tip

A false-positive in a drug test means that a person who is not using drugs tests positive for drug use.

7 A local educational authority wants to introduce random drug testing in secondary schools. It claims the tests have a very small false-positive rate of one half of one per cent.

   a Express one half of one per cent as a decimal.

   b The parents at a school with 800 students object to the test. They claim that four students could incorrectly test positive for drug use and that this could ruin their futures. Are the parents concerns valid? Explain why or why not.

   c There were 3 831 937 secondary school students under this authority in the year they wanted to do the drug testing. If they were all tested for drug use, how many of them would you expect to be incorrectly accused of being drug users?

8 Professional athletes are routinely tested for prohibited performance enhancing drugs. The testing authority estimates only 1% of the athletes tested are actually using prohibited drugs. If an athlete is using prohibited drugs, 90% of the time he or she will test positive in the test (in other words, fail the drug test). But, 10% of the athletes who are not using prohibited drugs will also test positive (in other words, fail the drug test even though they are not using drugs).

   a Complete this table to show how many athletes will pass or fail the drug test for every 1000 athletes tested.

Status	Test positive (i.e. fail drug test)	Test negative (i.e. pass drug test)	Total
Athletes who are using illegal substances			10
Athletes who are not using illegal substances			990
Total			1000

   b Represent the same information on a frequency tree.

   c If an athlete tests positive for the illegal substance, what is the chance that he or she is not actually using the substance? Give your answer as a percentage.

   d If an athlete tests negative for the substances is it certain that he or she is not using them? Explain your answer.

9 Eighty people are asked if they can tell the difference between butter and margarine. Of these, 37 say they can, 24 say no and 19 say they are not sure. The interviewer then carries out a blind taste test. Of those who said they could tell the difference, 14 got it wrong, of those who said no, 9 got it right and 14 of those who said they were not sure got it wrong. Draw a frequency tree to show the outcomes of this experiment.

10 Lee and Haroon want to know what the probability is of getting two heads when you flip two coins one after the other. Lee says it is 25% and Haroon says it is $33\frac{1}{3}\%$. They decide to do an experiment in which they flip five different sets (to be fair) of two coins and record the outcomes.

   a List all the possible outcomes when you flip two coins.

   b Complete this table to show the results of Lee and Haroon's experiment.

Set of coins	Number of flips	Number of times we got two heads	Running total of two heads	Percentage of two heads (running total)
Two 10p coins	25	6	6	$\frac{6}{25} \times 100 = 24$
Two 50p coins	25	8		
Two £1 coins	25	5		
Two £2 coins	25	7		
Two 20p coins	25	9		

   c What type of probability have they recorded?

   d What is the relative frequency of getting two heads according to their results?

   e Do their results settle their argument? Explain why or why not?

11 Lee finds a computer program that simulates coin flips. He does a trial flipping two coins 1000 times. The computer produces a graph (right) of his results.

   a What does the graph show?

   b What does the yellow line on the graph represent?

   c Why does the other line vary up and down?

   d What happens to the line showing the results as the number of tosses increases?

   e How does this graph help to settle the argument between Lee and Haroon?

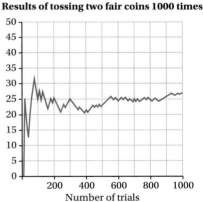

Results of tossing two fair coins 1000 times

Checklist of learning and understanding

Range of probabilities

- The probability scale ranges from 0 to 1. Impossible events have a probability of 0 and certain events have a probability of 1. It is not possible to have a negative probability (< 0) or a probability greater than 1.
- Probabilities between 0 and 1 can be expressed as fractions, decimals or percentages.

Find answers at: cambridge.org/ukschools/gcsemaths-studentbookanswers

Theoretical probability

- Probability of an event = $\dfrac{\text{number of favourable outcomes}}{\text{number of possible outcomes}}$

- The greater the number of trials, the closer the relative frequency is likely to be to the theoretical probability.

Sum of probabilities and complementary events

- The sum of probabilities will always total 1.
- The probability of an event not happening is equal to 1 minus the probability that the event will happen. So, P(Not E) = 1 − P(E).

Experimental probability and relative frequency

- Experimental probability tells you how often a favourable outcome occurs in an experiment.

  Experimental probability = $\dfrac{\text{relative frequency of a favourable outcome}}{\text{number of possible outcomes}}$

- Tables and frequency trees can be used to organise the outcomes of different experiments.
- Statistical data can also be used to give the relative frequency of particular events. The relative frequency of an event can be used to predict future outcomes.

Mutually exclusive events

- Mutually exclusive events cannot happen at the same time. For example, you cannot throw a 1 and a 5 at the same time when you roll a dice.

For additional questions on the topics in this chapter, visit GCSE Mathematics Online.

Chapter review

1. Choose the correct answer. Sharon is playing a game and she needs to roll a six to start. What is the probability of her rolling a six on her first go?

   A $\dfrac{1}{6}$ **B** $\dfrac{5}{6}$ **C** $\dfrac{6}{6}$ **D** 0

2. I have 30 red, 40 white, 2 brown and 8 green beads in a container. If I choose one at random, what is the probability that it will be:

   a white   **b** not green   **c** brown or green?

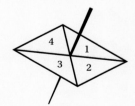

3. On the left is a four-sided spinner.

   The spinner is biased.

   The table shows the probabilities that the spinner will land on 1 or on 3

Number	1	2	3	4
Probability	0.2		0.1	

   The probability that the spinner will land on 2 is the same as the probability that the spinner will land on 4

a Work out the probability that the spinner will land on 4 *(3 marks)*

Shunya is going to spin the spinner 200 times.

b Work out an estimate for the number of times the spinner will land on 3 *(2 marks)*

©Pearson Education Ltd 2013

4 Nina rolled a dice 200 times and recorded her results in a table.

Result	1	2	3	4	5	6
Frequency	28	20	20	40	36	56
Experimental probability						

a Calculate the experimental probability of each result.

b What is the relative frequency of rolling an odd number?

c The theoretical probability of rolling a 4 is $\frac{1}{6}$. Compare this with Nina's data and suggest why the two probabilities might be different.

5 Jill interviews 64 people to get their opinions about sending texts when in company. 44 of those interviewed say it is rude to send texts in company. Jill then observes the people at a large event. Of those who said it was rude to send texts in company 13 sent texts when at the table with others. Of those who said it was acceptable, 9 did not send texts when in company. Complete the frequency tree (right) to show this data.

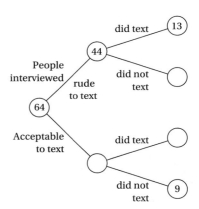

6 A company tests a new brand of soap and produces the following data.

Number of participants	Developed a rash	Did not develop a rash
2500	180	2320

a Use these results to determine the probability of using this soap and developing a rash.

b The company decides to print the following warning statement on the soap: '7% of people who use this soap may develop a rash.'

  i Is the statement correct?

  ii Why would the company use a percentage rather than giving the number of people who developed a rash?

  iii If 100 people used this soap, how many of them would you expect to develop a rash?

  iv Why is the word 'expect' used in part **iii** above?

21 Construction and loci

In this chapter you will learn how to …

- use a ruler, protractor and pair of compasses effectively.
- use a ruler and a pair of compasses to bisect lines and angles and construct perpendiculars.
- accurately construct geometrical figures.
- construct accurate diagrams to solve problems involving loci.

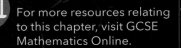

For more resources relating to this chapter, visit GCSE Mathematics Online.

Using mathematics: real-life applications

Draughtspeople and architects need to draw accurate scaled diagrams of the buildings and other structures they are working on.

Although the drawings are complicated, they still use ordinary mathematical instruments like pencils, rulers and pairs of compasses to draw them.

"I prepare technical drawings and plans that are given to me by an architect.

I use a CAD program, but I always start with a drawing board and plans that I draw using my ruler, set square and pair of compasses."

(Draughtsperson)

Before you start …

KS3	You need to be able to measure and draw angles accurately using a protractor.	1 Choose the correct measurement for each angle. a b 2 Use a ruler and a protractor to draw a reflex angle the same size as this one.

KS3	You should be able to convert between units of length.	3 Choose the correct answers.
		a 1 m is equivalent to:
		  A 10 mm B 100 mm C 1000 mm
		  D none of these measurements
		b Half of 8.7 cm is:
		  A 43 mm B 435 mm C 43.5 mm
		  D none of these measurements

Ch 2	You must know and be able to use the correct names for parts of shapes, including circles.	4 Match the letters **a** to **e** with the correct mathematical names from the box below. vertex centre radius side diameter

Assess your starting point using the Launchpad

STEP 1

1 Which of the following statements are true of this angle?

  A It is an acute angle.
  B It measures 120°.
  C It is called QRP.
  D If you extend arm QR, the size of the angle will increase.

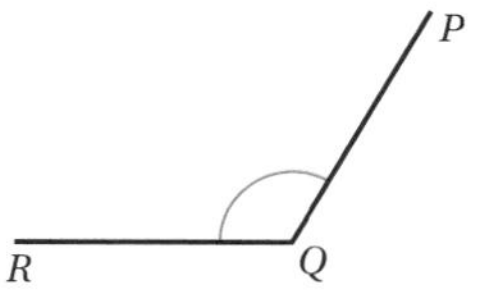

2 Which of the following statements are **not** true of this circle?

  A It has a radius of 5 cm.
  B It has a diameter of 5 cm.
  C $OC \perp AB$
  D $OC = \frac{1}{2}(AB)$

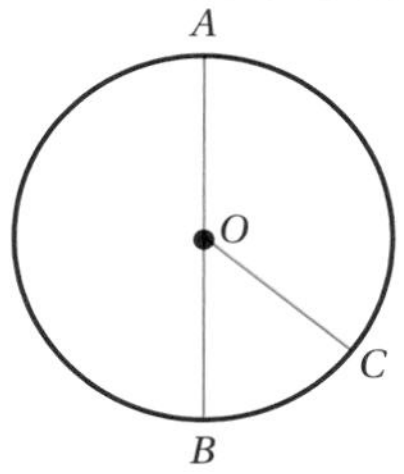

GO TO Section 1: Using geometrical instruments

STEP 2

3 Niresh did the construction shown here.

  a What do you call line BR?
  b What did Niresh do to produce points P and Q?
  c Given that angle ABC = 24°, state the size of angle ABR without measuring it.

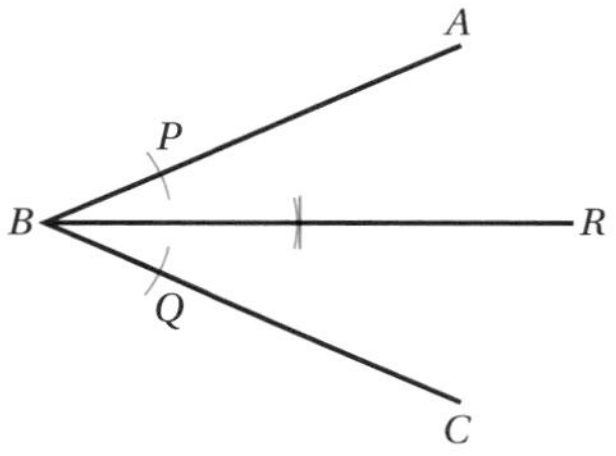

GO TO Section 2: Ruler and compass constructions

STEP 3

4 Two points, A and B, are four centimetres apart.
Find the locus of points that are equidistant from A and B.

GO TO Chapter review

GO TO Section 3: Loci

Section 4: Applying your skills

Find answers at: cambridge.org/ukschools/gcsemaths-studentbookanswers

Tip

Always measure and draw as accurately as you can. At this level you are expected to draw lengths correct to the nearest millimetre and angles correct to the nearest degree.

Tip

If the arms of the angle are too short to read the scale correctly use a ruler and a pencil to extend them. This doesn't change the size of the angle but it allows you to read the measurement more accurately. If you cannot draw on the angle (because it is in a book) you can extend the arm with the straight edge of a sheet of paper.

Section 1: Using geometrical instruments

Measuring and drawing angles

You use a protractor to measure and draw angles.

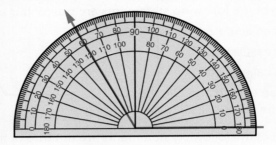

Protractors like the one above have two scales.

To avoid reading the wrong scale, estimate the size of the angle before you measure.

Use your knowledge of acute, right and obtuse angles to estimate as accurately as possible.

WORKED EXAMPLE 1

Estimate and then measure the size of each red angle.

a

b

a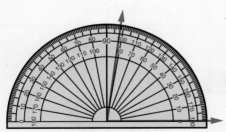

This is an acute angle but it is close to 90°. Estimate about 80°.

Use the inner scale to measure because this is the scale that has 0 on the arm of the angle.

$ABC = 82°$

b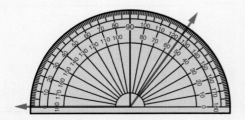

This is an obtuse angle. It is about one-third bigger than a right angle.

Estimate about 120°

Use the outer scale to measure because now this is the scale that has 0 on the arm of the angle.

$DEF = 125°$

21 Construction and loci

WORKED EXAMPLE 2

Use your protractor to draw angle $ABC = 76°$.

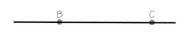

Draw a straight line using your ruler and mark the points B and C.
B is the vertex in angle ABC.

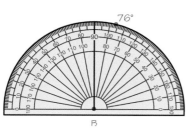

Place your protractor with its centre on B and baseline on the line you drew.

Measure and mark 76°.

Remove the protractor.
Draw a line from B through the 76° marking.
Label the angle correctly.

76° is an acute angle. Look at your angle and check that it looks less than 90°. If it does not, you have used the wrong scale on your protractor.

Using a pair of compasses

A pair of compasses (sometimes just called compasses) is useful for marking accurate line lengths in figures, for drawing circles, and for constructing some angles without measuring them with a protractor.

Make sure your pencil point is sharp and that the compasses are not too loose.

Parts of a circle

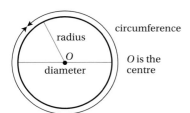 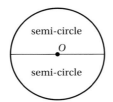 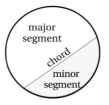

Key vocabulary

radius (plural radii): distance from the centre to the circumference of a circle. One radius is half of the **diameter** of the circle.

circumference: distance round the outside of a circle.

semi-circle: half of a circle.

chord: a straight line that joins one point on the circumference of a circle to another point on its circumference. The diameter is a chord that goes through the centre of the circle.

Find answers at: cambridge.org/ukschools/gcsemaths-studentbookanswers

Tip

Using a pair of compasses effectively takes practice.

WORKED EXAMPLE 3

Draw a circle with a **radius** of 4.5 cm.

Place the pair of compasses alongside a ruler and open it to 4.5 cm.

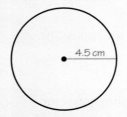

Draw a circle.

WORKED EXAMPLE 4

Make an accurate copy of this figure.

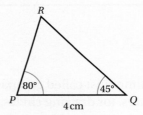

First draw the base line of 4 cm and label this PQ.

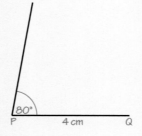

Use a protractor to measure the angle 80° from point P and draw a line.

Tip

Draw the lines in faintly first. Then go over them to show the final triangle. Never rub out your construction lines.

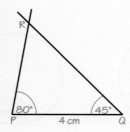

From point Q measure the angle 45° and draw a line from Q extending out so that it crosses the other line.

Where the two lines cross is point R. This is the apex of the triangle.

WORKED EXAMPLE 5

Construct an equilateral triangle with side lengths 6 cm.

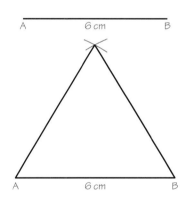

First draw the base line of 6 cm with a ruler and label it *AB*.

Then set your compasses to 6 cm and draw an arc above that line, setting the point of your compasses at *A*.

Repeat this from the other side at point *B*.

Where the two arcs join is the apex of the triangle. Use this point to complete the triangle.

You have constructed a triangle with three equal side lengths, so this is an equilateral triangle with three equal angles of 60°.

Tip

Leave the construction markings (arcs made using your pair of compasses) on your diagrams as this shows the method you used to construct them.

This method can be used for triangles with sides of different lengths by setting your pair of compasses to whatever the lengths of the sides are.

EXERCISE 21A

1 Use a ruler and protractor to draw and label the following angles.

   a $PQR = 25°$ **b** $DEF = 149°$ **c** $XYZ = 90°$

2 How could you use a protractor marked from 0° to 180° to measure an angle of 238°?

3 **a** Draw line *MN* which is 8.4 cm long.

   At *M*, measure and draw angle $NMR = 45°$.

   At *N*, measure and draw angle $RNM = 98°$ so that it forms a triangle *MNR*.

   b Explain why it is not necessary to know the lengths of *MR* and *NR* in order to construct the triangle accurately.

4 Use a pair of compasses to construct:

  a a circle of radius 4 cm

  b a circle of diameter 12 cm

  c a circle of diameter 2 cm which shares a centre, O, with another circle of radius 5 cm.

5 Draw a line AB which is 70 mm long. Construct the circle for which this line is the diameter.

6 **a** Draw a circle of radius 3.5 cm and centre O.

  Use a ruler to draw any two radii of the circle. Label them OA and OB.

  Join point A to point B to form triangle AOB.

  Measure angles AOB, OBA and BAO. Write the measurements on your diagram.

  b What type of triangle is triangle AOB?

7 Make accurate drawings of the shapes shown in these diagrams.

  a 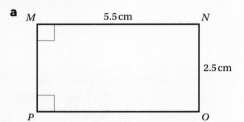 **b**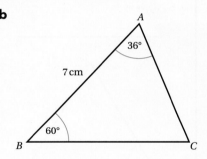

8 Prepare a step-by-step set of instructions for using **only** a ruler and a pair of compasses to construct:

  a an equilateral triangle ABC with sides of 6.4 cm.

  b a semi-circle with a radius of 30 mm.

Section 2: Ruler and compass constructions

Bisecting a line

> **Key vocabulary**
>
> **bisect**: divide exactly into two equal halves.

You can use a ruler and a pair of compasses to **bisect** any line **without** measuring it.

WORKED EXAMPLE 6

Bisect line AB by construction.

A————B

Open the pair of compasses to any width that is greater than half the line.

Continues on next page ...

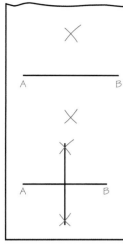

Place the point of the compasses on A and draw arcs above and below the line.

Keep the compasses open to the same width and place the point on B.

Draw arcs above and below the line so that they cut the first set of arcs.

Use a ruler to join the points where the arcs intersect.

The point where the constructed line cuts AB is called the **midpoint** of AB.

The distance from A to this point is equal to the distance from B to this point.

The constructed line is perpendicular to AB, so it is called the **perpendicular bisector** of AB. All points along the constructed line are an equal distance from both point A and point B.

Tip

Remember perpendicular means 'at right angles to'.

Key vocabulary

midpoint: the centre of a line; the point that divides the line into two equal halves.
perpendicular bisector: a line perpendicular to another that also cuts it in half.

Constructing perpendiculars

Construct a perpendicular at a given point on a line

WORKED EXAMPLE 7

Construct $XY \perp AB$ at point X.

Tip

Remember, $\perp$ is the symbol which means 'perpendicular to'.

Draw and label a line, AB, about 12 cm long and mark on it point X, about 5 cm from A.

Open your compasses to a width of about 4 cm.

Place the point of your compasses on X.

Draw two arcs to cut AB on either side of X.

Continues on next page ...

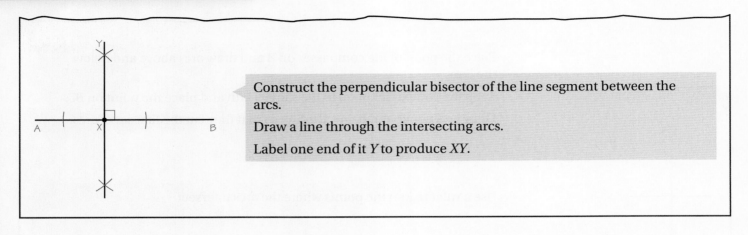

Construct the perpendicular bisector of the line segment between the arcs.

Draw a line through the intersecting arcs.

Label one end of it Y to produce XY.

Construct a perpendicular from a point to a line

WORKED EXAMPLE 8

Construct PX perpendicular to line AB from point P.

Tip

The shortest distance from any point to a line is a perpendicular from that point to the line.

Draw and label a line, AB, about 12 cm long and mark on it point P, about 5 cm above the line.

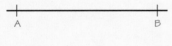

Place the point of your compasses on P.

Draw an arc that cuts AB in two places.

Label these places C and D.

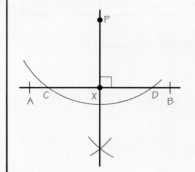

Open the pair of compasses to a width more than half the distance between C and D.

Place the point on C and draw an arc on the opposite side of the line to point P.

Place the point on D and draw an arc which intersects the one you just drew.

Draw a line from the intersecting arcs to P.

Label PX and mark the perpendicular.

Bisecting an angle

An angle bisector divides any angle into two equal halves.

> **Tip**
> Remember, if you bisect a 90° angle you will then have two angles of 45°.

WORKED EXAMPLE 9

Use a protractor to draw an angle PQR of 70°.

Construct the angle bisector of this angle without measuring.

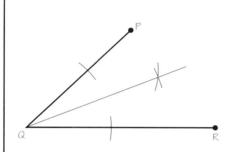

Place the pair of compasses on the vertex of the angle, Q.

Open your compasses a few centimetres and draw arcs that cut each arm of the angle.

Place the point of your compasses on each arc where it cuts the arm of the angle in turn and, keeping the width the same, draw arcs between the arms of the angle.

Draw a line from the vertex of the angle through the intersection of the arcs.

This is the angle bisector.

EXERCISE 21B

1 Measure and draw the following line segments. Find the midpoint of each by construction.

   a $AB = 9$ cm **b** $MN = 48$ mm **c** $PQ = 6.5$ cm

2 **a** Draw any three acute angles. Bisect each angle without measuring.

   b How could you check the accuracy of your constructions?

3 Draw the angles shown and then, using only a ruler and pair of compasses, bisect each angle.

   a   **b**   **c**

4 Draw any triangle ABC.

   a Construct the perpendicular bisector of each side of the triangle.

   b Use the point where the perpendicular bisectors meet as the centre and vertex A as a radius and draw a circle.

   c What do you notice about this circle?

5 Construct equilateral triangle DEF with sides of 7 cm.

   a Bisect each angle of the triangle by construction. Label the point where the angle bisectors meet as O.

   b Measure DO, EO and FO. What do you notice?

6 Draw $MN = 80$ mm. Insert any point A above MN.

   a Construct $AX \perp MN$.   **b** Draw $AB \parallel MN$.

Find answers at: cambridge.org/ukschools/gcsemaths-studentbookanswers

Section 3: Loci

Key vocabulary

locus (plural **loci**): a set of points that satisfy the same rule.

A **locus** is a set of points that obey a certain rule.

You can think of a locus as the path that shows all the possible positions for a point.

Some of the rules for loci produce shapes and lines (paths) that you are already familiar with from your work on constructions.

WORKED EXAMPLE 10

A tap is located at point X.

Draw the locus of points that are exactly 50 metres from the tap.

(Your diagram does not need to be to scale.)

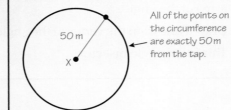

All of the points on the circumference are exactly 50 m from the tap.

Use compasses to draw a circle. Label the centre X and mark the radius.

Tip

The locus of points at a fixed distance from a given point is a circle with the given point at its centre and the fixed distance being the radius.

Using Worked example 10, the locus of points that are less than 50 m from the tap is the region inside the circle.

If you were asked to construct this locus, you would shade the interior of the circle to show that all the points inside the circumference meet the conditions of the locus.

You would show the circumference as a broken line to indicate that it is **not** included in the locus.

Tip

If a line is included in the locus you draw it as a solid line. If the line is not included, but just shows the edge of the locus you draw it as a broken, or dashed, line.

WORKED EXAMPLE 11

Anna lives at point A. Josie lives at point B.

They want to meet exactly midway between their homes.

Draw a diagram to show where they could meet.

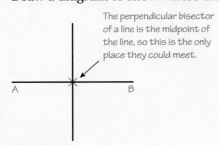

The perpendicular bisector of a line is the midpoint of the line, so this is the only place they could meet.

Draw a line to join the points.
Construct the perpendicular bisector of this line.
The point where the lines cross is exactly midway between their homes.

In Worked example 11 you can take any point on the perpendicular bisector and it will be the same distance from A and B.

If the girls wanted to meet at any point that was the same distance from their home, they could meet anywhere along that line.

However, the question asked you to find the point exactly midway between A and B, and the midpoint of line AB is the only point that meets that condition.

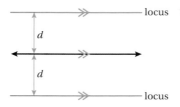

A line segment has a fixed length.

The locus of points equidistant from a line segment has to be the same distance from the line and also the same distance from its end point.

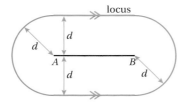

This produces an oval 'racing track' shape with all points the same distance (d) from the line segment.

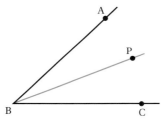

Tip

The locus of points equidistant from two fixed points is the perpendicular bisector of the line joining the two points. The locus of points a fixed distance (d) from a line is the pair of parallel lines which are d cm away from the given line. The locus of points at a fixed distance (d) from a line segment is a pair of parallel lines d cm away from the line segment along with the semi-circles of radius d cm at the ends of the line segment.

Tip

Remember a line continues to infinity in both directions, so it has no end points.

Tip

The locus of points an equal distance from two intersecting lines is their angle bisector.

EXERCISE 21C

1 Loci are common in architecture and also in the line markings on sports fields. Try to identify and describe some of the loci in the two photographs below and on the right.

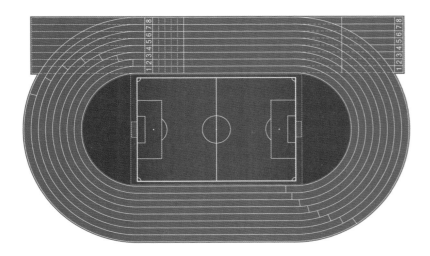

2 Without drawing them, describe the point, path or area that each locus will produce.
   a Points that are 200 km from a shop at point X.
   b Points that are more than 2 km but less than 3 km from a straight fence 1 km long.
   c Points that are equidistant from the two baselines of a tennis court.
   d Points that are equidistant from the four corners of a soccer field.
   e Points that are within 1 km of a railway line.

3 Accurately construct the locus of points 4 cm from a point D.

4 Draw angle MNO = 50°.
   Accurately construct the locus of points equidistant from MN and NO.

5 Draw PQ 4 cm long.
   Construct the locus of points 1 cm from PQ.

6 PQ is a line segment of 5 cm.
   X is a point exactly 4 cm from P and exactly 2.5 cm from Q.
   Show by construction the possible locations of point X.

7 Draw a rectangle ABCD with AB = 6 cm and BC = 4 cm.
   a Construct the locus of points that are equidistant from AB and BC.
   b Shade the locus of points that are less than 1 cm from the centre of the rectangle.
   c Construct the locus of points that are exactly 1 cm outside the perimeter of the rectangle.

Section 4: Applying your skills

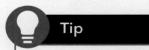

Tip

Scale was covered in Chapter 10.

If you are not given a scale, always state the scale that you have used.

Questions that require the accurate drawing of shapes and loci might be given in a real-world context.

You will need to decide which construction technique to use.

You might be asked to draw scaled diagrams to solve these problems.

Problem-solving framework

A, B and C represent three towns.

A mobile phone tower is to be erected in the area.

The tower is to be equidistant from towns A and B and within 30 km of town C.

Show by accurate construction on a scale diagram all possible sites for the tower.

Use a scale of 1 cm : 10 km.

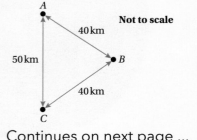

Continues on next page ...

Steps for solving problems	What you would do for this example
Step 1: Work out what you have to do. Start by reading the question carefully.	Although it doesn't say so, this is a locus problem. You have to find the locus of points equidistant from A and B and the locus of points that are less than 30 km from C. The solution is where these loci overlap.
Step 2: What information do you need? Have you got it all?	You have to draw a scale diagram. The distances and the scale are given. The conditions for the loci are given.
Step 3: Decide what maths you can do.	First work out the lengths you have to construct using the scale. The scale is 1 cm : 10 km. So: $\dfrac{40 \text{ km}}{10} = 4 \text{ cm}$ $\dfrac{50 \text{ km}}{10} = 5 \text{ cm}$ Use these lengths to construct a triangle using your ruler and pair of compasses. Once you have the triangle you can find the loci by construction.
Step 4: Set out your solution clearly. Check your working and that your answer is reasonable.	![diagram] The tower could be built at any position along the thick red line.
Step 5: Check that you've answered the question.	You have shown the overlapping loci and written a statement to answer the question.

EXERCISE 21D

1 Draw line $AB = 5.2$ cm.

Construct DE, the perpendicular bisector of AB. Draw DE so that it is 44 mm long.

Mark the intersection point of line AB and DE as C.

Mark point F on DE such that $DF = FC = 11$ mm.

Construct $MN \parallel AB$ and passing through point F.

2. Construct a parallelogram with sides of 46 mm and 28 mm and a longest diagonal of length 60 mm. Measure and write in the length of the other diagonal.

3. Accurately construct a square of side 45 mm.

4. Construct quadrilateral ABCD such that ABC = 90°, AB = DC = 2.2 cm and AD = BC = 5 cm.

    What kind of quadrilateral is this?

5. On a treasure map, the position of buried treasure is known to be 10 metres from the castle at point Y and 12 metres from the cave at point Z.

    Y and Z are 15 metres apart.

    Draw a scale diagram and mark with an X all the places where the treasure might be buried.

    Use a scale of 1 cm to 2 m.

6. A monkey is in a rectangular enclosure which is 10 m by 17.5 m.

    The monkey is able to stretch through the fence around its enclosure and reach a distance of 25 cm.

    a. Draw a scale diagram to show the locus of points that the monkey can reach outside its enclosure.

    b. Show on your diagram where you would place a safety barrier to make sure that people cannot touch the monkey.

    Give a reason for your choice.

Checklist of learning and understanding

Geometry constructions
- A protractor is used to measure and draw angles.
- You can use a ruler and pair of compasses to construct perpendicular lines and to bisect lines and angles.
- The perpendicular bisector of any line cuts the line at its midpoint.
- The shortest distance from a point to a line is always the perpendicular distance.

Loci
- A locus is a set of points that obey a certain rule.
- The locus of points can be a single point, a line, a curve or a shaded area.
- Loci can be used to solve problems involving equal distances and overlapping areas.

21 Construction and loci

Chapter review

For additional questions on the topics in this chapter, visit GCSE Mathematics Online.

1 **a** Use a protractor to measure angles a and b on this clock face.

   b Draw two angles which are the same size as a and b.

   c Bisect the two angles you have drawn by construction.

2 Draw a line AB of length 6.5 cm and find its midpoint by construction.

Indicate the locus of points that are equidistant from A and B on your diagram.

3 Town X is due north of Town Y and they are 20 km apart.

Town Z is 25 km from Town X and 35 km from Town Y and lies to the east of both towns.

   a Draw a scale diagram to show the location of Town Z in relation to the other two towns.

   Use a scale of 1 cm : 5 km.

   b A railway runs between towns X and Y such that it is equidistant from both towns.

   Indicate the position of the railway on your drawing.

   c The electricity supply from town X is carried on a cable that is the same distance from XZ and XY along its length.

   Indicate where this cable would be.

4 Here is a scale drawing of a rectangular garden $ABCD$.

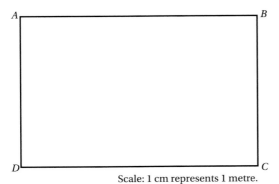

Scale: 1 cm represents 1 metre.

Jane wants to plant a tree in the garden
   at least 5 m from point C,
   nearer to AB than to AD
and less than 3 m from DC.

On the diagram, shade the region where Jane can plant the tree.

(4 marks)

©Pearson Education Ltd 2013

Find answers at: cambridge.org/ukschools/gcsemaths-studentbookanswers

333

22 Vectors

In this chapter you will learn how to ...
- represent vectors as a diagram or column vector.
- add and subtract vectors.
- multiply vectors by a scalar.

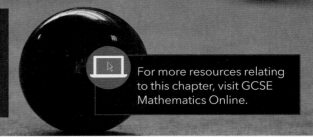

For more resources relating to this chapter, visit GCSE Mathematics Online.

Using mathematics: real-life applications

Vectors have huge applications in the physical world. For example, mathematical modelling of objects sliding down slopes with varying amounts of friction, working out how far objects can tilt before they tip over and making sure two ships don't crash in the night. All these problems involve the use of vectors.

"When landing at any airport I have to consider how the wind will blow me off course. Over a set amount of time I expect to travel through a particular vector but I have to add on the effect the wind has on my flight path. If I don't do this accurately I would struggle to land the plane safely." *(Pilot)*

Before you start ...

KS3	You need to be able to plot coordinates in all four quadrants.	1	Draw a set of axes going from −6 to 6 in both directions. Plot the points $A(2, 3)$, $B(-3, 4)$ and $C(-2, -3)$.
KS3	You need to be able to add, subtract and multiply negative numbers.	2	Calculate. a $3 - 7$ b $-4 + 11$ c $-5 - 18$ d -4×7 e -3×-9
Ch 14	You need to be able to solve simple linear equations.	3	Solve. a $12 = 4m - 36$ b $2k + 15 = 7$ c $-6 + 5d = -41$
Ch 14	You need to be able to solve simultaneous linear equations.	4	Solve. $3x + 2y = 8$ and $4x - 3y = 5$

Assess your starting point using the Launchpad

STEP 1

1 Give the column vector for $\overrightarrow{HG}$.

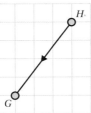

2 Draw the triangle ABC where $\overrightarrow{AB} = \begin{pmatrix} 3 \\ -5 \end{pmatrix}$ and $\overrightarrow{CA} = \begin{pmatrix} 2 \\ 7 \end{pmatrix}$.

**GO TO
Section 1**: Vector notation and representation

STEP 2

3 $\mathbf{j} = \begin{pmatrix} -1 \\ 3 \end{pmatrix}$ $\mathbf{k} = \begin{pmatrix} 2 \\ 1 \end{pmatrix}$ $\mathbf{l} = \begin{pmatrix} -4 \\ -2 \end{pmatrix}$

Write the following as single vectors.

a $\mathbf{j} + \mathbf{k}$ **b** $2\mathbf{k} - \mathbf{l}$

4 Find the values of f and g.

$\begin{pmatrix} 10 \\ g \end{pmatrix} - 4\begin{pmatrix} f \\ -3 \end{pmatrix} = \begin{pmatrix} -2 \\ 18 \end{pmatrix}$.

5 In the diagram below

$\overrightarrow{AC} = \begin{pmatrix} 14 \\ 2 \end{pmatrix}$ and $\overrightarrow{AB} = \begin{pmatrix} 9 \\ 12 \end{pmatrix}$.

Find:

a $\overrightarrow{CA}$
b $\overrightarrow{CA} + \overrightarrow{AB}$.

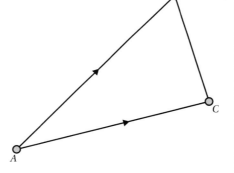

6 Which of these vectors are parallel?

$\begin{pmatrix} -3 \\ 4 \end{pmatrix} \begin{pmatrix} 9 \\ 16 \end{pmatrix} \begin{pmatrix} 15 \\ 20 \end{pmatrix} \begin{pmatrix} -3 \\ 2 \end{pmatrix}$

**GO TO
Section 2**: Vector arithmetic

**GO TO
Section 3**: Mixed practice

**GO TO
Chapter review**

Find answers at: cambridge.org/ukschools/gcsemaths-studentbookanswers

Section 1: Vector notation and representation

A **vector** describes movement from one point to another, it has a direction and a magnitude (size).

Vectors can be used to describe many different kinds of movement. For example: **displacement** of a shape following translation, displacement of a boat during its journey, the velocity of an object, and the acceleration of an object.

A vector that describes the movement from A to B can be represented by:

- an arrow in a diagram
- $\overrightarrow{AB}$ (arrow indicates direction)
- **a** (if handwritten this would be underlined, a)
- a column vector $\begin{pmatrix} x \\ y \end{pmatrix}$

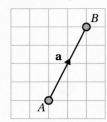

If you were to travel along this vector in the opposite direction, from B to A, you would represent this vector as:

- $\overrightarrow{BA}$
- **−a**
- $\begin{pmatrix} -2 \\ -4 \end{pmatrix}$

> **Key vocabulary**
>
> **vector**: a quantity that has both magnitude and direction.
>
> **displacement**: a change in position.

> **Tip**
>
> You include the negative sign because you are now moving in the opposite direction.

Column vectors

In a column vector x represents the **horizontal** movement; y represents the **vertical** movement.

	Movement	
	x	y
positive	right	up
negative	left	down

In the diagram, $\overrightarrow{AB} = \begin{pmatrix} 2 \\ 4 \end{pmatrix}$.

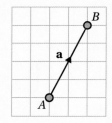

EXERCISE 22A

1 Match up equivalent representations of the vectors.

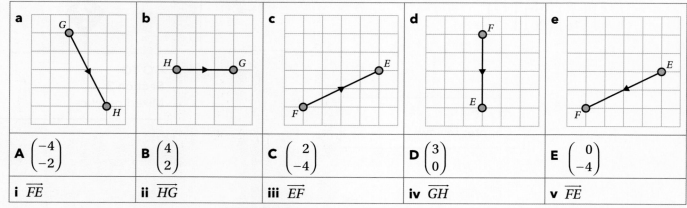

2 Use the diagram (right) to answer the following questions.

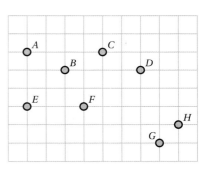

Find each of the column vectors.
- **a** $\overrightarrow{AB}$
- **b** $\overrightarrow{DC}$
- **c** $\overrightarrow{BC}$
- **d** $\overrightarrow{DF}$
- **e** $\overrightarrow{HF}$
- **f** $\overrightarrow{BH}$
- **g** What do you notice about $\overrightarrow{AB}$ and $\overrightarrow{DC}$?
- **h** What do you notice about $\overrightarrow{AB}$ and $\overrightarrow{BH}$?

3 Draw a pair of axes where x and y vary from -8 to 8.

Plot the point A (2, −1).

Then plot points B, C, D, E, F and G where:

$\overrightarrow{AB} = \begin{pmatrix} 2 \\ 7 \end{pmatrix}$ $\overrightarrow{AC} = \begin{pmatrix} -3 \\ 7 \end{pmatrix}$ $\overrightarrow{AD} = \begin{pmatrix} -6 \\ 3 \end{pmatrix}$

$\overrightarrow{AE} = \begin{pmatrix} 5 \\ 3 \end{pmatrix}$ $\overrightarrow{AF} = \begin{pmatrix} -3 \\ -1 \end{pmatrix}$ $\overrightarrow{AG} = \begin{pmatrix} 2 \\ -1 \end{pmatrix}$

4 a Find the vector from point A with coordinates (3, −4) to point B with coordinates (−1, 2).

b Give the coordinates of two more points E and F where the vector from E to F is the same as $\overrightarrow{AB}$.

5 a Find the vector from point K with coordinates (−2, −1) to point L with coordinates (−8, 9).

b Use your answer to find the coordinates of the midpoint of KL.

> **Tip**
>
> The midpoint of the line KL is halfway along the line from K to L.

6 These vectors describe how to move between points A, B, C and D.

$\overrightarrow{AB} = \begin{pmatrix} 2 \\ 1 \end{pmatrix}$ $\overrightarrow{BC} = \begin{pmatrix} 1 \\ 0 \end{pmatrix}$ $\overrightarrow{DA} = \begin{pmatrix} -1 \\ 2 \end{pmatrix}$

Draw a diagram showing how the points are positioned to form the quadrilateral $ABCD$.

7 The vector $\begin{pmatrix} 12 \\ -8 \end{pmatrix}$ describes the displacement from point A to point B.

a What is the vector from point B to point A?

b Point A has coordinates (3, 5). What are the coordinates of point B?

Section 2: Vector arithmetic

Addition and subtraction

The diagram shows $\overrightarrow{AB} = \begin{pmatrix} 2 \\ 4 \end{pmatrix}$, $\overrightarrow{BC} = \begin{pmatrix} 4 \\ -2 \end{pmatrix}$, and $\overrightarrow{AC} = \begin{pmatrix} 6 \\ 2 \end{pmatrix}$

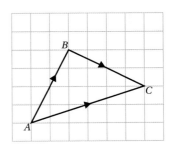

Moving from A to B and then from B to C is the same as moving directly from A to C. In other words, you can take a 'shortcut' from A to C by adding together $\overrightarrow{AB}$ and $\overrightarrow{BC}$.

$\overrightarrow{AC}$ is known as the **resultant** of $\overrightarrow{AB}$ and $\overrightarrow{BC}$.

$$\overrightarrow{AB} + \overrightarrow{BC} = \overrightarrow{AC}$$

$$\begin{pmatrix} 2 \\ 4 \end{pmatrix} + \begin{pmatrix} 4 \\ -2 \end{pmatrix} = \begin{pmatrix} 6 \\ 2 \end{pmatrix}$$

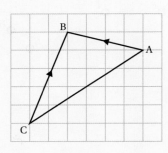

The diagram shows $\overrightarrow{AB}$ and $\overrightarrow{CB}$.

To find $\overrightarrow{AC}$ you need to travel along $\overrightarrow{CB}$ in the opposite direction.

So, **subtract** $\overrightarrow{CB}$.

$$\overrightarrow{AB} - \overrightarrow{CB} = \overrightarrow{AC}$$

$$\begin{pmatrix} -4 \\ 1 \end{pmatrix} - \begin{pmatrix} 2 \\ 5 \end{pmatrix} = \begin{pmatrix} -4-2 \\ 1-5 \end{pmatrix} = \begin{pmatrix} -6 \\ -4 \end{pmatrix}$$

Multiplying by a scalar

Key vocabulary

scalar: a numerical quantity (it has no direction).

Multiplying a vector by a **scalar** results in repeated addition.

This is the same as multiplying the x-component by the scalar, k, and the y-component by the same scalar, k.

$$\overrightarrow{AB} = \begin{pmatrix} 4 \\ -1 \end{pmatrix} \text{ and } \overrightarrow{CD} = \begin{pmatrix} 12 \\ -3 \end{pmatrix}$$

$CD = 3\overrightarrow{AB}$

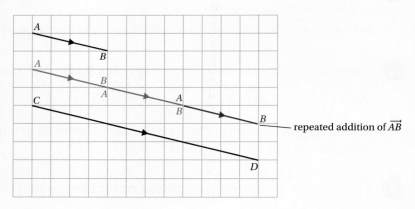

repeated addition of $\overrightarrow{AB}$

$$3 \times \begin{pmatrix} 4 \\ -1 \end{pmatrix} = \begin{pmatrix} 4 \\ -1 \end{pmatrix} + \begin{pmatrix} 4 \\ -1 \end{pmatrix} + \begin{pmatrix} 4 \\ -1 \end{pmatrix} \text{ —— repeated addition}$$

$$= \begin{pmatrix} 3 \times 4 \\ 3 \times -1 \end{pmatrix} \begin{array}{l} \text{—— multiplying the }x\text{-component by the scalar }k \\ \text{—— multiplying the }y\text{-component by the scalar }k \end{array}$$

$$= \begin{pmatrix} 12 \\ -3 \end{pmatrix}$$

Key vocabulary

parallel vectors: occur when one vector is a multiple of the other.

Multiplying a vector by a scalar, k, results in a **parallel vector** with a magnitude multiplied by k.

Vectors are parallel if one is a multiple of the other.

22 Vectors

WORK IT OUT 22.1

Which of the following vectors are parallel?

$\mathbf{a} = \begin{pmatrix} 3 \\ -1 \end{pmatrix}$ $\mathbf{b} = \begin{pmatrix} 4 \\ -3 \end{pmatrix}$ $\mathbf{c} = \begin{pmatrix} 9 \\ -3 \end{pmatrix}$ $\mathbf{d} = \begin{pmatrix} 6 \\ 2 \end{pmatrix}$ $\mathbf{e} = \begin{pmatrix} -6 \\ 2 \end{pmatrix}$

Option A	Option B	Option C
Vectors **a** and **c**	Vectors **d** and **e**	Vectors **a**, **c** and **e**

EXERCISE 22B

1 $\mathbf{p} = \begin{pmatrix} -3 \\ 2 \end{pmatrix}$ $\mathbf{q} = \begin{pmatrix} 5 \\ -1 \end{pmatrix}$ $\mathbf{r} = \begin{pmatrix} -3 \\ -2 \end{pmatrix}$ $\mathbf{s} = \begin{pmatrix} 4 \\ -7 \end{pmatrix}$

Write each of these as a single vector.

a $\mathbf{p} + \mathbf{q}$
b $\mathbf{s} - \mathbf{r}$
c $4\mathbf{p}$
d $-3\mathbf{s}$
e $\mathbf{p} + \mathbf{q} + \mathbf{r}$
f $2\mathbf{p} + \mathbf{q} - 2\mathbf{s}$
g Which of the results from parts **a** to **f** are parallel to the vector $\begin{pmatrix} 3 \\ -2 \end{pmatrix}$?

2 Give three vectors parallel to $\begin{pmatrix} 2 \\ -3 \end{pmatrix}$.

3 Find x, y, z and t in each of the following vector calculations.

a $\begin{pmatrix} x \\ 3 \end{pmatrix} + \begin{pmatrix} 5 \\ y \end{pmatrix} = \begin{pmatrix} 9 \\ 3 \end{pmatrix}$
b $\begin{pmatrix} 10 \\ y \end{pmatrix} - \begin{pmatrix} x \\ -3 \end{pmatrix} = \begin{pmatrix} -2 \\ 8 \end{pmatrix}$

c $\begin{pmatrix} x \\ -3 \end{pmatrix} + \begin{pmatrix} -6 \\ y \end{pmatrix} = \begin{pmatrix} 11 \\ -8 \end{pmatrix}$
d $z\begin{pmatrix} x \\ 12 \end{pmatrix} = \begin{pmatrix} 7 \\ -24 \end{pmatrix}$

e $z\begin{pmatrix} 12 \\ y \end{pmatrix} = \begin{pmatrix} 3 \\ 8 \end{pmatrix}$
f $\begin{pmatrix} 2 \\ -4 \end{pmatrix} + z\begin{pmatrix} 5 \\ y \end{pmatrix} = \begin{pmatrix} 17 \\ 14 \end{pmatrix}$

g $\begin{pmatrix} x \\ -4 \end{pmatrix} - z\begin{pmatrix} -5 \\ -3 \end{pmatrix} = \begin{pmatrix} 20 \\ 5 \end{pmatrix}$
h $z\begin{pmatrix} 3 \\ 4 \end{pmatrix} + t\begin{pmatrix} 2 \\ -2 \end{pmatrix} = \begin{pmatrix} 18 \\ 10 \end{pmatrix}$

4 In the diagram, $\overrightarrow{AB} = \begin{pmatrix} 20 \\ 16 \end{pmatrix}$. C lies on the line AB.

The ratio of $AC : CB$ is $1 : 3$.

a Find $\overrightarrow{AC}$.
b Find $\overrightarrow{BC}$.

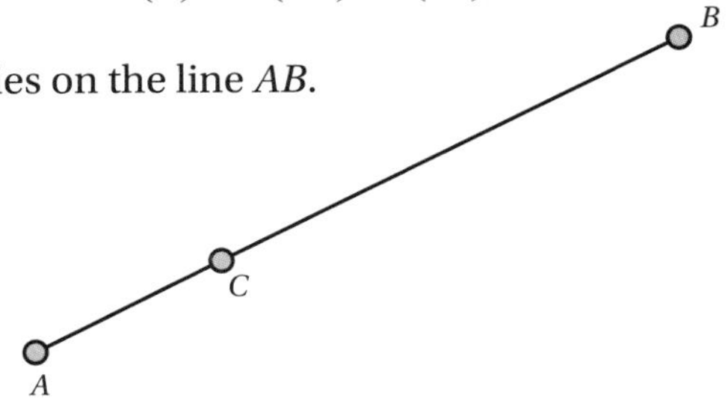

> **Tip**
>
> You can multiply a vector by a fractional scalar if you need to divide. If you need a reminder on fractions see Chapter 6: if you need a reminder of how to calculate ratios, see Chapter 19.

5 These vectors describe how to move between points E, F, G and H, which are four vertices of a quadrilateral.

$\overrightarrow{EF} = \begin{pmatrix} 3 \\ -1 \end{pmatrix}$ $\overrightarrow{HG} = \begin{pmatrix} 6 \\ -2 \end{pmatrix}$ $\overrightarrow{EH} = \begin{pmatrix} 0 \\ 1 \end{pmatrix}$

a What can you say about sides EF and HG?
b Predict what kind of quadrilateral $EFGH$ is.
c Draw the quadrilateral and find $\overrightarrow{GF}$.

Find answers at: cambridge.org/ukschools/gcsemaths-studentbookanswers

6 *ABCD* is a quadrilateral. $\overrightarrow{AB} = \overrightarrow{DC}$ and $\overrightarrow{DA} = \overrightarrow{CB}$.

What kind of quadrilateral is *ABCD*? How do you know this?

Section 3: Mixed practice

It is important that you understand what calculations are needed when given a problem involving vectors.

Test your knowledge using the exercise.

EXERCISE 22C

1 In the diagram, $\overrightarrow{AC} = \begin{pmatrix} 10 \\ 2 \end{pmatrix}$ and $\overrightarrow{AB} = \begin{pmatrix} 8 \\ 14 \end{pmatrix}$.

M is the midpoint of BC.

Find:

a $\overrightarrow{CA}$

b $\overrightarrow{CA} + \overrightarrow{AB}$

c $\overrightarrow{CM}$.

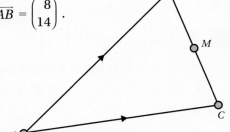

> **Tip**
>
> Remember that the line *CM* is half the length of the line *CB*. What does this mean about the journey from *C* to *M*?

2 Two triangles have vertices *ABC* and *DEF*.

The coordinates of the vertices are $A(0, 0)$, $B(3, 2)$, $C(2, 5)$ and $D(1, 1)$, $E(7, 5)$, $F(5, 11)$.

Compare the vectors:

a $\overrightarrow{AB}$ and $\overrightarrow{DE}$

b $\overrightarrow{AC}$ and $\overrightarrow{DF}$.

c What does this tell you about the triangles *ABC* and *DEF*?

3 In a game of chess different pieces move in different ways.

- A king can move one square in any direction (including diagonals).
- A knight moves two squares horizontally and one square vertically or two squares vertically and one horizontally.

A chessboard is eight squares wide and eight squares long.

What vectors can the following pieces move?

a King

b Knight

4 A ship travels 8 km east and 10 km north. What vector has it travelled?

5 The vector from *E* to *F* is $\begin{pmatrix} -6 \\ 2 \end{pmatrix}$ and the vector from *F* to *G* is $\begin{pmatrix} 5 \\ 1 \end{pmatrix}$.

What is the vector from:

a *E* to *G*

b *G* to *F*

c *E* to the midpoint of *EF*

d *G* to the midpoint of *EF*?

6 The vector from A to B is $\begin{pmatrix} 1 \\ -2 \end{pmatrix}$.

The vector joining C to D is parallel to AB, D is five times the distance from C as B is from A.

What is the vector from C to D?

Checklist of learning and understanding

Notation
- Vectors can be written in a variety of ways: $\overrightarrow{AB}$, **a**, $\begin{pmatrix} 1 \\ 2 \end{pmatrix}$.

Addition and subtraction
- To add or subtract vectors simply add or subtract the x- and y-components.

$$\begin{pmatrix} 3 \\ 2 \end{pmatrix} + \begin{pmatrix} 2 \\ -4 \end{pmatrix} = \begin{pmatrix} 5 \\ -2 \end{pmatrix} \qquad \begin{pmatrix} -1 \\ 4 \end{pmatrix} - \begin{pmatrix} 5 \\ -6 \end{pmatrix} = \begin{pmatrix} -6 \\ 10 \end{pmatrix}$$

Multiplication by a scalar
- To multiply by a scalar you can use repeated addition, or multiply the x-component by the scalar and the y-component by the scalar.

$$3\begin{pmatrix} -2 \\ 1 \end{pmatrix} = \begin{pmatrix} -2 \\ 1 \end{pmatrix} + \begin{pmatrix} -2 \\ 1 \end{pmatrix} + \begin{pmatrix} -2 \\ 1 \end{pmatrix} = \begin{pmatrix} -6 \\ 3 \end{pmatrix}$$

$$3\begin{pmatrix} -2 \\ 1 \end{pmatrix} = \begin{pmatrix} -6 \\ 3 \end{pmatrix}$$

- Multiplying a vector by a scalar quantity produces a parallel vector; you can identify that vectors are parallel if one vector is a multiple of the other. Parallel vectors can be part of the same line and described using a ratio.

Chapter review

For additional questions on the topics in this chapter, visit GCSE Mathematics Online.

1 What is the difference between coordinate (−2, 3) and vector $\begin{pmatrix} -2 \\ 3 \end{pmatrix}$?

2 Match the parallel vectors.

$\mathbf{a} = \begin{pmatrix} -6 \\ 2 \end{pmatrix} \qquad \mathbf{b} = \begin{pmatrix} 1 \\ 3 \end{pmatrix} \qquad \mathbf{c} = \begin{pmatrix} 3 \\ -1 \end{pmatrix} \qquad \mathbf{d} = \begin{pmatrix} 7 \\ 21 \end{pmatrix}$

$\mathbf{e} = \begin{pmatrix} -2 \\ 4 \end{pmatrix} \qquad \mathbf{f} = \begin{pmatrix} -6 \\ 12 \end{pmatrix} \qquad \mathbf{g} = \begin{pmatrix} -1 \\ 2 \end{pmatrix}$

3 Calculate.

a $\begin{pmatrix} 1 \\ -2 \end{pmatrix} + \begin{pmatrix} -2 \\ -1 \end{pmatrix}$ **b** $\begin{pmatrix} 0 \\ -3 \end{pmatrix} - \begin{pmatrix} -2 \\ 4 \end{pmatrix}$ **c** $-3\begin{pmatrix} 2 \\ -1 \end{pmatrix}$

Find answers at: cambridge.org/ukschools/gcsemaths-studentbookanswers

4 B lies on the line AC. The ratio of $\overrightarrow{AB} : \overrightarrow{BC}$ is 2 : 3.

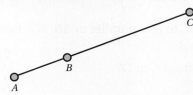

If $\overrightarrow{AB} = \begin{pmatrix} 6 \\ 4 \end{pmatrix}$, what is the column vector of

a $\overrightarrow{BC}$ **b** $\overrightarrow{AC}$?

5 In the diagram, M is the midpoint of AB.

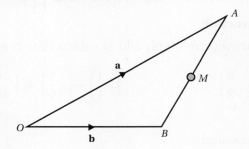

Find:

a $\overrightarrow{AB}$
b $\overrightarrow{AM}$
c $\overrightarrow{MO}$

6 PQRS is a parallelogram.

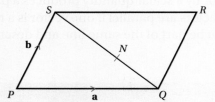

N is the point on SQ such that SN : NQ = 3 : 2

$\overrightarrow{PQ} = \mathbf{a}$

$\overrightarrow{PS} = \mathbf{b}$

a Write down, in terms of **a** and **b**, an expression for $\overrightarrow{SQ}$ *(1 mark)*

b Express $\overrightarrow{NR}$ in terms of **a** and **b**. *(3 marks)*

©Pearson Education Ltd 2013

23 Straight-line graphs

In this chapter you will learn how to …

- use a table of values to plot graphs of linear functions.
- find the equation of a line parallel to a given line.
- identify the main features of straight-line graphs.
- sketch graphs from linear equations in the form $y = mx + c$.
- find the equation of a straight line using the gradient and points on the line.

For more resources relating to this chapter, visit GCSE Mathematics Online.

Using mathematics: real-life applications

This is a photograph of a building nicknamed "The Gherkin", in London. The curves and lines of the building were designed using complex equations and their graphs. Architecture is just one of many professions in which people plot and use graphs in their work.

"When designing a new building, I use graphs to help identify and describe the structural properties the building needs to have."

(Architect)

Before you start …

Ch 15	You should remember how to generate terms in a sequence using a rule.	**1** Use the rule $T(n) = 3n - 2$ to complete this table.
		Term number | 1 | 3 | 5 | 10 **Term**
KS3	You should be able to give the coordinates of points on a grid.	**2** Use the grid. **a** Write down the coordinates of points A, D and E. **b** What point has the following coordinates? **i** $(-2, 2)$ **ii** $(0, -6)$ **c** What is the name given to the point $(0, 0)$? 
Ch 14	You must be able to manipulate and solve equations.	**3** Solve for x. **a** $4 - 3x = 13$ **b** $\dfrac{x}{7} = 6$ **c** $-3(5x + 2) = 0$ **4** If $y = 2x + 5$: **a** find y when $x = -2$ **b** find x when $y = 8$
Ch 16	You should remember how to change the subject of a formula.	**5** Make y the subject of each equation. **a** $-2x - y + 1 = 0$ **b** $2x + 3y = 6$ **c** $x - 2y = -2$

Find answers at: cambridge.org/ukschools/gcsemaths-studentbookanswers

Assess your starting point using the Launchpad

STEP 1

1 Complete the table of values for each function.

a $x - y = 2$

x	−2	−1	0	1
y				

b $x + y = 4$

x	−2	0	1	2
y				

c $2x + y + 2 = 0$

x	−3	−2	0	1
y				

d $x - 2y + 2 = 0$

x	−2	0	2	4
y				

2 The graphs of two of the functions from Question **1** are shown here.

Match each graph to its equation.

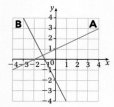

GO TO
Section 1: Plotting graphs

STEP 2

3 a Sketch the graph of $y = 2x + 4$ without plotting a table of values.
b Find the gradient and y-intercept of the resulting straight line.

4 Find the equation of the straight line that passes through the points (1, 4) and (3, 7).

5 A line cuts the x-axis at 4 and the y-axis at 5. What is its gradient?

GO TO
Section 2: Gradient and intercepts of straight-line graphs

STEP 3

6 Which of these lines are parallel to each other?
a $y = -3x + 3$
b $y = 7 - 3x$
c $y = 3x + 7$
d $y = \frac{1}{3}x + 3$
e $y = 7 - 2x$

7 A line is parallel to the line $y = \frac{1}{2}x$ and passes through the point (2, 4). What is its equation?

GO TO
Section 3: Parallel lines
Section 4: Working with straight-line graphs

GO TO Chapter review

Section 1: Plotting graphs

You can draw straight-line and curved graphs to show relationships.

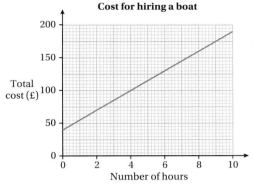

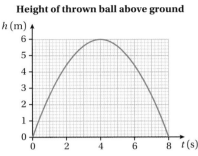

Key vocabulary

function: a set of instructions for changing one number (the input) into another number (the output).

coordinates: an ordered pair (x, y) identifying a position on a grid.

A graph is a picture of a **function**. The graph shows what happens when you apply a rule to x to get a value of y.

Every value of x on the graph has a corresponding value of y. Each pair of values form the **coordinates** (x, y) of a point on a line or a curve.

Functions that produce straight lines when you plot matching x- and y-values are called linear functions.

Tip

When you worked with functions and sequences in Chapter 16 you used a rule to find the terms in a pattern or sequence. You will apply these skills again in this section.

Plotting linear functions

You can plot the graph of a function by drawing up a table of values. Choose three or four values of x.

Apply the function to each value of x to find the corresponding value of y. Then plot the (x, y) coordinates on a set of axes to draw the graph.

Tip

It is usually simpler to use small values of x such as -1, 0, 1 and 2, but make sure the points are spaced out and not three points close together to achieve an accurate graph.

WORKED EXAMPLE 1

Draw up a table of values and plot the graph of $y = 2x + 1$.

x	-1	0	1	2
y	-1	1	3	5

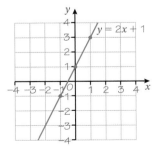

Choose some values for x.

Substitute each x-value into the equation to find the matching y-value.

Plot at least three points using the coordinates.

Points $(-1, -1)$, $(0, 1)$ and $(1, 3)$ have been plotted here.

Draw a straight line through the points.

Label the graph with the equation.

Tip

When you draw a graph, continue the line in both directions through the points. Don't just join the three plotted points together; they are just three of the infinite number of points on the line.

Tip

If the points you have plotted do not line up, then check your calculations for the y-values and/or check that you have plotted the points correctly.

EXERCISE 23A

1 Complete a table of values for each function. Plot the graphs.

   a $y = x$ **b** $y = x + 2$ **c** $y = 3x - 5$

   d $y = 6 - x$ **e** $y = 2x + 1$ **f** $y = x - 1$

   g $y = -2x + 3$ **h** $y = 4 - x$ **i** $y = 3x - 2$

2 What is the minimum number of points you need to plot a straight line accurately? Why?

Section 2: Gradient and intercepts of straight-line graphs

You can use the characteristics of straight-line graphs and their equations to sketch graphs without drawing up a table of values.

The main characteristics of a straight-line graph are:

- the **gradient**, or slope of the graph
- the *x*-intercept (where it crosses the *x*-axis)
- the *y*-intercept (where it crosses the *y*-axis)

Key vocabulary

gradient: a measure of the steepness of a line.

gradient = $\dfrac{\text{change in } y}{\text{change in } x}$

***x*-intercept**: the point where a line crosses the *x*-axis when $y = 0$.

***y*-intercept**: the point where a line crosses the *y*-axis when $x = 0$.

Gradient

On a graph, the gradient is the vertical distance travelled (difference between the *y*-coordinates) divided by the horizontal distance (difference between the *x*-coordinates):

Gradient of a line = $\dfrac{\text{difference in } y\text{-values}}{\text{difference in } x\text{-values}}$

Or, more simply, gradient = $\dfrac{\text{vertical rise}}{\text{horizontal run}}$

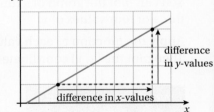

On the graph of $y = 2x + 4$, a right-angled triangle has been drawn on the graph so you can measure the gradient.

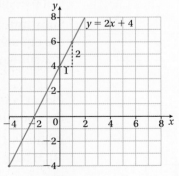

The graph of the linear equation $y = 2x + 4$ shown above slopes up to the right. This is described as a **positive** gradient.

Gradient = $\dfrac{\text{vertical rise}}{\text{horizontal run}} = \dfrac{2}{1} = 2$

23 Straight-line graphs

WORKED EXAMPLE 2

Calculate the gradient of each line.

a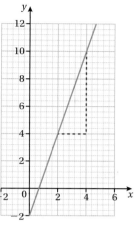
Movement up is positive

b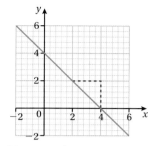
Movement down is negative

a Gradient = $\dfrac{\text{change in } y}{\text{change in } x} = \dfrac{6}{2} = 3$

b Gradient = $\dfrac{\text{change in } y}{\text{change in } x} = \dfrac{-2}{2} = -1$

> This line slopes from top left to bottom right.
> This is described as a **negative** gradient.

You don't need to draw the graph to find the gradient of a line.

You can calculate the gradient if you know the coordinates of any two points on the line.

WORKED EXAMPLE 3

Calculate the gradient of the line that passes through the points (1, 4) and (3, 8).

$\dfrac{\text{difference in } y\text{-values}}{\text{difference in } x\text{-values}} = \dfrac{(y_2 - y_1)}{(x_2 - x_1)}$

$= \dfrac{(8 - 4)}{(3 - 1)}$

$= \dfrac{4}{2}$

$= 2$

> Let the coordinates be (x_1, y_1) and (x_2, y_2) and substitute the values to find the gradient.

The gradient of the line that passes through the points (1, 4) and (3, 8) is 2.

Find answers at: cambridge.org/ukschools/gcsemaths-studentbookanswers

EXERCISE 23B

1 Calculate the gradient of each line. Leave your answer as a fraction in lowest terms if necessary.

a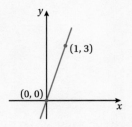
b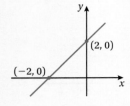
c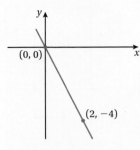

d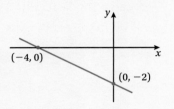
e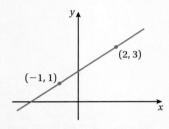
f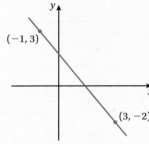

2 Find the gradient of the line that passes through points A and B in each case.

a $A(1, 2)$ and $B(3, 8)$
b $A(0, 6)$ and $B(3, 9)$
c $A(-1, -4)$ and $B(-3, 2)$
d $A(3, 5)$ and $B(7, 12)$

The x-intercept and y-intercept

All points on the x-axis have a y-value of 0.
All points on the y-axis have an x-value of 0.

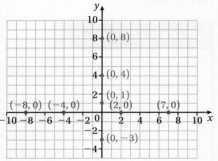

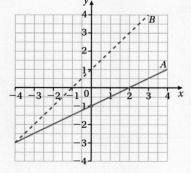

Consider these two lines.

Line A has an x-intercept at $(2, 0)$ and a y-intercept at $(0, -1)$.

Line B has an x-intercept at $(-1, 0)$ and a y-intercept at $(0, 1)$.

You will use the intercepts to sketch graphs and to find the equation of a graph.

23 Straight-line graphs

Using the gradient and y-intercept to sketch graphs

This is the graph of $y = -x + 2$.

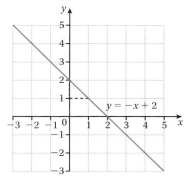

The graph slopes down to the right so it has a **negative** gradient. For every one unit it moves right it goes one unit down. The gradient of the line is $\frac{-1}{1} = -1$.

The line crosses the y-axis at the point $(0, 2)$. This is the y-intercept.

In the equation $y = -x + 2$, the **coefficient** of x is -1 and the **constant** is 2.

The general form of a linear equation is $y = mx + c$.

In this form:
- The value of the coefficient m is the gradient of the graph.
- The value of the constant c is the y-intercept.

Tip

You learned about coefficients and constants in Chapter 13.

Tip

$y = mx + c$, it is known as the **gradient-intercept** form of a straight line.

WORK IT OUT 23.1

Without plotting a graph, find the gradient and y-intercept of the linear function $y = -2x + 4$.

In which direction does the line move across the page?

Which of the answers below is correct?

Option A	Option B	Option C
$y = -2x + 4$	$y = -2x + 4$	$y = -2x + 4$
2 is the coefficient of x.	-2 is the coefficient of x.	-2 is the coefficient of x.
4 is the constant.	4 is the constant.	4 is the constant.
So, the gradient is 2 and the y-intercept is 4.	So, the gradient is -2 and the y-intercept is 4.	So, the gradient is -2 and the y-intercept is 4.
The graph goes up to the right.	The graph goes down to the right.	The graph goes down to the left.

You can use the values of m and c to sketch a graph.

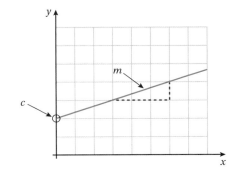

Tip

When the equation is in the form $y = mx + c$ you can identify the gradient and y-intercept without plotting the graph. If a linear equation is not written in this form, you can rearrange it so that it is.

Find answers at: cambridge.org/ukschools/gcsemaths-studentbookanswers

Using the *x*-intercept and *y*-intercept to sketch a graph

If you know where the graph cuts the axes you can sketch the graph of the line.

You can find the *x*-intercept by substituting $y = 0$ into the equation and you can find the *y*-intercept by substituting $x = 0$ into the equation.

WORKED EXAMPLE 4

Find the *x*- and *y*-intercepts and use them to sketch the graph of $y + 2x = 6$.

When $x = 0$
$y + 2(0) = 6$
The *y*-intercept is $(0, 6)$.

When $y = 0$
$0 + 2x = 6$
$2x = 6$
$x = 3$
The *x*-intercept is $(3, 0)$.

First find the intercepts.

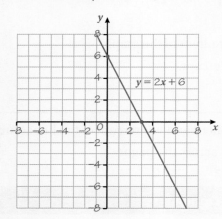

Plot the two points and join them to draw the graph. Label the line.

This method of sketching graphs is called the intercept-intercept method. It is useful when the equation is in the general form $ax + by = c$. You don't need to rearrange the equation to find the intercepts.

EXERCISE 23C

1 Plot the graph of each function. Write a description of each one, giving the *y*-intercept and gradient, stating whether this is positive or negative.

  a $y = 3x - 2$ **b** $y = -2x + 3$ **c** $y = \frac{1}{2}x - 1$ **d** $y = x - 1$

2 Rearrange each equation so it is in the gradient-intercept form ($y = mx + c$). Sketch each of the lines.

  a $3x - 2y = 6$ **b** $6x + 2y + 10 = 0$ **c** $3y - 6x + 12 = 0$
  d $2y - x + 18 = 0$ **e** $6y - 2x + 18 = 0$ **f** $2x - 3y + 12 = 0$

3 Match each graph to the correct linear equation.

a $y = x + 1$
b $y = 3 - x$
c $y = 9 - 3x$
d $y = x + 4$
e $y = -2x + 20$

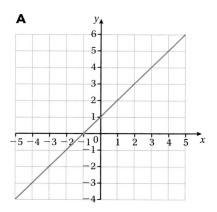
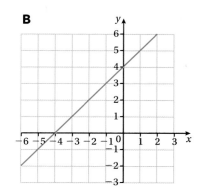
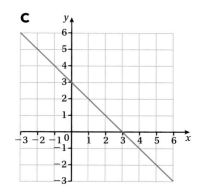
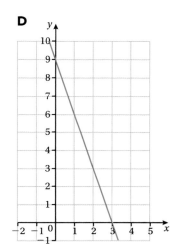
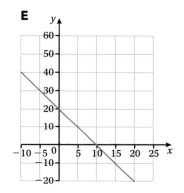

4 Sketch the graph of each line by calculating the coordinates of the *x*- and *y*-intercepts. Write down the gradient of each graph.

a $2x + y = 4$
b $3x + 4y = 12$
c $x + 2y = 1$
d $3x + y = 2$
e $x - y = 4$
f $x - y = 1$
g $4x - 2y = 8$
h $3x - 4y = 12$

Finding the equation of a line using two points on the line

If you have the coordinates of two points on a line you can use them to find the gradient.

Once you have the gradient, you can find the *y*-intercept by substituting the (*x*, *y*) values of one of the points on the line into the equation and solving it to find *c*.

You can then write the equation of the line in the gradient-intercept form: $y = mx + c$

WORKED EXAMPLE 5

Find the equation of the line passing through points (3, 11) and (6, 7).

Gradient = $\dfrac{\text{change in } y}{\text{change in } x} = \dfrac{y_2 - y_1}{x_2 - x_1} = \dfrac{7 - 11}{6 - 3} = -\dfrac{4}{3}$

Find the gradient first.
Let the points be (x_1, y_1) and (x_2, y_2).

$y = -\dfrac{4}{3}x + c$

So, $7 = -\dfrac{4}{3} \times 6 + c$

Use either of the points on the line to calculate c. Here we substitute the point (6, 7).

$7 = -8 + c$
$15 = c$

The equation is $y = -\dfrac{4}{3}x + 15$

This equation could be written $3y = -4x + 45$

WORKED EXAMPLE 6

Find the equation of a line that has the same gradient as the line $y = \dfrac{1}{2}x - 3$ and passes through the point (−1, 2).

$y = mx + c$
$y = \dfrac{1}{2}x + c$

$2 = (\dfrac{1}{2} \times -1) + c$

Substitute the values you know into the equation to find the value of c.

$c = \dfrac{5}{2}$

The equation is $y = \dfrac{1}{2}x + \dfrac{5}{2}$

This equation could be written $2y = x + 5$

EXERCISE 23D

1. For each equation, find c if the given point is on the line.

   a $y = 3x + c$ (1, 5)
   b $y = 6x + c$ (1, 2)
   c $y = -2x + c$ (−3, −3)
   d $y = \dfrac{3}{4}x + c$ (4, −5)

2. Find the equation of the line passing through each pair of points.

   a (0, 0) and (6, −2)
   b (0, 0) and (−2, −3)
   c (−2, −5) and (−4, −1)
   d (−2, 9) and (3, −1)

3. a A line passes through the point (2, 4) and has gradient 2. Find the y-coordinate of the point on the line when $x = 3$.

   b A line passes through the point (4, 8) and has gradient $\dfrac{1}{2}$. Find the y-coordinate of the point on the line when $x = 8$.

   c A line passes through the point (−1, 6) and has gradient −1. Find the y-coordinate of the point on the line when $x = 4$.

Tip

Find the equation of the line before you try to find the coordinates of points on it.

Section 3: Parallel lines

These three graphs are parallel to each other.

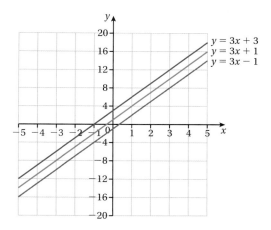

Look at the equations of each line, you can see that they all have the same gradient. Lines with equal gradients are parallel to each other.

These three graphs are members of a family of parallel lines with a gradient of positive 3. Each line in the family can be defined by the equation $y = 3x + c$.

Tip

You learned about parallel lines in Chapter 2.

WORK IT OUT 23.2

In which option have parallel lines been correctly grouped together?

Option A	Option B	Option C
$y = 4 - 2x$	$y = \frac{1}{3}x + 1$	$y = x - 1$
$y + 2x = 5$	$3y + x = 1$	$y + x = 1$
$y = -2x + 1$	$y = 3x + 1$	$y = 1 - x$
$y = 3x + 1$	$2y = x + 1$	$x = y + 1$
$y - 3x = -1$	$2y - x = 3$	$x - y = 1$
$y = 2 + 3x$	$y = \frac{1}{2}x - 1$	$x = 1 - y$

In mathematics one way to prove a statement is true is to follow a series of statements that end in a valid conclusion. You can use this method to prove that lines are parallel.

Steps for approaching a proof	What would you do to prove lines are parallel?
Step 1: Draw a diagram?	A diagram might take too long, especially if you are working with many different lines.
Step 2: Identify what you have to do.	Compare the gradients to see if they are the same or not.
Step 3: Test the problem with what you know.	If lines are parallel they will have an identical value for the gradient. The gradient is the coefficient of x in the equation (the value of m).
Step 4: What maths can you do?	Rearrange the equations so they are all in the form $y = mx + c$ to compare the gradients (m). If there are identical values for m then you have proved that the lines are parallel lines.

EXERCISE 23E

1. Identify the parallel lines in each set.

   a $y = 3x - 5$ $y = x - 5$ $3x + 7 = y$ $6x - 2y = -1$

   b $y = 2x + 3$ $y = 3x + 2$ $2x - y = -6$ $2y = x + 3$ $y = 2x - 3$

   c $y + 3 = x$ $x + y = 3$ $y = 3x - 1$ $y + x = 8$

2. a If $y = (2a - 3)x + 1$ is parallel to $y = 3x - 4$, find the value of a.

   b If $y = (3a + 2)x - 1$ is parallel to $y = ax - 4$, find the value of a.

3. Find the equation of the blue line.

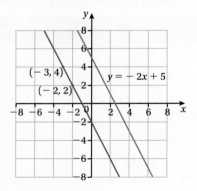

Tip

You learned about the properties of quadrilaterals in Chapter 2.

4. The vertices (corners) of a quadrilateral have coordinates $A(1, 6)$, $B(3, 14)$, $C(15, 16)$ and $D(13, 8)$.

   a Find the gradient of the line AB.

   b Find the equation of the line AB.

   c Prove that $ABCD$ is a parallelogram.

5. Investigate lines that are parallel to the axes. How are the equations for these graphs different to those for slanted graphs? Why?

Section 4: Working with straight-line graphs

You need to be able to interpret straight-line graphs. This means you can:
- determine the equation of the graph
- calculate the gradient of a line using given information
- use graphs to model and solve problems, including solving simultaneous equations.

The point of intersection of any two graphs is the solution to the simultaneous equations (of the graphs).

WORKED EXAMPLE 7

a Sketch the graphs of $x + 3y = 6$ and $y = 2x - 5$.
b What are the coordinates of the point of intersection of the two graphs?
c Show by substitution that these values are the simultaneous solution to the equations $x + 3y = 6$ and $y = 2x - 5$.

a Sketch the graphs.

$x + 3y = 6$
Let $x = 0$
$3y = 6$
$y = 2$

Let $y = 0$
$x = 6$
Plot $(0, 2)$ and $(6, 0)$

Use the intercept-intercept method for $x + 3y = 6$

$y = 2x - 5$
y-intercept $= -5$
gradient $= 2$

Use the gradient-intercept method for $y = 2x - 5$.

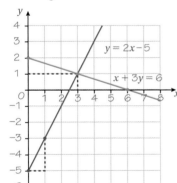

b The point of intersection is $(3, 1)$.

Read this from the graph.

c Let $x = 3$ and $y = 1$
Substitute in $x + 3y = 6$:
$3 + 3(1) = 6$
Substitute in $y = 2x - 5$:
$1 = 2(3) - 5$
$1 = 6 - 5$
$1 = 1$

The solutions work for both equations.

EXERCISE 23F

1 This table of values has been generated from a function.

x	−2	−1	0	1	2	3
y	−4	−3	−2	−1	0	1

a Which of these functions would produce the values in the table?

$y = -x + 2$ $y = 2x - 1$ $y = -2x + 4$ $y = x - 2$

b Plot the graph of this function.
c Draw a line parallel to your graph which crosses the y-axis at $(0, 3)$ and write its equation.

 Find answers at: cambridge.org/ukschools/gcsemaths-studentbookanswers

2. Find and write in gradient-intercept form $y = mx + c$ equations which satisfy the following statements.
   a A linear equation that does not pass through the first quadrant.
   b Two lines whose gradients differ by 2.
   c An equation of a straight line that passes through (2, 3) and has a gradient of 3.

3. Calculate the gradient of the line shown in the diagram and write down the equation of the line.

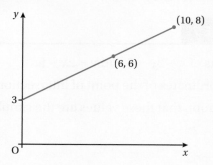

4. Sketch each of the following graphs.
   a $y = 3x + 2$ b $y = -2x - 1$ c $2y = x + 8$
   d $x - y = -3$ e $y + 4 = x$ f $3x + 4y = 12$

Tip

Sketch means draw a basic diagram to represent each equation showing the direction and intercept on the y-axis.

5. Find the equation of the line that passes through each pair of points.
   a (5, 6) and (−4, 10)   b (3, 4) and (−2, 8)   c (−2, 6) and (1, 10)

6. a Find the equation of the line with gradient −4 that passes through the point (0, −6).
   b Find the equation of the line with gradient −4 that passes through the point (3, 8).
   c Find the equation of the line that passes through the points (−4, 8) and (−6, −2).

7. The line passing through the points (−1, 6) and (4, b) has gradient −2. Find the value of b.

8. a Is the gradient of this straight line 2 or −2? Write down the equation of the line.

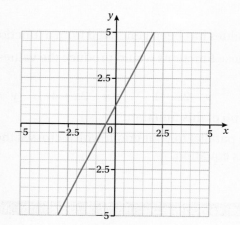

b Write the equation of this line.

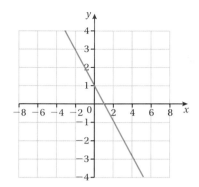

9 Find the equations of the four straight lines that make this square.

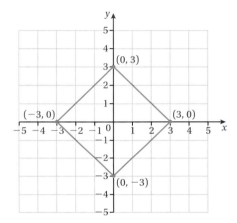

 Checklist of learning and understanding

Plotting straight-line graphs

- You can use the equation of a line to generate a table of x- and y-values.
- Choose any three (x, y) values, plot them and join the points to draw the graph.

Characteristics of straight-line graphs

- The general form of a straight-line graph is represented by $y = mx + c$, where m is the gradient and c is the point where the line cuts the y-axis (y-intercept).
- You can find the equation of a straight line if you have two points on it or one point and the gradient.
- You can sketch graphs using the gradient and y-intercept or using the x- and y-intercepts.

Parallel graphs

- Parallel lines have the same gradient so the values of m are equal when the equations are written in the form $y = mx + c$.

Find answers at: cambridge.org/ukschools/gcsemaths-studentbookanswers

Chapter review

1. On the grid, draw the graph of $y = \frac{1}{2}x + 5$ for values of x from -2 to 4. *(3 marks)*

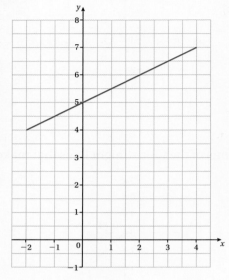

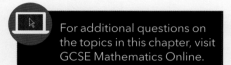

©*Pearson Education Ltd 2013*

2. Draw these lines on the same grid.

   a $y = x + 1$ **b** $y = 2x + 5$ **c** $y + 2 = 4x$

3. Write the equation of each line.

   a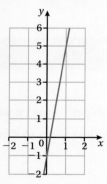

   b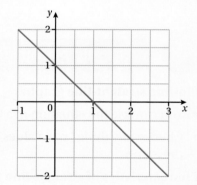

   c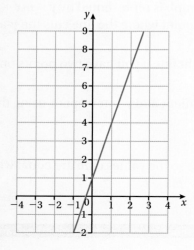

   d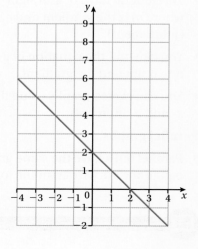

4 Which of the following lines are parallel to each other?

   a $y = \frac{1}{2}x + 1$ **b** $2y - x = 4$ **c** $y = 2x - 5$ **d** $y = 0.5x + 3$

5 Find the equation of a line parallel to $y = \frac{1}{2}x + 1$ passing through the point $(-1, 2)$.

6 Find the equation of the line that passes through the two points $(2, 4)$ and $(6, -12)$.

7 **a** Draw a graph of the two lines $y = 3x - 2$ and $y + 2x = 3$ and find their point of intersection.

   b Show by substitution that this is the simultaneous solution to the two equations.

24 Graphs of functions and equations

In this chapter you will learn how to ...
- plot and sketch graphs of linear and quadratic functions.
- identify the main features of graphs of quadratic functions and equations.
- plot and sketch other polynomials and reciprocal functions.

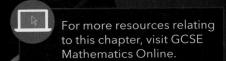

For more resources relating to this chapter, visit GCSE Mathematics Online.

Using mathematics: real-life applications

Graphs are used to process information, make predictions and generalise patterns from sets of data. The nature of the data and the relationship between values determines the shape and form of the graph.

"I study the Earth using gravity, magnetic, electrical, and seismic methods. I used this graph in a study of the Pacific and Atlantic oceans. I need to be able to understand equations and recognise the features of graphs to understand and interpret it." *(Geophysicist)*

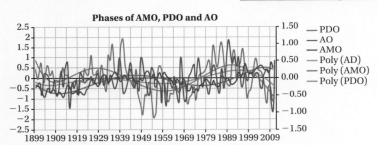

Before you start ...

Ch 23	You should be able to interpret equations of linear graphs.	1	For the graph $y = 3x + 1$: a Identify the gradient of the graph. b Give the coordinates of the y-intercept. c Find the value of x when $y = -14$. d Show that it is parallel to the graph $2y - 6x = -4$.
Ch 23	You must be able to generate a table of values from a function.	2	Given $y = 3x^2 + 1$, complete the table of values. | x | -2 | -1 | 0 | 1 | 2 | |---|---|---|---|---|---| | y | | | | | |
Ch 14	You need to be able to find the roots of a quadratic equation algebraically.	3	What are the roots of: a $x^2 + 2x - 8 = 0$? b $x^2 + 5x = -4$?
Ch 8	You should be able to work with cubed numbers and cube roots.	4	Evaluate. a 2^3 b 4^3 c $\sqrt[3]{125}$ d $\sqrt[3]{-27}$

Assess your starting point using the Launchpad

STEP 1

1 How many points do you need to calculate to plot the graph of a linear equation?

2 Sketch the graph of the linear equation $y = 2x + 1$.

GO TO
Section 1: Review of linear graphs

STEP 2

3 **a** Is this the graph of $y = x^2 + 1$ or $y = -x^2 + 1$?
b How can you tell?

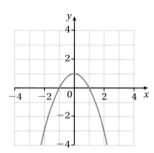

c Is the turning point a maximum or a minimum?
d What are the coordinates of the vertex?
e For what values of x is $y = 0$?

GO TO
Section 2: Quadratic functions

STEP 3

4 **a** What type of equation is $y = x^3$?
b How many points do you need to calculate to plot the graph of $y = x^3$?

5 Given $y = \dfrac{1}{x}$:
a explain what happens when $x = 0$.
b what happens to the value of y as the value of x increases?
c when $x = 60$ what is the value of y?

GO TO
Section 3: Other polynomials and reciprocals

Section 4: Plotting, sketching and recognising graphs

GO TO Chapter review

Find answers at: cambridge.org/ukschools/gcsemaths-studentbookanswers

Section 1: Review of linear graphs

Graphs in the form of $y = mx$

In the general form of a linear function, $y = mx + c$, m is the gradient of the line and the value of c tells you where the line cuts the y-axis. When there is no value of c in the equation you get a graph in the form of $y = mx$.

These graphs pass through the origin (0, 0) because when $y = 0$, x must be 0.

The graphs of $y = mx$ and $y = -mx$ are shown on the left.

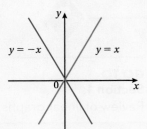

If $m = 1$ the equation is written as $y = x$. If $m = -1$, the equation is written as $y = -x$.

$y = x$ is the line that passes through the origin going up from left to right at an angle of 45°.

$y = -x$ is the line that passes through the origin going down from left to right making an angle of 45°.

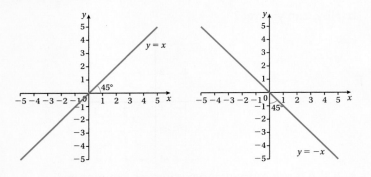

Sketching $y = ax$ ...	Examples	Notes
if a is greater than 1.	$y = 3x$ $y = 7x$	The line still passes through the origin but is steeper than $y = x$.
if a is a value between 0 and 1.	$y = \frac{1}{2}x$ $y = \frac{1}{5}x$	The line still passes through the origin but is less steep than $y = x$.
if a is a negative value.	$y = -3x$ $y = -\frac{1}{2}x$	The line still passes through the origin, but will go down from left to right like $y = -x$.

Vertical and horizontal lines

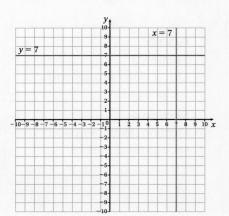

Consider the equation $x = 7$.

The graph of this equation passes through the point (7, 0). It also passes through all points on the grid with an x-coordinate of 7. For example: (7, −4), (7, −1), (7, 3), (7, 50).

Now consider the equation $y = 7$.

This graph would need to pass through the point $(0, 7)$ and all other points with a y-coordinate of 7. For example: $(3, 7)$, $(-2, 7)$, $(7, 7)$, $(50, 7)$.

- Any graph in the form of $x = a$ is parallel to the y-axis and passes through a on the x-axis.
- Any graph in the form of $y = b$ is parallel to the x-axis and passes through b on the y-axis.
- The values of a and b can be positive or negative.

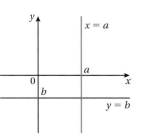

The axes themselves are straight lines and can be described using equations.

The x-axis is the line $y = 0$ and the y-axis is the line $x = 0$.

Tip

You should be able to recognise and sketch the graphs of any line in the form $x = a$ or $y = b$.

EXERCISE 24A

1 Write down which of the graphs on the grid (right) can be described in each of the following ways.

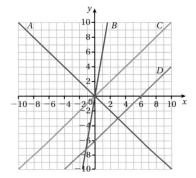

  a The x-coordinate of each point is equal to the y-coordinate.
  b The gradient is negative.
  c The general form of the graph is $y = mx$.
  d The y-coordinate is 6 times the x-coordinate.
  e The y-coordinate is 6 less than the x-coordinate.

2 Give the equation of each line on the grid in question **1**.

3 a Write down the equation of each line A to F shown on the grid (right).

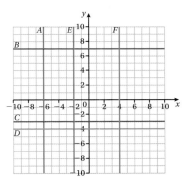

  b Determine the equation of the line parallel to D which passes through point $(0, 2)$.

4 Draw the following graphs on a grid numbered from -6 to 6 on each axis.

  a $x = -3$ b $y = 5$ c $y = -3$ d $x = 5$

5 What type of quadrilateral is enclosed by the four lines you drew in question **4**? Give reasons for your answer.

6 a Write the equations of two lines that would divide the quadrilateral formed in question **4** into two identical rectangles.

  b What is the mathematical name for lines that divide shapes into two identical halves?

c It is possible to draw two other lines that divide the quadrilateral into two equal halves.

   i Draw these lines on your diagram.

   ii Determine the equation of each line.

Section 2: Quadratic functions

A **quadratic** expression has the form $ax^2 + bx + c$, where $a \neq 0$.

$2x^2 + 4x - 1$ and $3x^2 + 7$ are quadratic expressions.

Key vocabulary

quadratic: an expression with a variable to the power of 2 but no higher power.

Tip

$2x - 1$ is a linear expression as it has no x^2 terms.

Tip

In Chapter 14 you learned how to solve quadratic equations by writing them as the product of two factors equal to 0. Revise that chapter if you have forgotten how to do this.

The path of moving objects such as this basketball can be modelled as a curved graph called a parabola.

The graph of a quadratic function is a curve called a **parabola**.

The simplest equation of a parabola is $y = x^2$.

You can plot the graph of a quadratic function by drawing up a table of values.

Key vocabulary

parabola: the symmetrical curve produced by the graph of a quadratic function.

Tip

The shape of a parabola can extend downwards to a minimum, $y = x^2$, or upwards to a maximum, $y = -x^2$. The equation of the graph tells you which of these shapes it will be.

WORKED EXAMPLE 1

Plot the graph of the function $y = x^2 + 2$.

Tip

Remember that the square of a negative number is positive.
For example, when $x = -3$: $(-3)^2 + 2 = 9 + 2 = 11$

x	−3	−2	−1	0	1	2	3
y	11	6	3	2	3	6	11

Draw up a table of x- and y-values that satisfy the equation $y = x^2 + 2$.

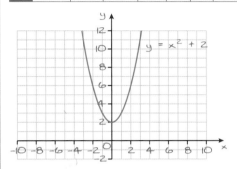

Plot all the points on a grid and join them to produce a smooth curved graph that extends through and beyond the points you have plotted.

Label the graph.

EXERCISE 24B

1 **a** Complete a table of values of x from −3 to 3 for each equation. Draw the graphs on the same grid.

   i $y = x^2$ **ii** $y = -x^2$

   b How are the graphs different?

   c Look at the equation of each graph. How does the sign of the x^2 term affect the shape of the graph?

2 **a** Complete a table of values of x from −3 to 3 for each function. Draw the graphs on the same grid.

   i $y = 3x^2$ **ii** $y = \frac{1}{3}x^2$ **iii** $y = 4x^2 + 1$ **iv** $y = -2x^2 + 3$

   b Look at the equation of each graph. What does the constant value in the equations of graphs **iii** and **iv** tell you about the graphs?

Features of parabolas

Quadratic graphs have characteristics that you can use to sketch and interpret them.

- The axis of symmetry – a line which divides the parabola into two symmetrical halves.
- The y-intercept for graphs of the general form $y = ax^2 + bx + c$ is where the curve cuts the y-axis.
- The turning point or vertex of the graph – this is either the minimum or maximum point of the graph.
- The x-intercepts where the graph cuts the x-axis – a parabola can have 0, 1 or 2 x-intercepts depending on the position of the graph.

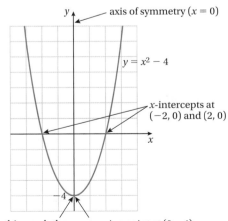

in this graph the turning point is a minimum (lowest point on graph)

turning point at $(0, -4)$ which is also the y-intercept

The graph either has a minimum turning point or a maximum turning point.

If the x^2 term is positive, the turning point will be a minimum (this is the lowest point of the graph).

If the x^2 term is negative, the turning point will be a maximum (this is the highest point of the graph).

WORKED EXAMPLE 2

For each parabola determine:

i the turning point and whether it is a minimum or maximum

ii the axis of symmetry

iii the y-intercept

iv the x-intercepts.

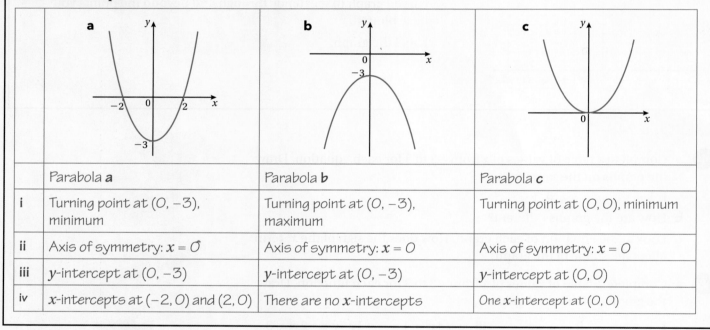

	Parabola **a**	Parabola **b**	Parabola **c**
i	Turning point at $(0, -3)$, minimum	Turning point at $(0, -3)$, maximum	Turning point at $(0, 0)$, minimum
ii	Axis of symmetry: $x = 0$	Axis of symmetry: $x = 0$	Axis of symmetry: $x = 0$
iii	y-intercept at $(0, -3)$	y-intercept at $(0, -3)$	y-intercept at $(0, 0)$
iv	x-intercepts at $(-2, 0)$ and $(2, 0)$	There are no x-intercepts	One x-intercept at $(0, 0)$

x-intercepts and roots of a quadratic equation

The x-intercepts of a parabola are the roots of the quadratic equation that defines the graph.

You can find the roots graphically by reading their value off the graph. These are the values of x where the curve cuts the x-axis.

Tip

Solving a quadratic equation to find its roots is useful when you have to sketch the graph.

EXERCISE 24C

 For each parabola determine:

i the turning point and whether it is a minimum or maximum.

ii the axis of symmetry.

iii the y-intercept.

iv the x-intercepts.

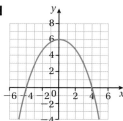
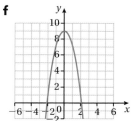

2. Graphs A to E are shown on the same grid. Use the diagram to determine whether each statement is true or false.

   a Graph A has two x-intercepts.
   b Graphs A and B have minimum turning points.
   c Graph B has y-intercepts at (−2, 0) and (2, 0).
   d Graphs B and C have the same x-intercepts.
   e The equation of graph D will have a positive x^2 term.
   f Graph E has a maximum turning point at (0, −2).
   g The equation of graph D has a constant of 2. This tells you the y-intercept is (0, 2).
   h All of these graphs are symmetrical about $x = 0$.

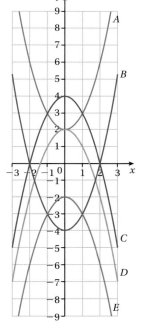

Sketching quadratic graphs

You can use the characteristics of a parabola to sketch graphs without drawing up a table of values. Remember, a sketch shows the general features of a graph but it does not have to be drawn on graph paper.

To sketch a parabola of the general form $y = ax^2 + c$ (where a is the coefficient of x^2 and c is a constant):

- Check the sign of a to determine whether the graph goes up to a maximum turning point or down to a minimum turning point.
- Work out the y-intercept (this is given by c in the equation).
- Calculate the x-intercepts by substituting $y = 0$ and solving for x. If there are no x-intercepts or only one when the graph only touches the x-axis you can find the coordinates of one point on the graph to help you draw an accurate sketch.
- Mark the y-intercept and x-intercepts (if they exist) and use the shape of the graph as a guide to draw a smooth curve.
- Label your graph.

WORKED EXAMPLE 3

Sketch the graph of $y = 3x^2$

- 3 is positive so the graph goes down to a minimum turning point.
- There is no constant, so the graph goes through the origin (0, 0).
- There is only one x-intercept (0, 0), so find one point on the curve.
  When $x = 1$
  $y = 3(1)^2 = 3$
  So, (1, 3) is a point on the curve.
- Sketch and label the graph.

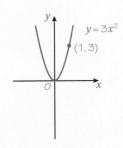

WORKED EXAMPLE 4

Sketch the graph of $y = -x^2 + 4$

- Coefficient of x is -1, so graph goes up to a maximum turning point.
- Constant is 4, so y-intercept is (0, 4).
- x-intercepts when $y = 0$
  $0 = -x^2 + 4$
  $\therefore x^2 - 4 = 0$
  This is a difference of squares
  $(x + 2)(x - 2) = 0$
  $x + 2 = 0$ or $x - 2 = 0$
  $x = -2$ or $x = 2$
  So, intercepts are $(-2, 0)$ and $(2, 0)$.
- Sketch and label the graph.

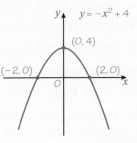

Tip

Draw a smooth curve to join the points and try to make your graph as symmetrical as possible.

EXERCISE 24D

1. Sketch and label these graphs on the same grid.

   a $y = x^2$ **b** $y = 2x^2$ **c** $y = \frac{1}{2}x^2$ **d** $y = -2x^2$ **e** $y = -\frac{1}{2}x^2$

2. How does the value of the coefficient of x^2 affect the wideness of the parabola's shape?

3. Solve each equation algebraically to find the x-intercepts of each graph. Do not draw the graphs.

   a $y = x^2 - 4$ **b** $y = x^2 - 9$

4. Sketch the following graphs on the same grid.

   a $y = x^2 - 3$ **b** $y = 3x^2 - 4$

   c $y = -x^2 - 2$ **d** $y = -3x^2 + 12$

5 Noor sketched these quadratic graphs but she didn't write the equations on them. Use the features of each graph to work out what the correct equations are.

a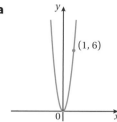
b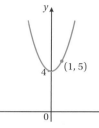
c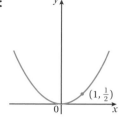
d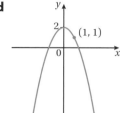
e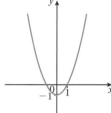
f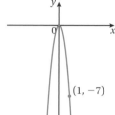

Section 3: Other polynomials and reciprocals

A **polynomial** is an expression with many terms. If the highest power of x is 3, the expression is called a cubic expression. For example, $2x^3$ and $2x^3 + x^2 + 3$ are both cubics.

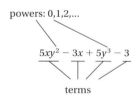

a polynomial

not polynomials

Key vocabulary

polynomial: an expression made up of many terms with positive powers for the variables.

You can use a table of values to plot the graph of a cubic equation. The simplest equation of a cubic graph is $y = x^3$.

The diagram shows the basic shape of cubic graphs in the form of $y = ax^3$.

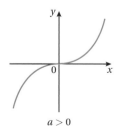
$a > 0$

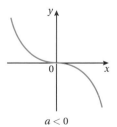
$a < 0$

The shape on the left occurs when a is positive. This is an increasing curve.

The shape on the right occurs when a is negative. This is a decreasing curve.

Tip

Remember that when you cube a negative number you will get a negative result.

For example,
$(-1)^3 = -1 \times -1 \times -1$
$= (-1 \times -1) \times -1$
$= +1 \times -1 = -1$

The larger the value of a, the steeper the curve.

This is a table of values for $y = x^3$ for values of x from -3 to 3.

x	-3	-2	-1	0	1	2	3
y	-27	-8	-1	0	1	8	27

To draw an accurate graph you plot all of the calculated points and draw a smooth curve through and beyond them.

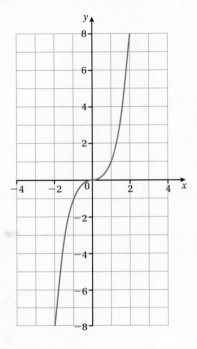

When you sketch a cubic graph you need to show its general shape and its important features. You still use a table of values, but you don't need to work out as many points as you need for an accurate graph.

EXERCISE 24E

1 Complete a table of values for whole number values of x from -3 to 3 for $y = -x^3$. Draw the graph $y = -x^3$. How does this graph differ from the graph of $y = x^3$?

2 Complete a table for whole number values from -3 to 3 to sketch the following pairs of cubic graphs (use separate grids for parts a and b).

  a $y = -2x^3$ and $y = 2x^3$
  b $y = \frac{1}{2}x^3$ and $y = -\frac{1}{2}x^3$

3 Work with a partner to compare the pairs of graphs you drew in question **2**. Discuss how you could sketch the graph of $y = -4x^3$ if you were given the graph of $y = 4x^3$.

4 Complete a table of values for whole number values of x from -3 to 3 for these equations and draw a graph of each curve.

  a $y = x^3 + 1$
  b $y = x^3 - 2$

5 Look at the graphs and their equations in question **4**. What information does the constant give you about the graph?

6 The red line on the diagram (right) is the graph $y = x^3$. What are the equations of graphs *A* and *B*?

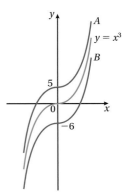

Reciprocal functions

The product of a number and its **reciprocal** is 1.

$8 \times \dfrac{1}{8} = 1$ $\dfrac{2}{5} \times \dfrac{5}{2} = 1$

Every number has a reciprocal except for 0, as $\dfrac{1}{0}$ cannot be defined.

The reciprocal of a is $\dfrac{1}{a}$: $a \times \dfrac{1}{a} = 1$.

The general equation of a reciprocal function is $y = \dfrac{a}{x}$, where a is a constant value. This equation can be re-arranged to give $xy = a$.

> **Tip**
>
> You learned about reciprocals in Chapter 6.

The graphs of reciprocal functions have a characteristic shape. Each graph is made of two curves which are mirror images in diagonally opposite quadrants of the grid.

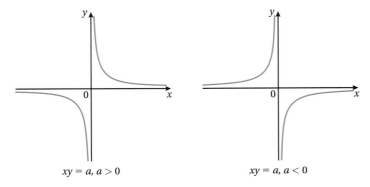

$xy = a, a > 0$ $xy = a, a < 0$

This is the table of values for the equation of $y = \dfrac{1}{x}$.

x	-3	-2	-1	$-\dfrac{1}{2}$	$-\dfrac{1}{3}$	0	$\dfrac{1}{2}$	$\dfrac{1}{3}$	1	2	3
y	$-\dfrac{1}{3}$	$-\dfrac{1}{2}$	-1	-2	-3	undefined	2	3	1	$\dfrac{1}{2}$	$\dfrac{1}{3}$

In order to draw a reciprocal graph accurately you need to work with some non-integer values of x.

Note that there is no y-value when $x = 0$ because division by 0 is undefined.

To draw the graph:

- Plot the (x, y) values from the table.
- Join the points with a smooth curve.
- Write the equation on both parts of the graph.

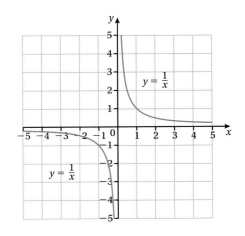

Note that as the value for x gets bigger the value for y gets closer and closer to 0 but the graph never actually meets the y-axis.

EXERCISE 24F

1 Copy and complete each table for the given values of x. Plot the graphs on the same grid.

a $y = \dfrac{2}{x}$

x	−4	−2	−1	1	2	4
y						

b $y = \dfrac{6}{x}$

x	−6	−3	−1	1	3	6
y						

c $xy = -12$

x	−10	−8	−6	−4	−2	2	4	6	8
y									

d $y = \dfrac{8}{x}$

x	−8	−6	−4	−2	1	2	4	6	8
y									

2 Compare the graphs that you have drawn for question **1**. How does the value of the constant in the equation affect the position of the graph?

3 Here are four reciprocal graphs.

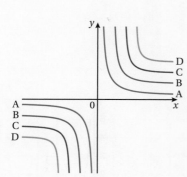

a Without doing any calculation, match each of these equations to a graph.

$y = \dfrac{2}{x}$ $y = \dfrac{4}{x}$ $y = \dfrac{8}{x}$ $y = \dfrac{10}{x}$

b Explain how you worked out which equation belonged with each graph.

4 Plot each of the following graphs on the same grid using x-values from −5 to 5.

a $y = \dfrac{1}{x}$ **b** $y = \dfrac{1}{x} + 1$ **c** $y = \dfrac{1}{x} + 3$

5 Use your graphs from question **4** to describe how the constant c in the equation $y = \dfrac{a}{x} + c$ changes the reciprocal graph of $y = \dfrac{a}{x}$.

6 Neo says that the line $y = x$ is a line of symmetry of the graph $y = \dfrac{1}{x}$. Is he correct? Explain your answer.

Section 4: Plotting, sketching and recognising graphs

You can use the characteristics of different types of graphs to work out what they will look like and sketch them.

- $y = mx + c$ will produce a straight line, or linear graph.
- $y = ax^2 + c$ will produce a quadratic graph called a parabola.
- $y = ax^3$ will produce a cubic curve.
- $y = \dfrac{a}{x}$ and $xy = a$ will produce a reciprocal curve.

EXERCISE 24G

1 Determine whether each statement is true or false. Correct any false statements.

  a The points (3, 7) (0, 1), (−1, −1) all lie on the line $y = 2x + 1$.

  b The graph of the equation $y = \dfrac{5}{x}$ cannot be evaluated for $x = 0$.

  c The diagram below shows a graph of a quadratic equation which has roots 1 and −1.

  d $xy = 5$ is the same function as $y = \dfrac{5}{x}$.

  e $y = -2x^2 + 4$ goes down to a minimum turning point with a y-intercept at (0, −2)

  f The line $y = -x$ is a line of symmetry of $xy = 4$.

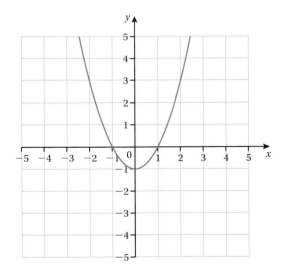

2 Sketch each of the following graphs.

  a $y = x + 2$ **b** $y = 3x - 4$ **c** $y = 7 - 3x$
  d $y = 3$ **e** $x = -5$

3 Draw up a table of values for x from -3 to 3, and plot each graph.

  a $y = 2x^2$ **b** $y = x^2 - 3$ **c** $y = -x^2 + 5$
  d $y = x^3 + 4$ **e** $y = \dfrac{1}{x}$ **f** $y = \dfrac{1}{x} + 1$

4 Solve the following equations algebraically and hence write down the x-intercepts of each graph.

  a $y = x^2 - 16$ **b** $y = x^2 - 2x$

5 If a quadratic equation has roots $x = -3$ and $x = 5$, give the coordinates of the points of intersection with the x-axis.

6 Match each of the following graphs with the appropriate equation from the list.

a 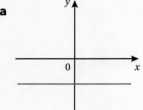 **b** 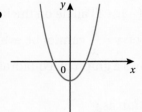 **c**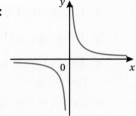

d 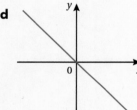 **e** 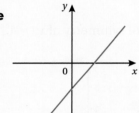 **f**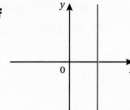

g 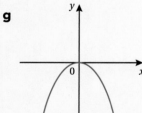 **h** 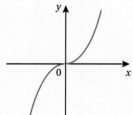 **i**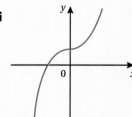

Equations:

$y = x^2 - 2$ $y = 3x - 4$ $xy = 1$

$y = -x$ $y = 2x^3$ $y = 3x^3 + 3$

$x = 4$ $y = -4$ $y = -x^2$

Checklist of learning and understanding

Linear functions
- Linear functions produce straight-line graphs.
- The general form of the linear function is $y = mx + c$.
- Graphs $x = a$ are vertical lines parallel to the y-axis.
- Graphs $y = a$ are horizontal lines parallel to the x-axis.
- Lines of the form $y = mx$ go through the origin.

Quadratic functions
- The graphs of quadratic equations such as $y = ax^2$ and $y = ax^2 + c$ are called parabolas.
- When a is positive the graph goes down to a minimum point. When a is negative the graph goes up to a maximum point. The y-intercept is given by c.
- A parabola has a turning point which can be a minimum or maximum depending on the shape of the graph.

Polynomials and reciprocals
- To draw graphs of polynomials first calculate a table of values that satisfy the equation for a range of values for x.
- $y = ax^3$ is an example of a cubic function.
- A reciprocal function is a graph made of two curves in diagonally opposite quadrants defined by $y = \dfrac{a}{x}$ or $xy = a$.

Chapter review

For additional questions on the topics in this chapter, visit GCSE Mathematics Online.

1 a Complete this table for $y = 2x + 1$ and plot the points to draw a straight-line graph.

x	-2	-1	0	1	2	3
y						

  b Use the graph to find:

  i the value of y when $x = -1.5$.   **ii** the value of x when $y = 6$.

2 a Complete this table for $y = x^2 - 3$ and plot the points to draw the curve of the parabola.

x	-2	-1	0	1	2	3
y			-3	-2		

  b Give the coordinates of the minimum point of the curve (the turning point).

  c What line is the axis of symmetry for this graph?

  d Estimate the roots of the equation $x^2 - 3 = 0$.

3

a Complete the table of values for $y = 2x^2 - 1$ *(2 marks)*

x	−2	−1	0	1	2
y	7			1	

b On the grid below, draw the graph of $y = 2x^2 - 1$ for values of x from $x = -2$ to $x = 2$ *(2 marks)*

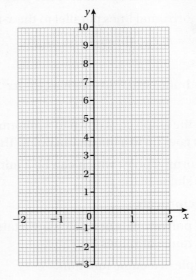

c Use your graph to write down estimates of the solutions of the equation $2x^2 - 1 = 0$ *(2 marks)*

©*Pearson Education Ltd 2012*

4 Match the equations $y = x^3$ and $y = \dfrac{1}{x}$ with the correct graph.

a 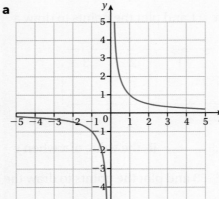   b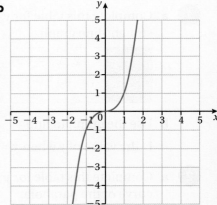

5 Draw a sketch diagram of $y = x^3 - 5$ and $y = \dfrac{1}{x} + 2$.

6 If the roots of the quadratic equation $x^2 + 2x - 3 = 0$ are $x = 1$ and $x = -3$, what are the coordinates of the points at which the equation $y = x^2 + 2x - 3$ cuts the x-axis?

25 Angles

In this chapter you will learn how to ...

- apply basic angle facts to find unknown angles.
- use the angles associated with parallel lines to find unknown angles in a range of figures.
- prove that the sum of angles in a triangle is 180°.
- use known angle facts to derive the sum of exterior and interior angles of polygons.
- use angle facts and properties of shapes to justify and prove results.

For more resources relating to this chapter, visit GCSE Mathematics Online.

Using mathematics: real-life applications

Many people rely on an understanding of angles and spatial relationships in their daily work. These include designers, architects, opticians and tree surgeons.

"I had to work quite carefully with the 360 degrees around the centre to place each of the 32 pods correctly on the London Eye." *(Structural engineer)*

Before you start ...

Ch 1	You should be able to use inverse operations to make 180 and 360.	**1** Copy and complete. **a** $180 - 96 = \square$ **b** $180 - 116 = \square$ **c** $360 - 173 = \square$ **d** $360 - 55 - 97 = \square$
Ch 2	You need to know and apply the basic properties of triangles and quadrilaterals.	**2** Use the marked properties to name each polygon as accurately as possible.  **3** What can you say about angles x and y in figure **c**? Why?
Ch 21	You need to know how to use a protractor to measure angles.	**4** Measure the following angles. **a** **b**

Find answers at: cambridge.org/ukschools/gcsemaths-studentbookanswers

Assess your starting point using the Launchpad

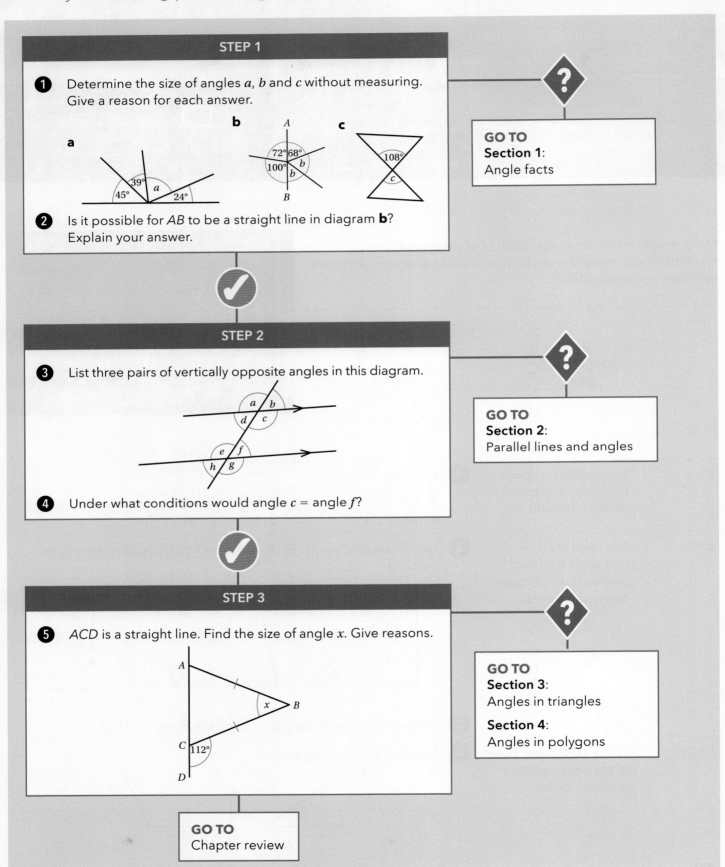

25 Angles

Section 1: Angle facts

Angles around a point

A 90° angle is one quarter of a turn, two 90° angles are half a turn and so on.

There are four quarter turns around a point.
$4 \times 90° = 360°$

The sum of angles around a point is 360°.

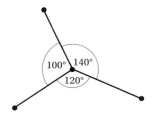

WORKED EXAMPLE 1

Find the size of x.

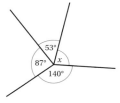

$87° + 53° + x + 140° = 360°$ — Use the angles round a point fact to form an equation in x.

$x = 360° - 140° - 53° - 87°$
$ = 360° - 280°$
$ = 80°$ — Solve the equation.

Angles on a straight line

Angles on a straight line add to 180°.

This is true for any number of angles which meet at a point on a line.

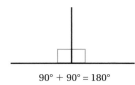
$90° + 90° = 180°$

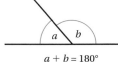
$a + b = 180°$

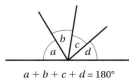
$a + b + c + d = 180°$

WORKED EXAMPLE 2

Determine the size of x.

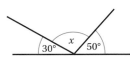

$30 + x + 50 = 180°$ — Use the angles on a straight line fact to form an equation in x.

$x = 180° - 30° - 50°$
$ = 100°$ — Solve the equation.

Tip

Angles that add to 180° are called supplementary angles.

Vertically opposite angles

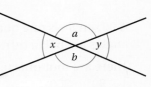

Key vocabulary

vertically opposite angles: angles that are opposite one another at an intersection of two straight lines. Vertical here means 'of the same vertex or point', not up and down.

When two lines cross, or intersect, they form four angles.

The angles a and b are vertically opposite each other, i.e. they share the same vertex, as do angles x and y.

Vertically opposite angles are equal.

$a = b$ and $x = y$

EXERCISE 25A

Tip

Angles less than 90° are **acute**.
Angles greater than 90° but less than 180° are **obtuse**.
Angles greater than 180° but less than 360° are **reflex angles**.

1 Calculate the size of each missing angle.

a 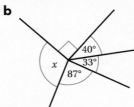 b 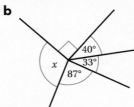 c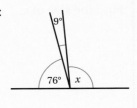

d What type of angle is x in each case?

2 Find the value of the lettered angles in each diagram.
Give reasons for any deductions you make.

a Find x and y.   **b** Find x.   **c** Find p.

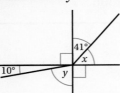 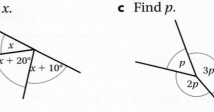

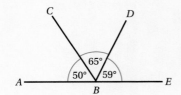

3 Explain why AE (left) cannot be a straight line.

4 Calculate the size of the marked angles in each of the two figures below.
The lines are straight lines but the diagrams are not to scale.
Show your working and give reasons for any deductions you make.

a 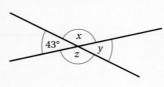 b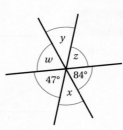

5 Given that $x = 50°$, find the size of angle z in the figure on the right.

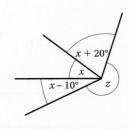

"When I'm designing and making clothes I need to be able to cut on the bias (at a given angle) and also bisect angles to add darts and fit sleeves."

(Fashion designer)

Section 2: Parallel lines and angles

A line intersecting two or more parallel lines is called a **transversal**.

When a transversal intersects with parallel lines, some pairs of angles have useful properties.

Vertically opposite angles are equal, so:

$a = c$ $\quad\quad$ $b = d$ $\quad\quad$ $e = g$ $\quad\quad$ $f = h$

Angles on a straight line add to 180° so:

$a + d = 180°$ $\quad$ $b + c = 180°$ $\quad$ $a + b = 180°$ $\quad$ $c + d = 180°$ and so on...

Key vocabulary

transversal: a straight line that crosses a pair of parallel lines.

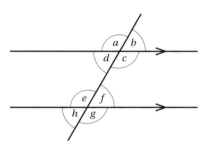

Corresponding angles

Corresponding angles, formed between a transversal and each parallel line, are equal.

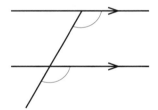

Key vocabulary

corresponding angles: angles that are created at the same point of the intersection when a transversal crosses a pair of parallel lines.

There are four pairs of corresponding angles.

The corresponding angle pairs are:

$a = b$
$c = d$
$e = f$
$g = h$

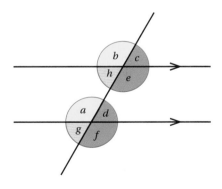

Alternate angles

Alternate angles, on opposite sides of the transversal and between the parallel lines, are equal.

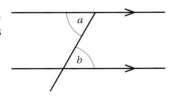 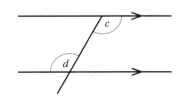

Key vocabulary

alternate angles: the angles on parallel lines on opposite sides of a transversal.

Co-interior angles

Key vocabulary

co-interior angles: the angles within the parallel lines on the same side of the transversal. Co-interior angles are sometimes referred to as 'allied angles'.

The angles inside the parallel lines and on the same side of the transversal are called **co-interior angles**. They add up to 180°.

EXERCISE 25B

1. Find the size of the missing angles a, b, c and d.

   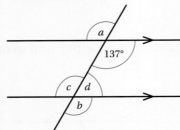

2. Given that the two poles are parallel to each other, find the angle x that the second pole makes with the incline upwards.

   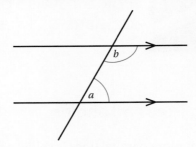

3. Find the size of the missing angles a, b and c.

   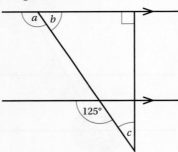

4. Find the size of angles x and y.

   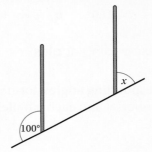

5. Find the size of each missing angle.

   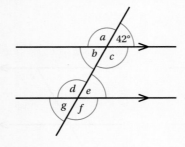

6. Find the size of $\angle CEG$.

   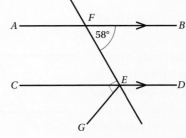

7 Find the size of ∠DCF.

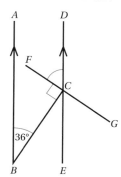

Section 3: Angles in triangles

Angle sum of a triangle

To prove that the angles in a triangle add up to 180° you have to construct a line parallel to one side of the triangle like this:

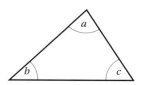 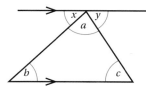

Once you have done that, you can prove that $a + b + c = 180°$ using mathematical principles.

$x + a + y = 180°$ Angles on a straight line sum to 180°.

But: $x = b$ and $y = c$ Alternate angles are equal.

Substitute b for x and c for y and you prove that $a + b + c = 180°$.

The exterior angle is equal to the sum of the opposite interior angles

> **Tip**
>
> Make sure you know the differences between an equilateral triangle, a right-angled triangle, an isosceles triangle and a scalene triangle. Look back at Chapter 2 if you need to revise the properties of these triangles.

WORKED EXAMPLE 3

Show that $a + b = x$, and hence prove that the exterior angle of any triangle is equal to the sum of the opposite interior angles. (see pages 385 and 386 for a definition of interior and exterior angles)

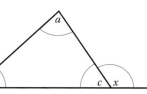

$c + x = 180°$ Angles on a straight line sum to 180°.
$\therefore c = 180° - x$

$a + b + c = 180°$ Angle sum of triangle.
$\therefore c = 180° - (a + b)$

But, $c = 180° - x$ From above.

So, $180° - (a + b) = 180° - x$
$\therefore a + b = x$

You can now combine basic angle facts, angle facts related to parallel lines and angle facts about triangles to find unknown angles in different figures.

Always give reasons for any statements you make based on known facts.

EXERCISE 25C

1 Calculate the size of each missing angle.

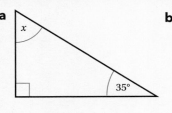
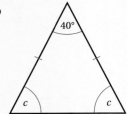

2 Find the size of angles x and y.

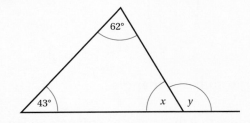

3 Find the size of angles a and b.

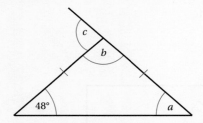

4 Find the size of angles x, y and z.

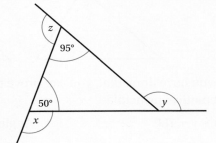

5 Work out the sizes of the angles marked a, b and c. Give reasons to justify your answers.

6 Calculate x.

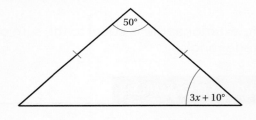

Section 4: Angles in polygons

Remember that:
- A polygon is a plane shape with three or more straight sides.
- Polygons are regular if all their sides and angles are equal.
- Polygons are named according to how many sides they have: triangle (3), quadrilateral (4), pentagon (5), hexagon (6), heptagon (7) and octagon (8).

These basalt columns are formed naturally when lava cools.

The end faces are almost perfectly hexagonal.

The angle sum of a polygon

You can divide any polygon into triangles by drawing in the diagonals from one vertex.

This allows you to use the angle sum of triangles to work out the sum of the interior angles in a polygon.

This regular hexagon has six **interior angles** which are all equal.

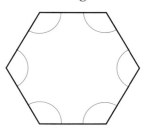

> **Key vocabulary**
>
> **interior angles**: angles inside a two-dimensional shape at the vertices or corners.

The diagonals divide the hexagon into four triangles.

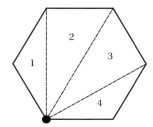

So the sum of the interior angles of a hexagon is $4 \times 180° = 720°$.

$720° \div 6 = 120°$. So, each interior angle is $120°$.

Work through the investigation in Exercise 25D to develop a rule for finding the angle sum of any polygon.

EXERCISE 25D

1 Draw the following polygons and divide them into triangles by drawing diagonals from one vertex as with the hexagon above.

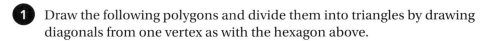

2 Predict how many triangles you could form if you did the same for a 10-sided and 20-sided polygon.

3 Copy and complete this table using your results from questions **1** and **2**.

Number of sides in polygon	3	4	5	6	7	8	10	20
Number of triangles	1			4				
Angle sum of interior angles	180°			720°				

Find answers at: cambridge.org/ukschools/gcsemaths-studentbookanswers

4. What is the relationship between the number of sides in a polygon and the number of triangles you can form in this way?

5. If a polygon has n sides, how many triangles can you form in this way?

6. Write a rule for finding the angle sum of a polygon:
   a in words
   b in general algebraic terms for a polygon of n sides.

7. Use your rule to find the angle sum of a polygon with 12 sides.

8. How could you find the size of each interior and exterior angle of a regular 12-sided polygon?

WORK IT OUT 25.1

Three students attempted to calculate the size of interior angles in a regular pentagon.

Which is the correct solution?

What mistakes have been made by the other students?

Option A	Option B	Option 3
There are three triangles within the pentagon.	There are five triangles in a pentagon.	There is a trapezium and a triangle inside the pentagon.
$3 \times 180° = 540°$	$5 \times 180° = 900°$	$360° + 180° = 540°$
Five angles in a pentagon.	$900° \div 5 = 180°$	There are five angles inside the pentagon including the central 360° which gives a total of 900°.
$540° \div 5 = 108°$	Interior angle = 180°.	
Each interior angle = 108°.		Interior angle = 900° ÷ 5 = 180°.

The sum of exterior angles of a polygon

Look at the regular hexagon again.

You know that each interior angle is 120°.

By extending each side, you can form six **exterior angles** as shown in the diagram on the right.

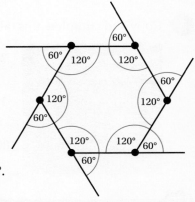

There are six sides, so this produces six pairs of angles (one interior and one exterior per pair) on straight lines.

The sum of the angles on a straight line is 180°.

Key vocabulary

exterior angles: angles produced by extending the sides of a polygon.

So, the sum of these six angle pairs is 6 × 180° = 1080°

But, you already know that the sum of just the interior angles is
180°(n − 2) = 180° × 4 = 720°.

So, the sum of the exterior angles of the hexagon is 1080° − 720° = 360°.

Now consider any polygon with n sides.

The sum of interior plus exterior angles can be found using 180n, where n is the number of sides.

The sum of the interior angles can be found using 180(n − 2) where n is the number of sides.

So, for a pentagon, the sum of the exterior and interior angles would be 180° × 5 = 900° because there are five sides.

The sum of interior angles is 180(5 − 2) = 180° × 3 = 540°.

The sum of the exterior angles
 = the sum of the interior and exterior angles − the sum of the interior angles
 = 900° − 540°
 = 360°

The sum of the exterior angles of any polygon is always 360°.

You can use these rules to find the angle sum of any polygon.

WORKED EXAMPLE 4

For a regular 10-sided polygon, find:

a the sum of the interior angles **b** the size of each interior angle.

a Angle sum = 180(n − 2) = 180(8)
  = 1440°   *In a 10-sided figure n = 10.*

b One interior angle = $\frac{1440}{10}$ = 144°   *There are 10 interior angles.*

If the polygon is regular, you can also calculate the size of each interior and exterior angle.

WORKED EXAMPLE 5

A regular polygon has an exterior angle of 18°. How many sides does it have?

Sum of exterior angles = 360

Number of angles = $\frac{360}{18}$ = 20

∴ Number of sides = 20   *The number of sides equals the number of angles.*

Find answers at: cambridge.org/ukschools/gcsemaths-studentbookanswers

EXERCISE 25E

1 Copy and complete the following table.

Regular polygon	Sum of interior angles	Size of interior angle	Size of exterior angle
triangle			120°
quadrilateral		90°	
pentagon			
		120°	60°
heptagon			

2 Calculate the sum of the interior angles of a polygon with:

   a 9 sides   **b** 12 sides   **c** 25 sides.

3 A regular polygon has 15 sides. Find:

   a the sum of the interior angles   **b** the sum of the exterior angles
   c the size of an interior angle   **d** the size of an exterior angle.

4 A regular polygon has an interior angle that is three times the size of the exterior angle.

   a What is the size of each exterior angle?
   b What is the size of each interior angle?
   c What is the name of the regular polygon?

5 Find the size of x in each polygon.

   a   **b**   **c**

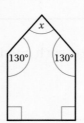  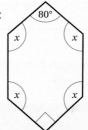

Checklist of learning and understanding

Basic angle facts
- Angles around a point sum to 360°.
- Angles on a straight line sum to 180°.
- Vertically opposite angles are equal.

Angles associated with parallel lines
- Corresponding angles are equal.
- Alternate angles are equal.
- Co-interior angles sum to 180°.

Geometric proofs

- Using properties of alternate and corresponding angles, you can show that the three interior angles of any triangle sum to 180°.
- Using the angle sum of triangles and properties of angles at a line and at a point, you can find the interior and exterior angles of any polygon.

Chapter review

For additional questions on the topics in this chapter, visit GCSE Mathematics Online.

1 Select from the list below the correct size for the marked angle in each diagram.

a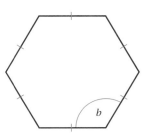

b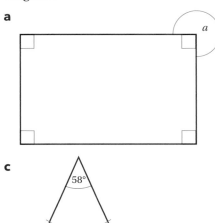

c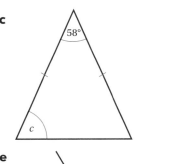

d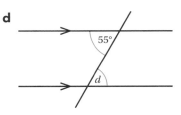

e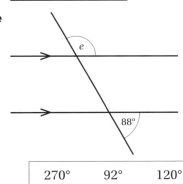

f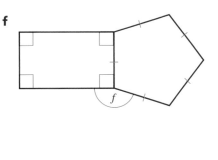

| 270° | 92° | 120° | 162° | 61° | 55° |

2 Triangle *ABC* is a right-angled triangle.

ADB is a straight line.

DA = *DC*

Angle *BCD* = 20°

Work out the size of the angle marked *x*.

You must give reasons for each stage of your working. *(4 marks)*

©Pearson Education Ltd 2013

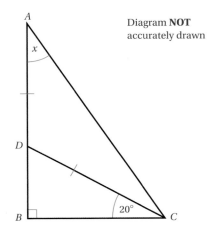

Diagram **NOT** accurately drawn

3. If the interior angle of a regular polygon is 108°, what type of polygon is it?

4. Calculate $p + q + r + s + t$.

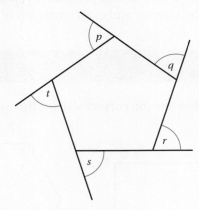

5. In this irregular hexagon, calculate the value of z.

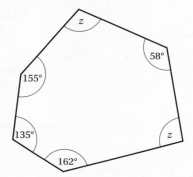

6. The diagram on the left shows part of a regular polygon. The interior angle is 144°.

   a Calculate the number of sides of the polygon.

   b What is the name given to this polygon?

7. Calculate the exterior angle of a regular nine-sided polygon.

8. Explain mathematically why the sum of the exterior angles of any polygon is 360°.

9. Find the values of x and y.

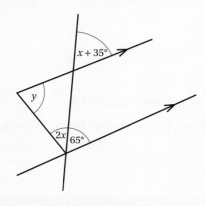

26 Probability – combined events

In this chapter you will learn how to ...

- use a range of diagrams to list outcomes of combined events.
- apply the addition rule.
- use various representations to solve probability problems.

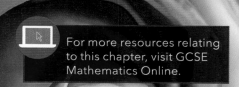

For more resources relating to this chapter, visit GCSE Mathematics Online.

Using mathematics: real-life applications

Medical researchers have developed a range of tests to detect drug use, blood-alcohol levels, disease markers and genetic and birth defects in unborn children. The probability that the test results are accurate is very high, but it is seldom 100%.

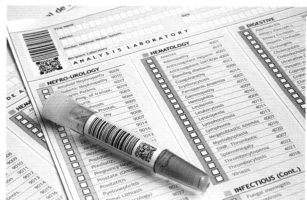

"An incorrect test result can be devastating. People can be convicted of drink-driving or more serious crimes, be expelled from competitive sports, risk surgery or decide to terminate a pregnancy based on test results, so it is really important to understand the probability of a good test giving a bad result." (Medical statistician)

Before you start ...

Chs 6 and 7	You'll need to be able to calculate effectively with fractions and decimals.	1	These calculations are all incorrect. What should the answers be? **a** $\dfrac{1}{8} + \dfrac{1}{4} = \dfrac{1}{12}$ **b** $\dfrac{2}{3} + \dfrac{1}{5} = \dfrac{2}{15}$ **c** $1 - \dfrac{3}{5} = -\dfrac{2}{5}$ **d** $\dfrac{2}{3} \times \dfrac{2}{5} = \dfrac{2}{15}$ **e** $0.3 \times 0.6 = 1.8$
Ch 20	Check that you can list all the possible outcomes of an experiment.	2	Complete each list of possible outcomes. **a** Two students are to be chosen at random from a group of males and females: FF, FM, ... **b** Two coins are to be flipped at the same time: HH, ... **c** Two cards are selected from a set of three cards labelled A, B and C and placed next to each other in the order they are drawn: AB, AC, ...

Find answers at: cambridge.org/ukschools/gcsemaths-studentbookanswers

Assess your starting point using the Launchpad

STEP 1

1 The Venn diagram shows the different sports chosen by students from one particular class. T represents students who play tennis and S represents those who take swimming.

   a How many students are in the class?
   b How many students play tennis?
   c How many students play tennis and swim?
   d If a student is chosen at random from the class, what is the probability that he or she will take swimming?

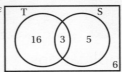

2 Copy and complete the tree diagram to show all the possible outcomes if you spin this spinner twice in a row. Add the probability of landing on each colour to the tree diagram.

GO TO
Section 1: Representing combined events

STEP 2

3 The diagram is used to find the probability of drawing hearts or twos at random from a normal pack of 52 playing cards.

What is the probability that a card drawn at random will be:

   a both a heart and a two
   b either a heart or a two
   c neither a heart nor a two?

4 A black (B) or white (W) counter is drawn at random from a box containing both black and white counters. The counter is replaced before a second counter is drawn. The possible outcomes and the probabilities of each outcome are shown on the right.

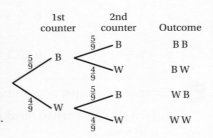

What is the probability of:

   a drawing a black counter on the first draw
   b drawing two counters the same colour
   c drawing a white counter first and a black counter second?

GO TO
Section 2: Theoretical probability of combined events

GO TO
Chapter review

Section 1: Representing combined events

Two or more events can happen at the same time. For example, if you flip a coin and roll a dice then you get heads (H) or tails (T) and a number from 1 to 6. These are called **combined events** because there is a combination of outcomes.

For combined events you need efficient ways of identifying all the possible outcomes so that you can find the probability of different combinations of outcomes.

The picture shows the **sample space** for flipping a coin and rolling a dice at the same time. The sample space must include all possible outcomes of the combined events.

Key vocabulary

combined events: one event followed by another event producing two or more outcomes.

sample space: a list or diagram that shows all possible outcomes from two or more events.

The sample space in the diagram can be represented as an ordered list like this:

H, 1   H, 2   H, 3   H, 4   H, 5   H, 6   T, 1   T, 2   T, 3   T, 4   T, 5   T, 6

Tables and grids

Listing the sample space can take a long time and you might make mistakes. Two-way tables and grids let you work faster and also let you see quickly whether you have left out, or repeated any outcomes.

WORKED EXAMPLE 1

Represent the sample space for flipping a coin and rolling a dice using:

a a table.   **b** a grid.

a Table

Dice / Coin	1	2	3	4	5	6
Heads	H1	H2	H3	H4	H5	H6
Tails	T1	T2	T3	T4	T5	T6

b Grid

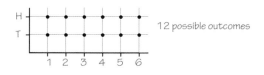

12 possible outcomes

Tip

You can check that you have recorded all the combinations. There are two possible outcomes for flipping a coin and six for rolling a dice. Combining these gives a total of $2 \times 6 = 12$.

Using tables to list probabilities of favourable outcomes

In some cases you don't have to list all the possible outcomes. For example, let's say you want to know how many ways there are to get a total score of 7 when you roll two ordinary dice. You can draw up a table like this one:

Number on dice	1	2	3	4	5	6
1	2	3	4	5	6	7
2	3	4	5	6	7	
3	4	5	6	7		
4	4	6	7			
5	6	7				
6	7					

> **Tip**
>
> Once you get to a sum of 7 you can stop because the next sum will be greater than that.

> **Tip**
>
> In all exercises in this chapter, assume that all dice and coins are 'fair', i.e. that there is an equal chance of landing on any of the sides of the dice, or either side of the coin.

The table shows there are six ways of getting a score of 7: (1, 6), (2, 5), (3, 4), (4, 3), (5, 2) and (6, 1).

Even though you haven't filled in the empty blocks, you can still see that there are 36 possible outcomes. So the probability of getting 7 is $\frac{6}{36}$ or $\frac{1}{6}$.

EXERCISE 26A

1 Use a grid to represent the sample space for:

   a flipping two coins

   b choosing a letter at random from the word DOG and flipping a coin

   c drawing a counter from each of two bags containing one red, one blue and one yellow counter.

2 **a** Draw a table to show:

   i all possible combinations of scores when you roll two dice

   ii the sample space for flipping a coin and spinning a spinner with equal sectors A, B, C and D.

   b For each table above, make up five probability questions that could be answered from the tables. Exchange questions with a partner and try to answer each other's questions.

3 Copy and complete this two-way table to show all the possible outcomes for drawing red or black cards if you draw two cards from a pack of 52 cards. The first-drawn card is replaced before drawing the second.

Second card \ First card	Red (diamond or heart)	Black (club or spade)
Red (diamond or heart)		
Black (club or spade)		

   a How many possible outcomes are there?

   b What is the probability of drawing two black cards?

   c What is the most likely outcome?

Venn diagrams

Venn diagrams show the mathematical relationships between sets of data. Different events (sets of outcomes) are represented by circles inside a rectangular frame which in turn represents the sample space (universal set).

Look at the Venn diagram below and read through the information to revise the main features of Venn diagrams.

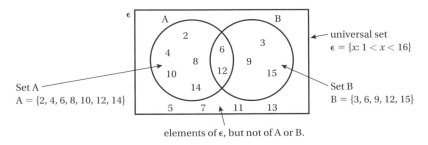

Tip

ε can also be written as:
$\varepsilon = \{$set of numbers between 1 and 16$\}$.

ε is the universal set. In this case it is the whole numbers between 1 and 16 (not 1 and not 16).

The circles A and B represent sets.

Set A is the set of even numbers between 1 and 16: A = {2, 4, 6, 8, 10, 12, 14}.

Set B is the set of multiples of three between 1 and 16: B = {3, 6, 9, 12, 15}.

There are seven elements in Set A. This can be written as n(A) = 7.

There are five elements in Set B, so n(B) = 5.

The numbers 5, 7, 11 and 13 are not elements of A or B but they are elements of the universal set, so they are written inside the rectangle, but outside the circles.

Tip

The curly brackets { } are used to show you are describing a set. The numbers between the brackets are elements of the set (for example 2 ∈ A). The symbol ∉ means 'not an element of a set'. 3 ∉ A

Intersection, union and complement of sets

Venn diagrams can also represent operations between sets.

Set A and Set B have two elements in common. The numbers 6 and 12 are written in the overlapping section of the circles.

This is the **intersection** between A and B.

$A \cap B = \{6, 12\}$

$n(A \cap B) = 2$

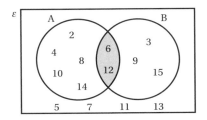

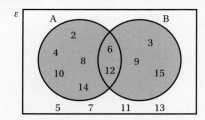

The shaded area in the diagram (left) represents the **union** of A and B. This is the combined elements of both sets with no elements repeated.

A ∪ B = {2, 3, 4, 6, 8, 9, 10, 12, 14, 15}

n(A ∪ B) = 10

The **complement** of a set refers to all the members of the universal set not in the given set. In the diagram below, the complement of Set A is shaded.

A' = {3, 5, 7, 9, 11, 13, 15}

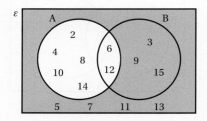

Set notation

Consider, A = {integers greater than zero but less than 20}

This is quite a lot to write out, so it makes sense to find a shorter notation.

A = {$x: x$ is an integer, $0 < x < 20$}

The element symbol (∈) might also appear in set notation. For example:
C = {$x: x \in$ primes, $10 < x < 20$}

You read this as C is the set of prime numbers greater than 10 but less than 20. This information allows you to work out that C = {11, 13, 17, 19}

WORKED EXAMPLE 2

Given that ε = {$x: x$ is a letter from a to h inclusive}, A = {a, b, c, e} and B = {c, d, e, f, g}, draw a Venn diagram to represent this information.

c and e are elements of A and B, so A ∩ B = {c, e}

> Look for elements that are common to both sets. These will go in the intersection.

h is not in A or B, so (A ∪ B)' = h

> Look for any elements which are in the universal set but which are not in A or B (the complement of A and B or (A ∪ B)').

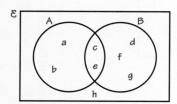

> Draw the diagram and label it correctly.

In some problems you might be given information and have to define the sets yourself. In some cases you can't list the separate elements of the sets so you write the number of elements in each set.

WORKED EXAMPLE 3

In a survey of 25 people it was found that they all liked either chocolate or ice cream or both. 15 people said they liked ice cream and 18 said they liked chocolate.

Draw a Venn diagram and use it to work out the probability that a person chosen at random from this group will like both chocolate and ice cream.

ε = {number of people surveyed}, so, $n(\varepsilon)$ = 25
C = {people who like chocolate}, so, $n(C)$ = 15
I = {people who like ice cream}, so, $n(I)$ = 18

Start by defining the sets and writing the information in set language.

$n(C) + n(I) = 15 + 18 = 33$,
$n(C \cap I) = 33 - 25 = 8$

But there were only 25 people surveyed, so eight people must have said they liked both chocolate and ice cream.

Use the figures to draw your Venn diagram.

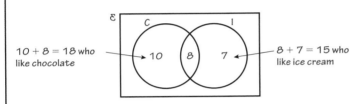

10 + 8 = 18 who like chocolate

8 + 7 = 15 who like ice cream

P(person likes both)
$= \dfrac{\text{number of people who like both}}{\text{number of people surveyed}} = \dfrac{8}{25} = 0.32$

Finally, calculate the probability.

EXERCISE 26B

1 ε = {integers from 1 to 20 inclusive}, A = {6, 7, 8, 9, 10, 11, 12} and B = {factors of 24}.

   a Draw a Venn diagram to show this information.

   b Use your Venn diagram to find:

   i $A \cap B$ **ii** $A \cup B$ **iii** $n(A)$ **iv** $n(A')$ **v** $n(B')$.

2 Nadia has 20 pairs of shoes. Six pairs are sports shoes, four pairs are red and only one of the pairs of sports shoes is red.

Draw a Venn diagram to show this information and work out the probability that a pair of shoes chosen at random from her shoe collection will be neither red nor sports shoes.

3 A factory employs 100 people. 47 of the employees have to work with moving machinery so if they have long hair they have to tie it back. 35 employees have long hair, and of these, some work with moving machinery. 23 employees neither have long hair nor work with moving machinery.

Draw a Venn diagram to show this information and use it to work out the probability of a random employee having to tie his or her hair back at work.

 Of the first 20 students to walk into a classroom, 13 were wearing headphones and 15 were sending texts. Four students were not wearing headphones or sending texts.

Represent this information on a Venn diagram and state how many students were wearing headphones while sending texts when they walked into class.

Tree diagrams

A tree diagram is a branching diagram that shows all the possible outcomes (sample space) of one or more activities.

To draw a tree diagram:
- Make a dot to represent the first activity.
- Draw branches from the dot to show all possible outcomes of that activity only.
- Write the outcomes at the end of each branch.
- Draw a dot at the end of each branch to represent the next activity.
- Draw branches from this point to show all possible outcomes of that activity.
- Write the outcomes at the end of the branches.

These two diagrams both show the possible outcomes for throwing a dice and flipping a coin at the same time. Both diagrams are correct.

Tip

Tree diagrams show probabilities, not actual responses. This is the main difference between them and the frequency trees you worked with in Chapter 20.

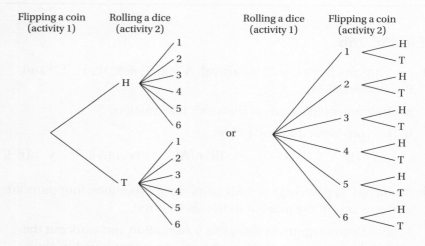

Once you've drawn a tree diagram you can list the possible outcomes by following the paths along the branches. Listing the combinations lets you work out the probability of different events.

WORKED EXAMPLE 4

Draw a tree diagram to find how many possible combinations of boys and girls there are in a three-child family if the probability of having a boy or a girl is equal.

Use your diagram to find:

a the probability of three girls **b** the probability of three children with the same gender.

```
1st      2nd      3rd      Possible
child    child    child    combinations
                     B      BBB
              B
                     G      BBG
         B
                     B      BGB
              G
                     G      BGG
                     B      GBB
              B
                     G      GBG
         G
                     B      GGB
              G
                     G      GGG
```

> Draw a dot for the first-born child.
> Draw and label two branches, one B and one G.
> Repeat this at the end of each branch for the second and third child.
> List the possible combinations.

There are eight combinations.

> As all the branches have equal probability, you can see from the diagram there are eight, equally likely, possible combinations of boys and girls.

a $P(3 \text{ girls}) = \dfrac{1}{8}$

> There is only one outcome that produces three girls.

b $P(\text{all the same gender}) = \dfrac{2}{8} = \dfrac{1}{4}$

> Same gender is all girls or all boys. There are two outcomes that satisfy this, GGG and BBB.

EXERCISE 26C

1 Here are two groups of jelly beans.

Draw a tree diagram to show the sample space for taking a jelly bean at random from each group of jelly beans.

2 Sandy has a bag containing a red, a blue and a green pen. Copy and complete the tree diagram (right) to show the sample space when she takes a pen from the bag at random, replaces it, and then takes another pen.

3 Draw a tree diagram to show the sample space when three coins are flipped one after the other.

How many ways are there to get two heads and a tail?

4 Andrew offers a gift-wrapping service in a large department store. Customers can choose striped, checked, metallic, spiral or plain brown wrapping paper and white, silver, black or pink ribbon. You should assume that customers are equally likely to choose any of the papers and any of the ribbons.

  a Draw a tree diagram to show all the possible combinations of paper and ribbon.

  b How many possible combinations are there?

Did you know?

The tree diagram in Worked example 4 assumes that a boy or a girl is equally likely for each pregnancy. In reality the probability of having a boy or a girl varies by family and by country.

c What is the probability that a customer will choose metallic paper and a silver ribbon?

d What is the probability of choosing metallic or brown paper with a black ribbon?

5 In a knockout quiz, the winner goes on to the next round. Naresh takes part in a four-round quiz and he estimates that he has an equal chance of winning or losing each round.

a Using W to represent win and L to represent loss, draw a tree diagram to show all possible outcomes for Naresh.

b How many possible outcomes are there?

c What is the probability that he will win the first round given his own estimate of his chances?

Section 2: Theoretical probability of combined events

Once you have identified all the possible equally likely outcomes you can mark the ones that are favourable and use these to find the probability of different events.

The formula, $P(\text{event happens}) = \dfrac{\text{number of favourable outcomes}}{\text{number of possible outcomes}}$, can still be used to work this out.

WORKED EXAMPLE 5

Jay has six cards with the numbers 0, 0, 2, 2, 3 and 7 on them. He picks a card, returns it and then picks another at random.

a Draw a grid to show the sample space

b Use the grid to find the probability that Jay will pick:
   i two numbers that are the same
   ii two numbers that add up to 7.

a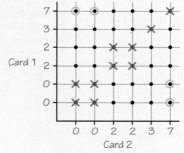

The grid shows there are 36 possible outcomes.

b i $P(\text{two numbers the same}) = \dfrac{10}{36}$

$= \dfrac{5}{18}$

or 0.28 (correct to 2 decimal places)

The successful outcomes are marked with a cross on the grid.

ii $P(\text{sum of 7}) = \dfrac{4}{36}$

$= \dfrac{1}{9}$

The successful outcomes are circled on the grid.

EXERCISE 26D

1 In a restaurant you can have your meal served with a choice of chips, baked potatoes or rice as well as a choice of green salad, coleslaw or mixed vegetables. (Assume each choice is equally likely.)

   a Draw up a tree diagram and complete it to show the sample space for the possible side dish combinations.

   b What is the probability of choosing rice and salad?

   c What is the probability of choosing rice and potatoes?

2 One box contains a red, a yellow and a blue marble and another box contains a red, a green and a purple marble. Assume marbles are chosen at random from each box.

   a Draw up a table to show the sample space if you choose one marble from each box.

   b What is the probability of choosing two red marbles?

   c What is the probability of choosing a red and a purple marble?

   d Is it possible to end up with one yellow marble and one blue marble? Why?

3 Linda and Annie each take a coin at random out of their pockets and add the values together to get the total. Linda has two £1 coins, a 50p coin, a £2 coin and three 20p coins in her pocket. Annie has three £2 coins, one £1 coin and three 50p coins.

   a Draw up a two-way table to show all the possible outcomes for the sum of the two coins.

   b What is the probability that the coins will add up to exactly £2.50?

   c What is the probability of the coins adding up to less than £2?

   d What is the probability that the coins will add up to £3 or more?

4 The diagram on the right shows picture cards from a normal pack. These cards are shuffled, a card is drawn, the type of card is noted and it is replaced before another card is drawn.

   a Draw a tree diagram to show the possible results (J, Q or K) when you draw three cards at random from this set.

   b What is the probability of drawing three cards with the same letter?

   c What is the probability that all three cards will have a different letter?

   d What is the probability of getting at least one king?

Different types of events

The type of event determines whether you add or multiply the probabilities.

Mutually exclusive events and the addition rule

Imagine you have a bag with 3 red, 2 yellow and 5 green sweets in it and you are allowed to choose one sweet at random. You cannot pick a red sweet and a yellow sweet at the same time, so the events P(red) and P(yellow) are mutually exclusive.

You can work out the probability of choosing *either* a red *or* a yellow sweet. There are 3 red and 2 yellow sweets, so $\frac{5}{10}$ of the sweets are either red or yellow.

$$P(\text{red or yellow}) = P(\text{red}) + P(\text{yellow}) = \frac{3}{10} + \frac{2}{10} = \frac{5}{10} = \frac{1}{2}$$

You can say that P(A or B) = P(A) + P(B) where A and B are mutually exclusive events.

This is called the addition rule for mutually exclusive events.

Events that are not mutually exclusive

When you list the elements in the union of sets you do not repeat shared elements (those in the intersection).

In set language, you can write this as $n(A \cup B) = n(A) + n(B) - n(A \cap B)$. The elements in the intersection of sets are not mutually exclusive and this affects your probability calculations.

WORKED EXAMPLE 6

The Venn diagram shows the possible outcomes when a six-sided dice is rolled. Set A = {prime numbers} and Set B = {odd numbers}. Use the diagram to find the probability of rolling a number that is either odd or prime.

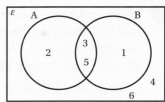

P(A or B) = P(A) + P(B) − P(A and B)

> You don't want to add the numbers that fall into the intersecting part twice, which is why you subtract $n(A \cap B)$ in the formula.

$P(A) = \frac{3}{6}$

$P(B) = \frac{3}{6}$

$P(A \text{ and } B) = \frac{2}{6}$

> The total number of outcomes is the denominator, and it's easier to add and subtract the fractions if you don't simplify the fractions first.

So, $P(A \text{ or } B) = \frac{3}{6} + \frac{3}{6} - \frac{2}{6} = \frac{4}{6} = \frac{2}{3}$

> You can see this is true by looking at the diagram. The combined elements of A and B are 1, 2, 3 and 5, giving you $\frac{4}{6}$ numbers falling into one or the other of these sets.

Tip

The word 'or' means it belongs in one set or the other, so you need to deal with the union of the sets.

26 Probability – combined events

Independent events

When the outcome of one event does not affect the outcome of the others, the events are **independent**.

Rolling a dice and flipping a coin are independent events. The score (outcome) on the dice doesn't affect whether you get heads or tails.

The probability of combined events on a tree diagram

This tree diagram shows the possible outcomes for throwing a dice and flipping a coin at the same time (H is used for head and T is used for tail).

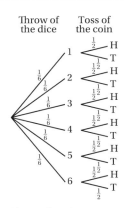

> **Key vocabulary**
>
> **independent events:** events that are not affected by what happened before.

To find the probability of one particular combination of outcomes multiply the probabilities on consecutive branches, for example, the probability of throwing a 5 and getting heads is $\frac{1}{6} \times \frac{1}{2} = \frac{1}{12}$.

This is called the multiplication rule and it works for independent events only.

$P(A \text{ and } B) = P(A) \times P(B)$

> **Tip**
>
> It can be helpful to use a colour to mark the route along the branches to show which events you are dealing with.

Combining the rules

To find the probability when there is more than one favourable combination or when the events are mutually exclusive:

- multiply the probabilities on consecutive branches
- add the probabilities (of each favourable combination) obtained by multiplication, for example, throwing 1 or 2 and getting an H

is $\left(\frac{1}{6} \times \frac{1}{2}\right) + \left(\frac{1}{6} \times \frac{1}{2}\right) = \frac{1}{12} + \frac{1}{12} = \frac{2}{12} = \frac{1}{6}$.

WORKED EXAMPLE 7

Two coins are flipped together. Draw a tree diagram to find the probability of getting:

a two tails.   **b** one head and one tail.

a $P(TT) = P(T \text{ on 1st flip}) \times P(T \text{ on 2nd flip})$

$= \frac{1}{2} \times \frac{1}{2} = \frac{1}{4}$

b $P(HT \text{ or } TH) = P(HT) + P(TH)$

$= \left(\frac{1}{2} \times \frac{1}{2}\right) + \left(\frac{1}{2} \times \frac{1}{2}\right) = \frac{1}{4} + \frac{1}{4} = \frac{1}{2}$

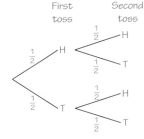
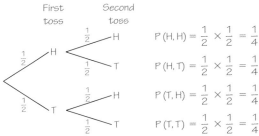

Find answers at: cambridge.org/ukschools/gcsemaths-studentbookanswers

Dependent events

When the outcome of one event affects the outcome of another the events are said to be **dependent**.

Key vocabulary

dependent events: events in which the outcome is affected by what happened before.

Here are 4 red and 2 yellow sweets.

Suppose you choose one sweet at random and eat it before you select a second sweet. What is the probability of the second sweet being red?

The answer to this depends on what colour the first sweet was. If the first sweet was red then the probability on the branch to the second red is $\frac{3}{5}$ because there are only 5 sweets left and only 3 of those are red.

1 red eaten

If the first sweet was yellow then the probability on the branch to the second one being red is $\frac{4}{5}$. There are still only five sweets left to choose from, but this time four of them are red.

1 yellow eaten

For dependent events you can find the probability by adapting the multiplication rule to accommodate the dependent event.

P(A *and then* B) = P(A) × P(B *given that* A has occurred)

Tree diagrams can be useful for showing dependent events.

WORKED EXAMPLE 8

A box contains three yellow, four red and two purple marbles. A marble is chosen at random and not replaced before choosing the next one. If three marbles are chosen (without replacement) what is the probability of choosing:

a three red marbles?

b a yellow, red and purple marble in that order?

a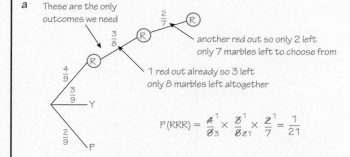

b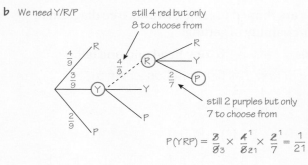

Problem-solving framework

When you use Venn diagrams to solve problems it is important to choose the correct operation to solve the problem. The wording of the problem normally gives you clues about which operation you need.

Given that Set A = {x: x is an even number} and Set B = {x: x is a multiple of 3}, you may be asked questions like the ones in the first column of the table.

Question: What is the probability of a number ...	Which operation is involved in finding the solution?	Why?
being even?	No operations	The probability will only involve the elements of one set.
being even and a multiple of three?	Intersection	The word 'and' tells you to look for numbers that are elements of both sets (i.e. those in the overlapping section).
being even or a multiple of three?	Union	The word 'or' tells you it can be in either of the two sets, so you need to include all the elements of both sets – without repeating any in the intersection.
not being even?	Complement	All the numbers that are outside the set of even numbers must be included. These are the numbers outside the circles and also the numbers in the multiples of three, but not in the overlap.
being neither even nor a multiple of three?	Complement	'Neither, nor' tells you that two sets have to be excluded, you are looking for the elements outside the circles in this case.

EXERCISE 26E

1 In a class of 28 students, 12 take physics, 15 take chemistry and 8 take neither physics nor chemistry.

  a Draw a Venn diagram to represent this information.

  b What is the probability that a student chosen at random from this class:

   i takes physics but not chemistry?

   ii takes physics or chemistry?

   iii takes physics and chemistry?

2 Nico is on a bus and he is bored. He passes the time by choosing a consonant and a vowel at random from the names of towns on road signs. The next road sign is DUNDEE.

  a Draw up a sample space diagram to list all the options that Nico has.

  b Calculate P(D and E).

  c Calculate P(D and E or U).

  d Calculate P(not N and not U).

3. A bag contains 3 red counters, 4 green counters, 2 yellow counters and 1 white counter. Two counters are drawn at random from the bag one after the other, without being replaced. Calculate:

   a P(2 red counters)
   b P(2 green counters)
   c P(2 yellow counters)
   d P(white and then red)
   e P(white or yellow, but not both)
   f P(white or red, but not both)
   g P(white or yellow first and then any other colour).

4. Mohammed has four Scrabble tiles with the letters A, B, C and D on them. He draws a letter at random and places it on the table, then he draws a second letter and a third, placing them down next to the previously drawn letter.

   a What is the probability that the letters he has drawn spell the words:
      i cad   ii bad   iii dad?
   b What is the probability that he will not draw the letter B?
   c What is Mohammed's chance of drawing the letters in alphabetical order?

5. In a standard pack of cards, A = {hearts} and B = {kings}. If a card is picked at random, determine:

   a P(A)   b P(B)   c P(A and B)   d P(A or B).

6. Maria has a bag containing 18 fruit drop sweets. 10 are apple flavoured and 8 are blackberry flavoured. She chooses a sweet at random and eats it. Then she chooses another sweet at random.

   a Calculate the probability that:
      i   both sweets are apple flavoured
      ii   both sweets are blackberry flavoured
      iii the first is apple and the second is blackberry
      iv   the first is blackberry and the second is apple.
   b Add up your answers from part a. Explain why you should get an answer of 1 if you worked out the probabilities correctly.

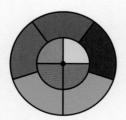

7. Amira has a colour wheel with four colours on the inside and five colours on the outside as shown. She turns the wheel at random to find possible colour combinations to use in her clothing designs.

   a Draw a sample space to show all the possible colour combinations on this wheel.
   b Determine P(blue and brown).
   c Determine P(yellow and orange).
   d Amira doesn't really like green or orange. If she picks a combination at random, what is the chance that she will not get a combination with those colours?

8 A cleaner accidentally knocked the name labels off three students' lockers. The labels say Raju, Sam and Kerry. The tree diagram shows the possible ways of replacing the labels.

   a Copy the diagram and write the probabilities next to each branch.

   b Are these events dependent or independent? Why?

   c How many correct ways are there to match the name labels to the lockers?

   d How many possible ways are there for the cleaner to label the lockers?

   e If the cleaner randomly stuck the names back onto the lockers, what is the chance of getting the names correct?

   f What is the probability of getting the labels on Lockers 2 and 3 correct given that the first one is correctly labelled Kerry?

Checklist of learning and understanding

Representing combined events

- The sample space of an event is all the possible outcomes of the event.
- When an event has two or more stages it is called a combined event.
- Lists, tables, grids, tree diagrams and Venn diagrams can be used to represent combined events.

Calculating probabilities for combined events

- For mutually exclusive events P(A or B) = P(A) + P(B).
- For independent events P(A and then B) = P(A) × P(B).
- When independent events are mutually exclusive, you need to add the probabilities obtained by multiplication.
- For dependent events P(A and then B) = P(A) × P(B given that A has happened).

Chapter review

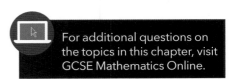

For additional questions on the topics in this chapter, visit GCSE Mathematics Online.

1 Choose the most appropriate method and use it to represent the sample space in each of the following.

   a A coin is flipped and an octagonal dice with faces numbered 0 to 7 is rolled at the same time.

   b Boxes A, B and C contain pink and yellow tickets. A box is selected at random and a ticket is drawn from it.

   c The number of ways in which three letters P, A and N can be arranged to form a three letter sequence.

   d In a class of 24 students, 10 take art, 12 take music and 5 take neither.

2 Two normal six-sided dice are rolled simultaneously. Draw a sample space for this information and hence calculate the probability of rolling:

   a double 2
   b at least one 4
   c a total greater than 9
   d a total of 6 or 7.

Find answers at: cambridge.org/ukschools/gcsemaths-studentbookanswers

3. The letters from the word MANCHESTER are written on cards and placed in a bag.

   a What is the probability of drawing a vowel if one letter is drawn at random?

   b Copy and complete the tree diagram (left) to show all probabilities for when a letter is drawn from the bag, noted and replaced and then another letter is drawn.

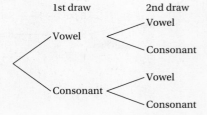

   c Use the tree diagram to determine the probability of drawing:

   i two vowels
   ii two consonants
   iii a vowel and a consonant
   iv at least one consonant.

   d Explain why drawing the letters can be considered independent events in this case.

   e How could you change the experiment to make the events dependent?

4. There are 50 students in a year group. 30 have brown eyes, nine have fair hair and three have both brown eyes and fair hair. Represent this information on a Venn diagram and use it to determine the probability that a student chosen at random from this group:

   a has neither brown eyes nor fair hair.

   b has brown eyes but not fair hair.

5. There are 10 socks in a drawer.

   7 of the socks are brown.

   3 of the socks are grey.

   Bevan takes at random two socks from the drawer at the same time.

   a Complete the probability tree diagram. *(2 marks)*

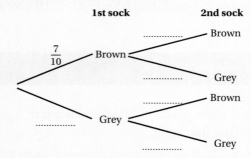

   b Work out the probability that Bevan takes two socks of the same colour. *(3 marks)*

   ©Pearson Education Ltd 2010

6. An unbiased cubical dice has six faces numbered 4, 6, 10, 12, 15 and 24. The dice is thrown twice and the highest common factor (HCF) of the scores is recorded.

   a Draw a sample space diagram to show the possible outcomes.

   b Calculate the probability that:

   i the HCF is 2
   ii the HCF is greater than 2
   iii the HCF is not 7
   iv the HCF is 3 or 5.

27 Standard form

In this chapter you will learn how to ...
- convert numbers to and from standard form.
- use a calculator to solve problems with numbers in standard form.
- apply the index laws to add, subtract, multiply and divide numbers in standard form without using a calculator.

For more resources relating to this chapter, visit GCSE Mathematics Online.

Using mathematics: real-life applications

The study of stars, moons and planets involves huge numbers. Astronomers use standard form to make it easier to write or type very large quantities, to make them easier to compare and to allow them to calculate with and without calculators. The Sun has a mass of 1.988×10^{30} kg. This is a number with 27 zeros and it would be clumsy and impractical to have to write it out each time you wanted to use it.

Calculator tip

Make sure you know how your calculator deals with powers and that you have it in the correct mode to do calculations involving powers.

"In astronomy we work with very large and very small numbers. There are 100 000 000 000 000 000 000 000 known stars alone! Imagine having to write this number out in full every time you wanted to use it! It is much easier to write 1×10^{23}."

(Astronomy student)

Before you start ...

Ch 7	You should be able to calculate efficiently with decimals.	**1** Evaluate these without using a calculator. **a** $2.9 + 5.8$ **b** $12.5 - 3.8$ **c** 4.5×1.5 **d** $4.5 \div 0.3$

Ch 9	You need to be able to round numbers to a given number of significant figures.	**2** Choose the correct answer.
		a 507 000 000 rounded to 2 significant figures.
		   A 50 700 **B** 510 000 000
		b 1.098 rounded to 3 significant figures.
		   A 1.10 **B** 1.09
		c 0.006 25 rounded to 1 significant figure.
		   A 0.6 **B** 0.006

Ch 8	You should know how to apply the laws of indices.	**3** State whether the following are true or false. **a** $3^2 \times 3^3 = 3^6$ **b** $x^5 \times x^3 = x^8$ **c** $x^{-3} \times x^4 = x$ **d** $\dfrac{x^4}{x^5} = x$ **e** $\dfrac{x^{-4}}{x^2} = x^{-6}$

Find answers at: cambridge.org/ukschools/gcsemaths-studentbookanswers

Assess your starting point using the Launchpad

STEP 1

In questions **1** to **4**, choose the correct answer.

1 2.4×10^7 written in full.
  A 0.00000024
  B 240000000
  C 0.000000024
  D 24000000
  E 2400000

2 7×10^{-3} written in full.
  A 0.0007
  B 7000
  C 0.007
  D 70000
  E 0.00007

3 23500 written in standard form.
  A 2.35×10^{-4}
  B 2.35×10^4
  C 235×10^2
  D 23.5×10^3
  E 2.35×10^{-2}

4 0.0000231 written in standard form.
  A 23.1×10^{-5}
  B 2.31×10^5
  C 2.31×10^{-4}
  D 2.31×10^4
  E 2.31×10^{-5}

GO TO
Section 1: Expressing numbers in standard form

STEP 2

5 Multiply 2400000×23000 giving your answer in standard form.

6 Calculate $\dfrac{4.6 \times 10^{-3}}{1.84 \times 10^4}$.

GO TO
Chapter review

GO TO
Section 2: Calculators and standard form
Section 3: Working in standard form

Section 1: Expressing numbers in standard form

Writing out very large or very small numbers takes time and you might make mistakes and skip zeros when you do calculations. You use standard form to write these numbers in a simpler way using powers of 10.

> **Remember**
> $10^2 = 100$ one hundred
> $10^3 = 1000$ one thousand
> $10^4 = 10\,000$ ten thousand
> $10^5 = 100\,000$ one hundred thousand
> $10^6 = 1\,000\,000$ one million

A number is in standard form when it is written as a product ($\times$) of a number and a power of 10. The number must be greater or equal to 1 and smaller than 10. For example, 2×10^2 and 1.2×10^{-2} are both in standard form. The numbers 0.4×10^5 and 12×10^3 are not in standard form because 0.4 is less than 1 and 12 is greater than 10.

A good understanding of place value makes standard form fairly easy to understand.

Look at the patterns in this table.

$1.5 \times 10^0 =$	1.5	Remember $n^0 = 1$	$1.5 \times 10^0 = 1.5$
$1.5 \times 10^1 =$	15		$1.5 \times 10^{-1} = 0.15$
$1.5 \times 10^2 =$	150		$1.5 \times 10^{-2} = 0.015$
$1.5 \times 10^3 =$	1500		$1.5 \times 10^{-3} = 0.0015$
$1.5 \times 10^4 =$	15 000		$1.5 \times 10^{-4} = 0.000\,15$
$1.5 \times 10^5 =$	150 000		$1.5 \times 10^{-5} = 0.000\,015$
$1.5 \times 10^6 = 1\,500\,000$			$1.5 \times 10^{-6} = 0.000\,001\,5$

The index (power of ten) gives you important information.

- Multiplying by 10^4 moves the digits 4 places to the left (or, if you prefer to think in terms of moving the decimal point, then move that 4 places to the right).
- Multiplying by 10^{-4} moves the digits 4 places to the right (the same as moving the decimal point 4 places to the left).

Tip

In the UK we use the term standard form for numbers in the form $A \times 10^n$, where $1 \leqslant A < 10$. This notation is also called scientific notation.

Writing a number in standard form

To write a number in standard form:

- Place the decimal point after the first non-zero digit.
- Find the power of 10 needed to move the first digit back to its original position. In other words, work out the number of places the first digit has moved and the direction in which it has moved.
- Write the number as a decimal multiplied by a power of 10.

WORKED EXAMPLE 1

Express these numbers in standard form.

a 416 000   **b** 0.0037

a 416000

4.16

4 1 6 0 0 0

> Write the number with the decimal point after the first non-zero digit, to get the number between 1 and 10.
>
> Work out how many decimal places the first digit needs to move to get back to its original place value.

$416000 = 4.16 \times 10^5$

> The first digit needs to move five places to the left. This gives a larger number, so the power is 5.

b 0.0037

3.7

0.0 0 3 7

> Write the number with the decimal point after the first non-zero digit, to get the number between 1 and 10.
>
> Work out how many decimal places the first digit needs to move to get back to its original place value.

$0.0037 = 3.7 \times 10^{-3}$

> The first digit needs to move three places to the right. This gives a smaller number, so the power is −3.

EXERCISE 27A

1 Express each of the following in standard form.

a 321 000	**b** 1340	**c** 3 010 000	**d** 0.08
e 0.0001	**f** 32 000 000	**g** 910 000	**h** 0.000 031 255
i 0.000 000 241 52	**j** 0.003 05	**k** 0.201	**l** 0.000 34
m 0.009	**n** 2.45	**o** 0.000 426	**p** 0.426

2 Express each of the following quantities in standard form.

a In 2011 the population of the Earth reached 7 000 000 000.

b The distance from the Earth to the Moon is approximately 240 000 miles.

c There are about 37 000 000 000 000 cells in your body.

d Some cells are about 0.000 000 2 metres in diameter.

e The surface area of the Earth's oceans is about 140 million square miles.

f An angstrom is a unit of measure. One angstrom is equivalent to 0.000 000 000 1 metre.

g Humans blink on average about 6 250 000 times per year.

h A dust particle has a mass of about 0.000 000 000 753 kg.

Converting from standard form to ordinary numbers

To convert numbers from standard form to ordinary numbers or decimals you need to look at the powers and move the digits the correct number of places to the left or right.

WORKED EXAMPLE 2

Write as ordinary numbers.

a 3.25×10^5 **b** 2.07×10^{-5}

a 3.25×10^5

3.2 5
3 2 5 _ _ _.

The digits need to move 5 decimal places to the left.

3 2 5 0 0 0
$3.25 \times 10^5 = 325\,000$

Fill in the correct number of zeros.

b 2.07×10^{-5}

2.07
_ . _ _ _ _ 2 0 7

The digits need to move 5 decimal places to the right.

0 . 0 0 0 0 2 0 7

Fill in the correct number of zeros.

$2.07 \times 10^{-5} = 0.000020\,7$

Remember to write the 0 before the decimal point as well.

EXERCISE 27B

1 Express each of the following as an ordinary number.

  a 1.4×10^2 **b** 4.8×10^4 **c** 2.9×10^3
  d 3.25×10^2 **e** 3.25×10^{-1} **f** 3.67×10^5
  g 4.5×10^7 **h** 2.13×10^{-2} **i** 3.209×10^4
  j 3.46×10^{-3} **k** 1.89×10^{-4} **l** 7×10^{-7}
  m 1.03×10^{-2} **n** 1.025×10^{-3} **o** 2.09×10^{-5}

2 Write each quantity out in full as an ordinary number.

  a The area of the Atlantic Ocean is 3.18×10^7 square miles.

  b The space between tracks on a DVD disk is 7.4×10^{-4} mm.

  c The diameter of the silk used to weave a spider's web is 1.24×10^{-6} mm.

d There are 3×10^9 possible ways to play the first four moves in a game of chess.

e A sheet of paper is about 1.2×10^{-4} m thick.

f The distance between the Sun and Jupiter is about 7.78×10^8 km.

g The Earth is about 1.5×10^{11} km from the Sun.

h The mass of an electron is about $9.109\,382\,2 \times 10^{-31}$ kg.

Section 2: Calculators and standard form

You can use a modern scientific calculator to enter calculations in standard form. The calculator will also give you an answer in standard form if it has too many digits to display on the screen.

Entering standard form calculations

You will need to use the ×10ˣ, Exp or EE button on your calculator.

These are known as the exponent keys and they all work in the same way, even though they might look different on different calculators.

When you are using the exponent function key of your calculator you don't enter the × 10 part of the calculation because the calculator function automatically includes that part.

Use your own calculator to work through this example. You should get the same result even if your function key is different to those in the example.

WORKED EXAMPLE 3

Enter these into your calculator.

a 2.134×10^4 b 3.124×10^{-6}

a [2] [.] [1] [3] [4] [×10ˣ] [4] [=]

Enter the figures using your calculator's key for the exponent.

21 340

You should get this. If not, check that you are using the exponent key correctly.

b [3] [.] [1] [2] [4] [Exp] [+/−] [6] [=]

Use the correct key for your calculator to enter the negative 6.

0.000 003 124

Calculator tip

Calculators work in different ways and you need to understand how your own calculator works. Make sure you know which buttons to use to enter standard form calculations, how to read and make sense of the display and how to convert your calculator answer into decimal form.

27 Standard form

Making sense of the calculator display

How your calculator displays an answer will depend on the calculator you use.

This is 5.98×10^{-6}

This is 2.56×10^{24}

To give the answer in standard form, read the display and write the answer correctly.

EXERCISE 27C

1 Enter each of these numbers into your calculator using the correct function key and write down what appears on the display.

    a 4.2×10^{12} **b** 1.8×10^{-5} **c** 2.7×10^{6}

    d 1.34×10^{-2} **e** 1.87×10^{-9} **f** 4.23×10^{7}

    g 3.102×10^{-4} **h** 3.098×10^{9} **i** 2.076×10^{-23}

2 Here are nine different calculator displays giving answers in exponential form. Write each answer correctly in standard form.

a 1.09 05   **b** 2.876 −06   **c** 4.012 09

d 1.89 07   **e** 3.123E13   **f** 2.876E−04

g 9.02E15   **h** 8.076E−12   **i** 8.124E−11

Significant figures

When you work with standard form, you will often be asked to give the answers in standard form correct to a given number of significant figures.

When working with decimal values, remember that none of the zeros before a non-zero digit are significant.

- 0.003   is correct to 1 significant figure.
- 0.01   is correct to 1 significant figure.
- 0.10   is correct to 2 significant figures.

A zero after a non-zero digit is significant.

> **Tip**
>
> You rounded numbers in Chapter 9.

EXERCISE 27D

Use your calculator to do these calculations. Give your answers in standard form correct to 3 significant figures.

1
 a 4216^6
 b $(0.00009)^4$
 c $0.0002 \div 2500^3$
 d $65\,000\,000 \div 0.000\,0045$

2
 a $(0.0029)^3 \times (0.00365)^5$
 b $(48 \times 987)^4$

3
 a $\dfrac{4525 \times 8760}{0.000\,020}$
 b $\dfrac{9500}{0.0005^4}$
 c $\sqrt{5.25} \times 10^8$
 d $3\sqrt{9.1} \times 10^{-8}$

Tip

Remember:
$n^x \times n^y = n^{x+y}$ $\dfrac{n^x}{n^y} = n^{x-y}$

Section 3: Working in standard form

Writing numbers in standard form allows you to use the laws of indices to calculate quickly without using a calculator.

Multiplying and dividing numbers in standard form

WORKED EXAMPLE 4

Do these calculations without using a calculator. Give your answers in standard form.

a $(3 \times 10^5) \times (2 \times 10^6)$
b $(2 \times 10^{-3}) \times (3 \times 10^{-7})$
c $(2 \times 10^3) \times (8 \times 10^7)$
d $\dfrac{2.8 \times 10^6}{1.4 \times 10^4}$
e $\dfrac{4 \times 10^8}{9 \times 10^5}$

a $(3 \times 10^5) \times (2 \times 10^6)$
This is the same as: $3 \times 2 \times 10^5 \times 10^6$
$= 6 \times 10^{5+6}$
$= 6 \times 10^{11}$

<- Simplify 3×2 and add the indices.

<- Write the answer in standard form.

b $(2 \times 10^{-3}) \times (3 \times 10^{-7})$
This is the same as: $2 \times 3 \times 10^{-3} \times 10^{-7}$
$= 6 \times 10^{-3+-7}$
$= 6 \times 10^{-10}$

c $(2 \times 10^3) \times (8 \times 10^7)$
This is the same as: $2 \times 8 \times 10^3 \times 10^7$
$= 16 \times 10^{3+7}$
$= 16 \times 10^{10}$

$= 1.6 \times 10 \times 10^{10}$
$= 1.6 \times 10^{11}$

<- But 16 is greater than 10 so this is not in standard form.

<- If you think of 16 as 1.6×10 you can change it to standard form.

Continues on next page ...

d $\dfrac{2.8 \times 10^6}{1.4 \times 10^4} = \dfrac{2.8}{1.4} \times \dfrac{10^6}{10^4}$

$= 2 \times 10^{6-4}$

$= 2 \times 10^2$

> Subtract the indices to divide the powers.

e $\dfrac{4 \times 10^8}{9 \times 10^5} = \dfrac{4}{9} \times \dfrac{10^8}{10^5}$

$= 0.44 \times 10^3$ (to 2 significant figures)

> 0.44 is smaller than 1 so this is not standard form.

$= 4.4 \times 10^{-1} \times 10^3$

$= 4.4 \times 10^2$ (to 2 significant figures)

> If you think of 0.44 as 4.4×10^{-1} you can change it to standard form.

You need to be able to solve problems involving numbers in standard form. Although the problems might seem complicated when you read them, they are usually simple once you work out what operation you need to do.

Problem-solving framework

The number of bacteria that can fit onto a piece of skin of area $1\,\text{mm}^2$ is approximately 1.5×10^{14}.

If your index fingertip has an approximate area of $3.75 \times 10^2\,\text{mm}$, how many bacteria could fit onto it?

Give your answer in standard form and as an ordinary number.

Steps for solving a problem	What you would do for this example
Step 1: Work out what you need to do.	Find the number of bacteria that will fit onto a larger area than the one given.
Step 2: Look for information that will help you.	The area of a fingertip is $3.75 \times 10^2\,\text{mm}$. The number of bacteria that can fit onto $1\,\text{mm}^2 = 1.5 \times 10^{14}$.
Step 3: What maths can you do?	Multiply the number of bacteria by the area: $1.5 \times 10^{14} \times 3.75 \times 10^2 = 1.5 \times 3.75 \times 10^{14} \times 10^2$ $= 5.625 \times 10^{16}$
Step 4: Set out the solution clearly, making sure you have answered the original question.	$5.625 \times 10^{16} = 56\,250\,000\,000\,000\,000$ bacteria

EXERCISE 27E

1 Simplify, giving the answers in standard form.

a $(2 \times 10^{13}) \times (4 \times 10^{17})$ **b** $(1.4 \times 10^8) \times (3 \times 10^4)$

c $(1.5 \times 10^{13}) \times (1.5 \times 10^{13})$ **d** $(0.2 \times 10^{17}) \times (0.7 \times 10^{16})$

e $(9 \times 10^{17}) \div (3 \times 10^{16})$ **f** $(8 \times 10^{17}) \div (4 \times 10^{16})$

g $(1.5 \times 10^8) \div (5 \times 10^4)$ **h** $(2.4 \times 10^{64}) \div (8 \times 10^{21})$

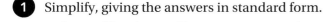
Find answers at: cambridge.org/ukschools/gcsemaths-studentbookanswers

2 Simplify, giving the answers in standard form.

a $(2 \times 10^{-4}) \times (4 \times 10^{-16})$
b $(1.6 \times 10^{-8}) \times (4 \times 10^{-4})$
c $(1.5 \times 10^{-6}) \times (2.1 \times 10^{-3})$
d $(11 \times 10^{-5}) \times (3 \times 10^2)$
e $(9 \times 10^{17}) \div (4.5 \times 10^{-16})$
f $(7 \times 10^{-21}) \div (1 \times 10^{16})$
g $(4.5 \times 10^8) \div (0.9 \times 10^{-4})$
h $(11 \times 10^{-5}) \times (3 \times 10^2) \div (2 \times 10^{-3})$

3 Carry out these calculations without using your calculator. Leave the answers in standard form.

a $(3 \times 10^{12}) \times (4 \times 10^{18})$
b $(1.5 \times 10^6) \times (3 \times 10^5)$
c $(1.5 \times 10^{12})^3$
d $(1.2 \times 10^{-5}) \times (1.1 \times 10^{-6})$
e $(0.4 \times 10^{15}) \times (0.5 \times 10^{12})$
f $(8 \times 10^{17}) \div (3 \times 10^{12})$
g $(1.44 \times 10^8) \div (1.2 \times 10^6)$
h $(8 \times 10^{-15}) \div (4 \times 10^{-12})$

4 The speed of light is approximately 3×10^8 metres per second. How far will the light travel in:

a 10 seconds?
b 20 seconds?
c 10^2 seconds?
d 2×10^3 seconds?

5 A human being blinks approximately 6.25×10^6 times per year.

a How often will you blink in 5 years? Give the answer in standard form and as an ordinary number.
b If there were 7.2×10^9 people on the planet, how many blinks would there be in a year?

6 A sheet of paper is 1.2×10^{-4} m thick.

a Work out the height (in metres) of a stack of 500 sheets of this paper, giving your answer in standard form and as an ordinary number.
b How many millimetres high is the stack of paper?

Calculator tip

If you are using a calculator you don't need to rewrite the numbers. You just need to enter the calculation correctly.

Adding and subtracting in standard form

When you have to add or subtract numbers in standard form you can rewrite them both as ordinary numbers to do the calculation. You then convert your answer back to standard form.

WORKED EXAMPLE 5

Calculate $6 \times 10^{-3} - 3 \times 10^{-4}$.

$6 \times 10^{-3} - 3 \times 10^{-4}$
$= 0.006 - 0.0003$
$= 0.0057$
$= 5.7 \times 10^{-3}$

Remember to write the numbers so the place values line up:

```
  0.0060
- 0.0003
  ──────
  0.0057
```

EXERCISE 27F

1 Carry out these calculations without using a calculator. Give your answers in standard form.

   a $(3 \times 10^8) + (2 \times 10^8)$ **b** $(3 \times 10^{-3}) - (1.5 \times 10^{-3})$

   c $(1.5 \times 10^5) + (3 \times 10^6)$ **d** $(6 \times 10^7) - (4 \times 10^6)$

   e $(4 \times 10^{-4}) + (3 \times 10^{-3})$ **f** $(5 \times 10^{-3}) - (2.5 \times 10^{-2})$

2 The Pacific Ocean has a surface area of approximately 1.65×10^8 km² and the Atlantic Ocean has a surface area of approximately 1.06×10^8 km².

   a Which ocean has the greater surface area?

   b How much larger is it?

   c If the total surface area of the world's oceans is 361 000 000 km², work out the combined surface area of the other three oceans (the Indian, Southern and Arctic), giving the answer in standard form.

3 The Earth is approximately 9.3×10^7 miles from the Sun and 2.4×10^5 miles from the Moon. How much further is it from the Earth to the Sun than from the Earth to the Moon?

Checklist of learning and understanding

Standard form

- Very large and very small numbers can be written in standard form by expressing them as the product of a value greater or equal to 1 and less than 10, and a power of 10.
- Positive powers of ten indicate large numbers and negative powers of ten indicate decimal fractions (small numbers).

Using a calculator

- The exponent function of the calculator allows you to enter calculations in standard form without entering the $\times 10$ part of the calculation.
- When a number has too many digits to display, the calculator will give the answer in exponent form.

Calculations in standard form

- You can multiply and divide numbers in standard form by applying the laws of indices.
- To add or subtract numbers written in standard form, you convert them to normal numbers, do the calculation, then convert your answer back to standard form.

For additional questions on the topics in this chapter, visit GCSE Mathematics Online.

Chapter review

1 Express the following numbers in standard form.

   a 45 000   **b** 80   **c** 2 345 000

   d 32 000 000 000   **e** 0.0065   **f** 0.009

   g 0.000 45   **h** 0.000 000 8   **i** 0.006 75

2 Write the following as ordinary numbers.

   a 2.5×10^3 **b** 3.9×10^4 **c** 4.265×10^5

   d 1.045×10^{-5} **e** 9.15×10^{-6} **f** 1.0×10^{-9}

   g 2.8×10^{-5} **h** 9.4×10^7 **i** 2.45×10^{-3}

3 Use a calculator and give the answers in standard form to 3 significant figures.

   a $5 \times 10^4 + 9 \times 10^6$ **b** $3.27 \times 10^{-3} \times 2.4 \times 10^2$

   c $5(8.1 \times 10^9 - 2 \times 10^7)$ **d** $(3.2 \times 10^{-1}) - (2.33 \times 10^{-3})$

4 Simplify the following without using a calculator and give the answers in standard form.

   a $(1.44 \times 10^7) + (4.3 \times 10^7)$ **b** $(4.9 \times 10^5) \times (3.6 \times 10^9)$

   c $(3 \times 10^4) + (4 \times 10^3)$ **d** $(4 \times 10^6) \div (3 \times 10^5)$

5

   a Write 8.2×10^5 as an ordinary number. *(1 mark)*

   b Write 0.000 376 in standard form. *(1 mark)*

   c Work out the value of $(2.3 \times 10^{12}) \div (4.6 \times 10^3)$

      Give your answer in standard form. *(2 marks)*

©*Pearson Education Ltd 2013*

6 The Sun has a mass of approximately 1.998×10^{27} tonnes. The planet Mercury has a mass of approximately 3.302×10^{20} tonnes.

   a Which has the greater mass?

   b How many times heavier is the greater than the smaller mass? Give your answer to 3 significant figures.

7 The UK has an approximate area of $2.4 \times 10^5 \, \text{km}^2$. Australia has an area of approximately $7.6 \times 10^6 \, \text{km}^2$.

   a What is the difference in the areas of the two countries? Give the answer in standard form.

   b What is the combined area of the two countries? Give the answer in standard form.

   c How many times bigger is the area of Australia than the area of the UK?

28 Similarity

In this chapter you will learn how to ...
- identify similar triangles and prove that two triangles are similar.
- apply the concept of similarity to calculate unknown lengths.
- work with scale factors to enlarge shapes on a grid.
- find the scale factor and centre of enlargement of a transformation.

For more resources relating to this chapter, visit GCSE Mathematics Online.

Using mathematics: real-life applications

When you enlarge a photo, project an image onto a screen or make scaled models you are dealing with similarity. Many toys and other objects are scaled, but similar, versions of larger objects.

"I work with scale drawings and scale models all the time. The models are mathematically similar to the real planes so the clients can see what they are buying. We made these scaled models to display at an international air show." *(Aircraft designer)*

Before you start ...

Ch 25	You need to be able to label angles correctly.	**1**	**a** Which angle is a right angle? **b** What size is the acute angle *BOA*? **c** What size is the obtuse angle *BOD*?
Ch 14	You need to know how to solve simple equations using inverse operations.	**2**	Solve **a** $3x = 24$ **b** $16 = \dfrac{h}{4}$ **c** $6.25 = 25k$
Ch 19	You need to be able to recognise numbers in equivalent ratios.	**3**	Which pairs of numbers are in the same ratio as $3:2$? **A** $6:5$ **B** $4:6$ **C** $0.15:0.1$
		4	Given that $\dfrac{x}{15} = \dfrac{4}{90}$, find x.

Find answers at: cambridge.org/ukschools/gcsemaths-studentbookanswers

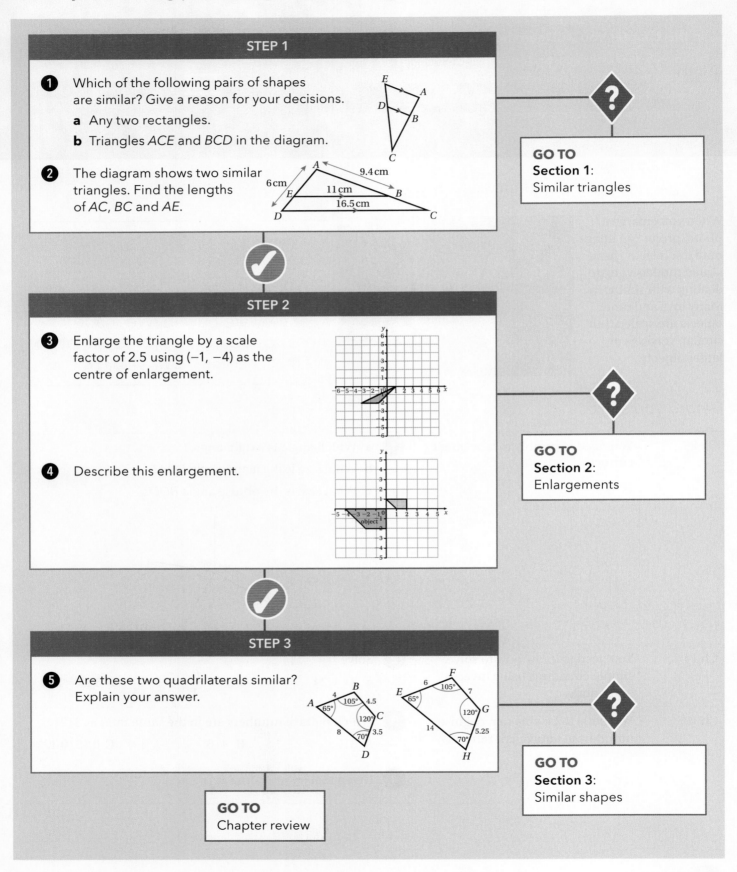

Section 1: Similar triangles

Two shapes are mathematically similar if they have the same shape and proportions but are different in size.

If the corresponding angles in two triangles are equal, then the corresponding sides will be in proportion, and the triangles will be similar.

To prove that two triangles are similar, you have to show that one of these statements is true:

- All the corresponding angles are equal.
- The three sides are in proportion.
- Two sides are in proportion and the included angle (between these two sides) are equal.

You must name triangles with the corresponding vertices in the correct order when you state facts about similarity.

WORKED EXAMPLE 1

Prove that the triangles ABC and RTS are similar.

State which angles are equal and which sides are in proportion.

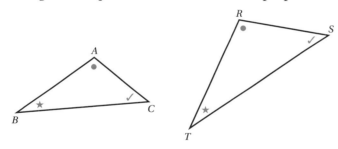

Triangle ABC is similar to triangle RTS because angle A = angle R, angle B = angle T and angle C = angle S.

The three sides are in proportion, so $\dfrac{AB}{RT} = \dfrac{AC}{RS} = \dfrac{BC}{TS}$.

WORKED EXAMPLE 2

Given that the following relationship exists between the sides of triangle PQR and triangle WXY, write down which angles are equal.

$$\dfrac{PQ}{WY} = \dfrac{PR}{WX} = \dfrac{QR}{YX}$$

Triangle PQR is similar to triangle WYX.
Therefore $\angle P = \angle W$; $\angle Q = \angle Y$ and $\angle R = \angle X$.

Finding unknown lengths using proportional sides

You can use the ratio of corresponding sides to find the lengths of unknown sides in similar figures.

Problem-solving framework

In the figure, triangle ABC is similar to triangle QRP. Find the length of x.

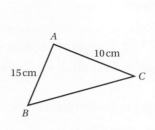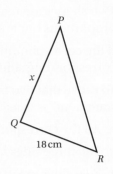

Steps for approaching a problem-solving question	What you would do for this example
Step 1: Identify the similar triangles in the problem and write down with the vertices in the correct order.	Triangle ABC is similar to triangle QRP.
Step 2: Write down what you know.	That the triangles are similar. Two sides of triangle ABC. One side of triangle QRP.
Step 3: Find the ratio between the sides.	AB corresponds to QR so the ratio is 15 : 18.
Step 4: Write a proportion with the unknown side.	$\dfrac{AB}{QR} = \dfrac{15}{18} = \dfrac{10}{x}$
Step 5: Solve the proportion equation.	$x = \dfrac{18 \times 10}{15} = 12\,\text{cm}$
Step 6: Have you answered the question?	$x = 12\,\text{cm}$

EXERCISE 28A

1 Each diagram below contains a pair of similar triangles. Identify the matching angles and the sides that are in proportion. Explain your reasoning using the correct angle vocabulary.

a 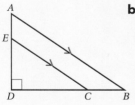   b 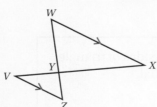   c 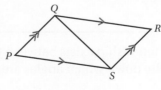   d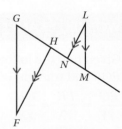

2 Are the following pairs of triangles similar? Explain your answers.

a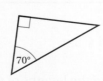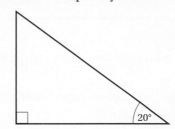

b

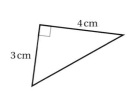
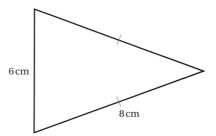

c

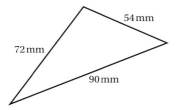
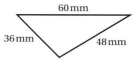

3 Are the statements below true or false? Explain your reasoning, give a counter-example for any statement you believe is false.

  a All isosceles triangles are similar.
  b All equilateral triangles are similar.
  c All right-angled triangles are similar.
  d All right-angled triangles with an angle of 30° are similar.
  e All right-angled isosceles triangles are similar.
  f No pair of scalene triangles are ever similar.

4 Each diagram below contains three similar triangles. Identify the matching angles and sides in each group of triangles, explaining your reasoning.

a
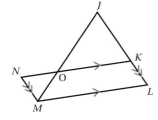

b
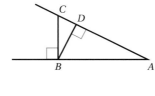

5 The two shapes below are similar. Find the missing lengths c and d.

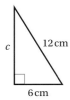
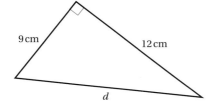

6 The two shapes below are similar. Find the missing lengths *e* and *f*.

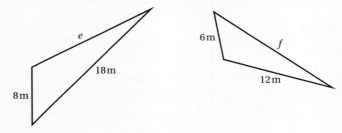

7 **a** Explain why triangles *EDC* and *ADB* are similar.
   b Find the lengths of *AE*, *CE* and *AB*.

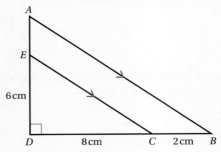

8 Find the lengths of *YZ* and *XY*.

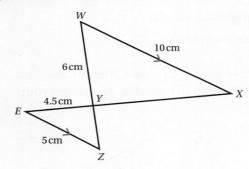

9 Arrange the lengths in the correct positions to complete the similar shapes.

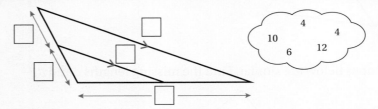

10 Arrange the lengths in the correct positions to complete the similar shapes.

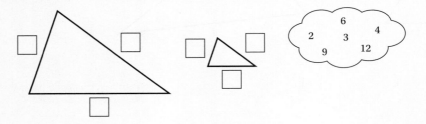

Section 2: Enlargements

An enlargement is a transformation that changes the size of a shape as well as its position. Under enlargement, an object and its image are similar shapes.

In mathematics, the word enlargement is used for all transformations that produce similar images even if the image is smaller than the original object. When you take a photo, the image you see on your screen is considered to be an enlargement of the scene in front of you even though it is smaller.

To construct an enlargement of a shape, you multiply the length of each side by the scale factor.

Before you enlarge a shape, consider its new dimensions.

Has it got bigger? | Has it stayed the same size? | Has it got smaller?

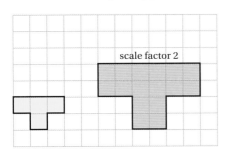
scale factor 2

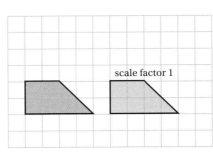
scale factor 1

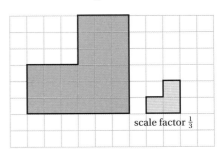
scale factor $\frac{1}{3}$

The object and its image are similar under enlargement. Sides are in the ratio $1:k$, where k is the scale factor.

WORKED EXAMPLE 3

Draw an enlargement of triangle ABC by a scale factor of 2.

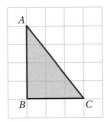

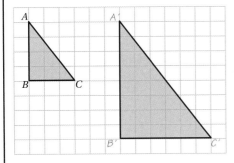

AB is 4 squares long so $A'B'$ is $2 \times 4 = 8$ squares long.

BC is 3 squares long so $B'C'$ is 6 squares long.

The angle at B is 90°.

Draw lines $A'B'$ and $B'C'$ at right angles to each other and join A' to C'.

Notice that triangle ABC is similar to triangle $A'B'C'$ and that the sides are in proportion.

EXERCISE 28B

1 Enlarge each shape as directed.

   a Enlarge A by a scale factor of 3.
   b Enlarge B by a scale factor of 0.5.
   c Enlarge C by a scale factor of $1\tfrac{1}{2}$.

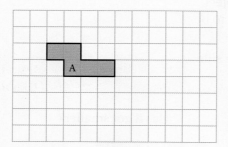
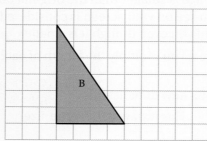
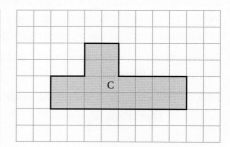

The centre of enlargement

You need two pieces of information to accurately draw an enlargement.

- The scale factor.
- The centre of enlargement.

The centre of enlargement is the point from where the enlargement is measured. In Worked example 3, you drew the enlargements at any convenient position on the grid. When you use a centre of enlargement, you draw the enlargement in a certain position in relation to the original object.

WORKED EXAMPLE 4

Enlarge the rectangle ABCD from the given centre by a scale factor 2.

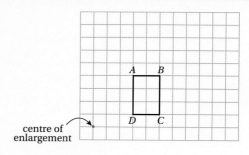

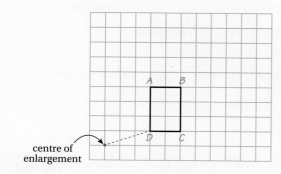

Find the distance from the centre of enlargement to a point on the object. You can draw a ray from the centre to the point.

Continues on next page …

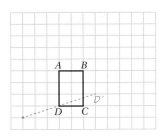

The original ray is three units to the right and one unit up. Double the distance of the ray to find the image of point D. Label it D'.

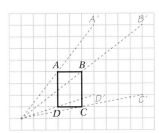

Follow the same process of drawing rays and extending them to find the images of all the vertices $ABCD$. Label them $A'B'C'D'$.

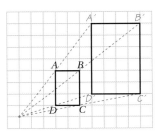

Draw in the image. Check that the lengths of the image are correct.

Note that lines from the corresponding vertices of the object and its image will meet at the centre of enlargement.

The procedure is the same for a centre of enlargement in any position, even for a centre of enlargement inside the shape itself.

The centre of enlargement can be anywhere: inside the object, on a vertex or side of the object or outside the object.

WORK IT OUT 28.1

This triangle is enlarged from centre $(-3, 4)$ scale factor 2. Draw its image.

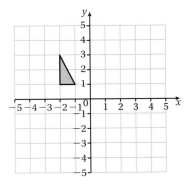

Which one of these answers is correct? Why are the others wrong?

How many marks would you give the incorrect answers if you were the teacher? Why?

Continues on next page …

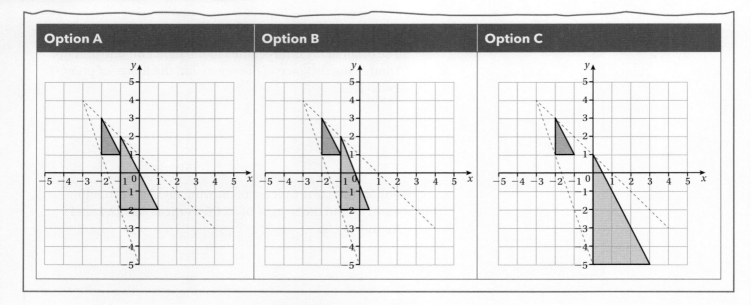

Tip

Sketch the new shape before you construct the enlargement. Draw one ray to identify the new position of the shape. After drawing the enlargement add in additional rays to check that it is in the correct position.

A fractional scale factor

If the scale factor is a fraction, the image will be smaller than the object.

A scale factor greater than 1 will enlarge the object. A scale factor smaller than 1 will reduce the size of the object although this is still called an enlargement.

WORKED EXAMPLE 5

Enlarge the triangle on the grid by a scale factor of $\frac{1}{2}$ through the given centre of enlargement.

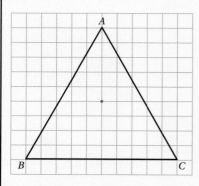
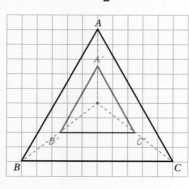

In this case, the rays from each vertex of the object are halved to find the position of the image.

EXERCISE 28C

1 Enlarge each shape as directed using the point C as the centre of enlargement.

  a Enlarge shape R by a scale factor of 3.
  b Enlarge shape S by a scale factor of 2.
  c Enlarge shape T by a scale factor of $\frac{1}{2}$.

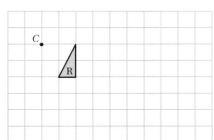
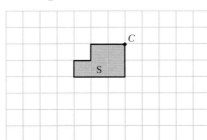
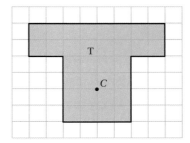

2 Enlarge the given shape by a scale factor of 3 using the origin as the centre of enlargement.

3 Enlarge the given shape by a scale factor of 2 using $(-4, 3)$ as the centre of enlargement.

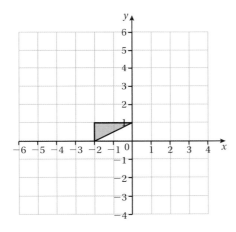
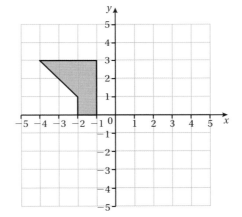

4 Enlarge the given shape by a scale factor of $\frac{1}{3}$ using $(-5, 2)$ as the centre of enlargement.

5 Enlarge the given shape by a scale factor of $1\frac{1}{2}$ using $(0, 1)$ as the centre of enlargement.

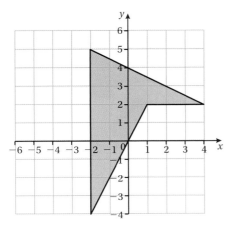
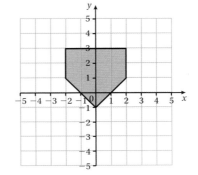

Describing enlargements

Tip

The ratio of sides gives the scale factor.

To describe an enlargement you need to give:
- the scale factor.
- the centre of enlargement.

To find the centre of enlargement you need to draw lines from corresponding vertices of the object and its image to find the point where they meet.

WORK IT OUT 28.2

What scale factors have been used to enlarge this shape?

Which one of these students' answers is correct?

What feedback would you give each student to make sure they don't make the same mistakes again?

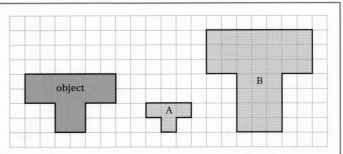

Ben	Ellie	Rosie
A scale factor of 2 has been used to produce shape A. The top of shape A is 3 squares across, multiply this by 2 to get 6, the length of the top of the object. B can't be an enlargement. Its top is 7 squares and the object is 6. You can't do that using multiplication. Maybe it's -1?	Since the sides have halved in length to get A, the scale factor is $\frac{1}{2}$. Shape B isn't an enlargement. The sides have been increased by different numbers of squares.	To get shape A you have to take away one square along the bottom and along each side edge. So the scale factor is -1. Shape B is a scale factor of $1\frac{1}{2}$ because 2 squares have become 3 squares.

EXERCISE 28D

1. Which of the photos below show an enlargement of the original? How can you tell?

Original

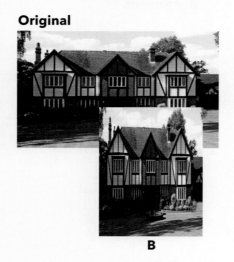

B

A

C

2 Which of the following houses are enlargements of house A? For each enlargement state the scale factor.

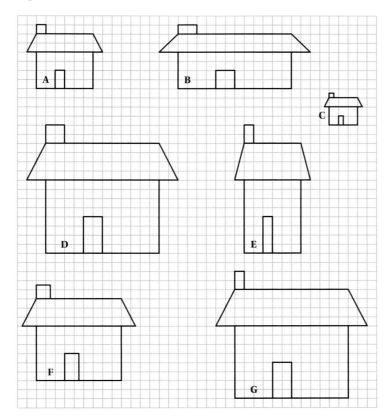

3 These diagrams each show an object and its image after an enlargement. Describe each of these enlargements by giving both the scale factor and the coordinates of the centre of enlargement. In each case the object is labelled.

a
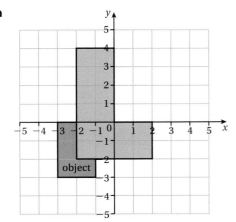

b
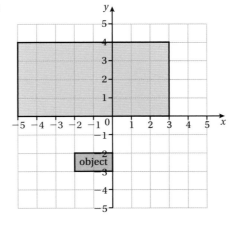

c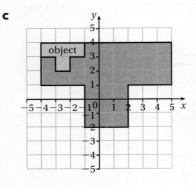

d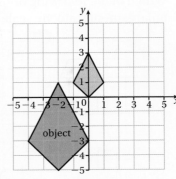

e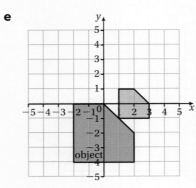

f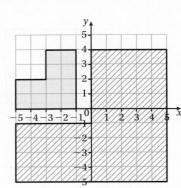

Section 3: Similar shapes

A polygon is similar to another polygon if it is an enlargement of the original polygon. So two polygons will be similar if the angles in one polygon are equal to the angles in the other polygon, **and** the ratio of the sides from the one polygon to the other are kept the same.

For polygons other than triangles, equal angles alone are not sufficient to prove similarity.

WORKED EXAMPLE 6

Compare each of the quadrilaterals B to D below to the first quadrilateral, A. Which of them are similar to A?

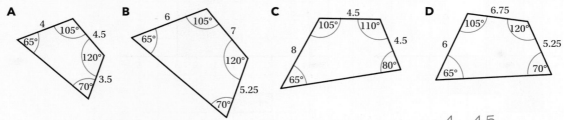

B has the same size angles as A, but the sides are not in proportion. For example, $\frac{4}{6} \neq \frac{4.5}{7}$. So B is not similar to A.

The angles in C are different to the angles in A, so C is not similar to A.

D has corresponding angles equal to those in A. Test to see whether the sides are in the same proportion:

$\frac{4}{6} = \frac{4.5}{6.75} = \frac{3.5}{5.25}$. So D is similar to A.

D is an enlargement of A with a scale factor of 1.5.

EXERCISE 28E

1 Decide whether each statement below is true or false. Explain your reasoning.

  a All squares are similar.

  b All hexagons are similar.

  c All rectangles are similar.

  d All regular octagons are similar.

2 Sketch the following pairs of shapes and decide if they are similar. Explain your reasoning.

  a Rectangle ABCD with AB = 5 cm and BC = 3 cm.

  Rectangle EFGH with EF = 10 cm and FG = 6 cm.

  b Rectangle ABCD with AB = 5 cm and BC = 3 cm.

  Rectangle EFGH with EF = 10 cm and FG = 9 cm.

  c Square ABCD with AB = 4 cm.

  Square EFGH with EF = 6 cm.

3 The two shapes below are similar. Find the missing lengths a and b.

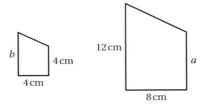

4 The two shapes below are similar. Find the lengths of the missing sides in the second shape.

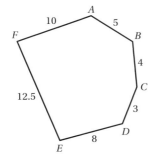
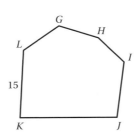

5 The first shape below has been enlarged by a scale factor of 1.5 to create the image GHIJKL. AB = 5 cm and BC = 7 cm. Find the lengths of sides JK and GL in the second shape.

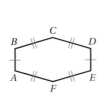
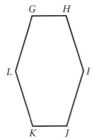

6 Emily drew the diagram below as a plan for a mass display at an athletics stadium. The actual size of each side will be 10 m.

5 cm

    a What is the scale factor of enlargement of the plan?

    b Write the scale as a ratio.

Checklist of learning and understanding

Similar triangles

- Two triangles are similar if all three corresponding angles are equal. Similar triangles are the same shape, and their corresponding sides are in proportion.
- The proportion between corresponding sides of similar triangles can be used to solve problems in geometry.

Enlargements

- An enlargement is a transformation that changes the position and size of a shape.
- Enlargements are described by a scale factor and centre of enlargement.

Similar shapes

- If shapes are enlarged, similar shapes are created with all their sides in proportion. The proportionality between lengths can be used to solve geometry problems.

For additional questions on the topics in this chapter, visit GCSE Mathematics Online.

Chapter review

1   **a** Prove that triangle *VWX* is similar to triangle *VYZ*.

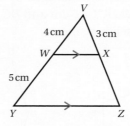

    b Find the length of *XZ*.

2. A tree which is 3 m high has a shadow length of 7.5 m. At the same time of day, a building casts a shadow that is 16.25 m long. Use similar triangles to calculate the height of the building.

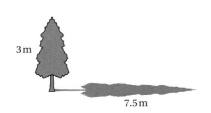
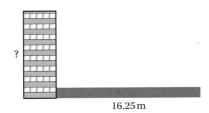

3. Draw an enlargement of ABCD by $\frac{1}{3}$, using the given point as the centre of enlargement.

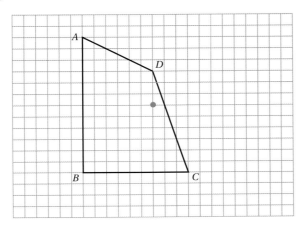

4. Draw an enlargement of this shape by 1.5, using the origin as the centre of enlargement.

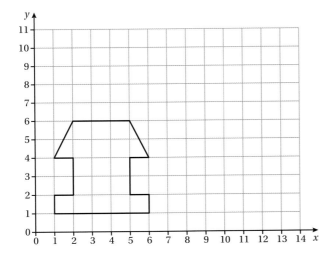

5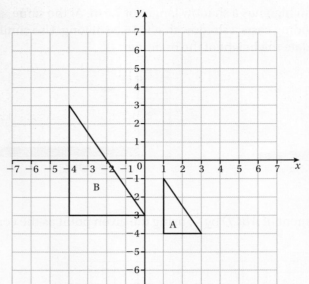

Describe fully the single transformation that maps triangle A onto triangle B. *(3 marks)*

©Pearson Education Ltd 2013

6 Are any two regular hexagons similar shapes? Explain.

7 Are any two rhombuses similar shapes? Explain.

29 Congruence

In this chapter you will learn how to ...
- prove that two triangles are congruent using the cases SSS, ASA, SAS, RHS.
- apply congruency in calculations and simple proofs.

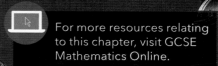

For more resources relating to this chapter, visit GCSE Mathematics Online.

Using mathematics: real-life applications

Congruent triangles are used in construction to reinforce structures that need to be strong and stable.

"When designing any bridge I have to allow for reinforcement. This ensures that the bridge stays strong and doesn't collapse under heavy traffic. Any bridge I design has many congruent triangles." *(Structural engineer)*

Before you start …

Ch 2	You need to know how to label angles and shapes that are equal.	**1** Here are two identical triangles. **a** Write down a pair of sides that are equal in length. **b** What angle is equal in size to angle *BAC*? **c** Write down another pair of angles that are equal in size.	
Chs 2 and 25	You need to know basic angle facts.	**2** Match up the correct statement with the correct diagram. **a** Vertically opposite angles are equal. **b** Alternate angles are equal. **c** Corresponding angles are equal.	
Chs 2 and 25	You should be able to apply angle facts to find angles in figures and to justify results in simple proofs.	**3** Decide whether each statement is true or false. **a** Angle *DBE* = 40° (alternate to angle *ADB*). **b** Angle *BEC* = 50° (complementary to angle *ADB*). **c** Triangle *ABD*, triangle *BDE* and triangle *BCE* are equilateral. **d** Angle *BDE* = angle *BED* = 70°.	
Ch 2	You need to know and be able to apply the properties of triangles and quadrilaterals.	**4** What is the value of *x*? Choose the correct answer. **A** 60° **B** 30° **C** 45° **D** 50°	

Find answers at: cambridge.org/ukschools/gcsemaths-studentbookanswers

Assess your starting point using the Launchpad

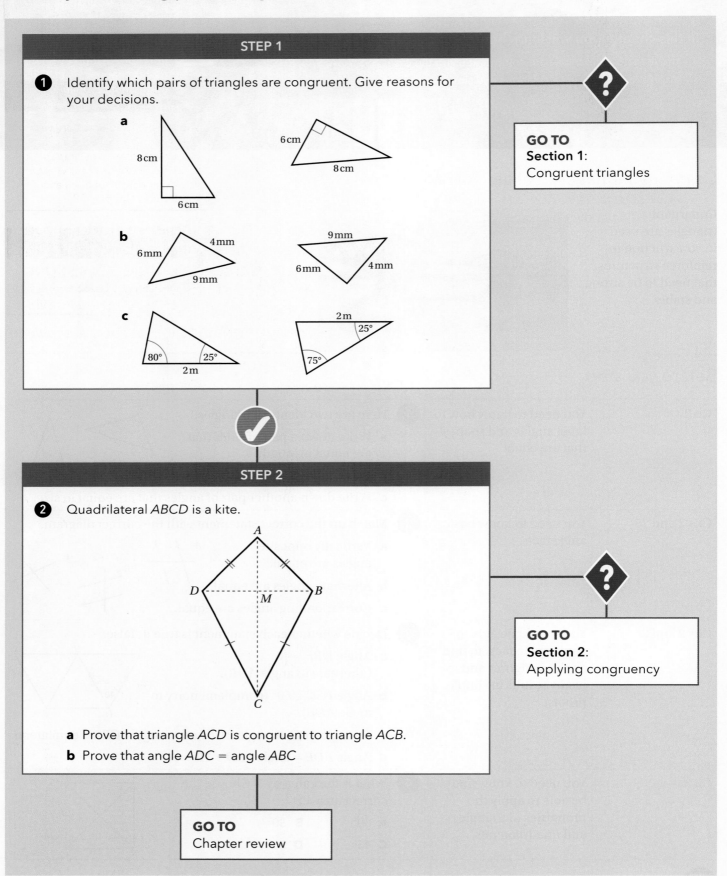

STEP 1

1. Identify which pairs of triangles are congruent. Give reasons for your decisions.

   a, b, c (see figures)

   GO TO Section 1: Congruent triangles

STEP 2

2. Quadrilateral ABCD is a kite.

   a Prove that triangle ACD is congruent to triangle ACB.
   b Prove that angle ADC = angle ABC

   GO TO Section 2: Applying congruency

GO TO Chapter review

Section 1: Congruent triangles

Congruent triangles are identical in shape and all corresponding measurements (lengths and angles) are equal.

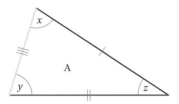 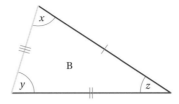

Congruent triangles can have different orientations.

When the triangles are in different orientations you need to think carefully about the corresponding sides and angles.

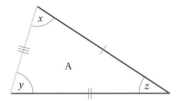 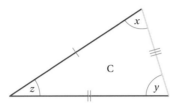

Key vocabulary

congruent: shapes that are identical in shape and size.

Tip

You learned about congruent shapes in Chapter 2. You will work with congruent shapes in different orientations when you deal with reflections and rotations in Chapter 38.

Tip

If you place two congruent triangles on top of each other the angles and sides will match up.

Two triangles are congruent if one of the following sets of conditions is true.

Side side side or **SSS**: the three sides of one triangle are equal in length to the three sides of the other triangle.	
Angle side angle or **ASA**: two angles and one side of one triangle are equal to the corresponding two angles and side of another triangle.	
Side angle side or **SAS**: two sides and the included angle of one triangle are equal to two sides and the included angle of the other triangle.	
Right angle hypotenuse side or **RHS**: the hypotenuse and one side of a right-angled triangle are equal to the hypotenuse and one other side of the other right-angled triangle.	

Find answers at: cambridge.org/ukschools/gcsemaths-studentbookanswers

Tip

It is important to write the letters of the vertices of the two triangles in the correct order.

When you write that triangle ABC is congruent to triangle DEF, it means that:

angle A = angle D, angle B = angle E, angle C = angle F

and

AB = DE; AC = DF and BC = EF.

The conditions in the table are the minimum conditions for proving that triangles are congruent. No other combinations of side and angle facts are sufficient to tell you whether a triangle is congruent or not.

For example:

Two triangles with all their angles equal can still be very different sizes.

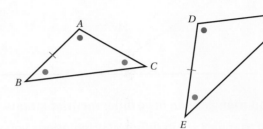

If you are given two triangles that have two equal sides and one equal angle, but where the equal angle is not included (between the two given sides), you do not know if they are congruent or not. The third side might have a different length in the two triangles.

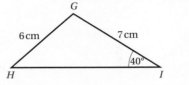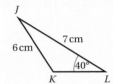

Although a pair of triangles with two sides and one angle matching might actually be congruent, there is not enough evidence to prove that they are.

WORK IT OUT 29.1

Here are three proofs for congruence for the pair of triangles.

Which one uses the correct reasoning?

Why are the others incorrect?

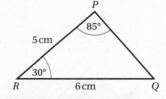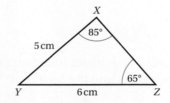

Option A	Option B	Option C
In triangle PQR and triangle XYZ:	In triangle PRQ and triangle XYZ:	In triangle PRQ and triangle XYZ:
PR = XY = 5 cm	In triangle PRQ, angle Q = 65° (sum of angles in a triangle)	PR = XY = 5 cm
angle P = angle X = 85°	angle Q = angle Z = 65°	RQ = YZ = 6 cm
RQ = YZ = 6 cm	RQ = YZ	In triangle XYZ, angle Y = 30° (sum of angles in a triangle)
so triangle PQR is congruent to triangle XYZ (SAS).	so the triangles are congruent.	so triangle PRQ is congruent to triangle XYZ (SAS).

EXERCISE 29A

1 Match up each of the congruency descriptions (SSS, ASA, SAS, RHS) with each pair of triangles below:

a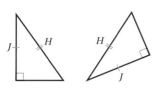

b

c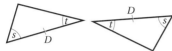

d

2 Which of the following figures show a pair of congruent triangles? In your answer, state whether the triangles are congruent or not, or whether there is insufficient information. Write the triangles with the vertices in the correct order and give the reasons as SSS, SAS, ASA or RHS.

a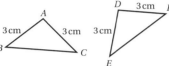

b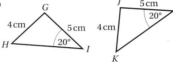

c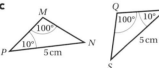

d

e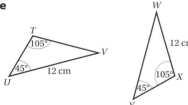

3 Prove that triangle ABC is congruent to triangle DCE.

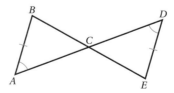

4 EFG is a straight line. Write down two different proofs for congruence of triangles DEF and DGF.

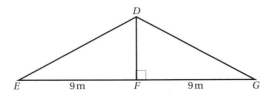

5. In the diagram, PQ is parallel to SR and QT = TR = 2 cm.

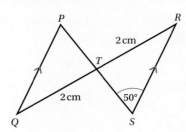

Prove that triangle PQT is congruent to triangle SRT.

6. Prove that triangles ABE and CBD in the figure are congruent, giving full reasons.

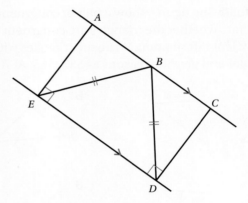

7. Triangle ABD is isosceles. AC is the perpendicular height. Prove that triangle ABC is congruent to triangle ADC.

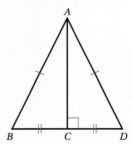

8. In the figure below, PR = SU and angle PUT = angle SRT. Prove that triangle PQR is congruent to triangle SQU.

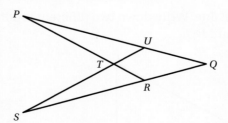

9 ABCD in the figure is a kite.

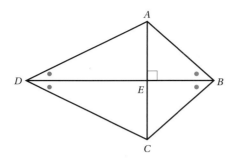

Prove that:

a triangle ADB is congruent to triangle CDB

b triangle AED is congruent to triangle CED.

10 Quadrilateral ABCD is a rhombus.

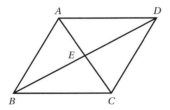

Prove that:

a triangle AED is congruent to triangle CEB

b triangle AEB is congruent to triangle CED.

Section 2: Applying congruency

Problem-solving framework

The following steps are useful for solving geometry problems:

In the diagram, AM = BM and PM = QM.

a Prove that triangle AMP is congruent to triangle BMQ.

b Prove that AP is parallel to BQ.

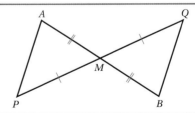

Steps for approaching a problem-solving question	What you would do for this example
Step 1: Read the question carefully to decide what you have to find.	**a** Prove that triangle AMP is congruent to triangle BMQ. **b** Prove that AP ∥ BQ.
Step 2: Write down any further information that might be useful.	The two triangles also have vertically opposite angles, which are equal.

Continues on next page ...

Step 3: Decide what method you'll use.	You are given two equal sides and you can see that the included angle is also equal, so use SAS to prove congruence.
Step 4: Set out your working clearly.	**a** In triangles AMP and BMQ: $AM = BM$ (given). $PM = QM$ (given). angle AMP = angle BMQ (vertically opposite angles at M). So triangle AMP is congruent to triangle BMQ (SAS). **b** Angle $APM = BQM$ (matching angles of congruent triangles). So $AP \parallel BQ$ (alternate angles are equal).

WORKED EXAMPLE 1

Triangle DEF is divided by GF into two smaller triangles. Prove that FG is perpendicular to DE.

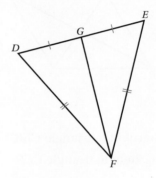

Tip

'Common' means included in both triangles.

In triangle FGD and triangle FGE:

Side FG is common to both triangles.

$DG = GE$

$DF = EF$

So the triangles are congruent (SSS).

Angle DGF = angle EGF and the two angles lie on a straight line.

So each angle = 90°, and FG is perpendicular to DE.

EXERCISE 29B

1 In the diagram, prove that $KL = ML$.

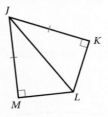

2 Use the facts given in the diagram to:
  a prove that angle *ABE* = angle *EDC*
  b prove that quadrilateral *ABCD* is a parallelogram.

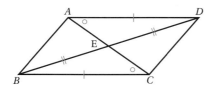

3 In the quadrilateral, *SP* = *SR*, *QP* = *QR* and *QP* ∥ *RS*. Angle *QRP* = 56°.

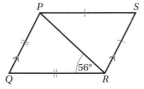

  a Calculate the size of angle *PSR*.
  b What does this tell you about quadrilateral *PQRS*?

4 In the diagram, *PQ* = *PT* and *QR* = *ST*. Prove that triangle *PRS* is isosceles.

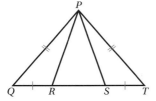

5 In the figure, prove that:
  a triangle *AEB* is congruent to triangle *CEB*
  b angle *EAD* = angle *ECD*.

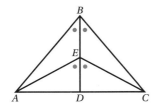

6 In quadrilateral *ABCD*, *AD* = *BC* and *AD* ∥ *BC*.
Prove that angle *B* = angle *D*.

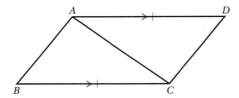

Checklist of learning and understanding

Congruent triangles

- You can prove that two triangles are congruent using one of the four cases of congruence:
  - Side side side or SSS: the three sides of one triangle are equal in length to the three sides of the other triangle.
  - Angle side angle or ASA: two angles and one side of one triangle are equal to the corresponding two angles and side of another triangle.
  - Side angle side or SAS: two sides and the **included** angle of one triangle are equal to two sides and the **included** angle of the other triangle.
  - Right angle hypotenuse side or RHS: the hypotenuse and one side of a right-angled triangle are equal to the hypotenuse and one other side of the other right-angled triangle.

 For additional questions on the topics in this chapter, visit GCSE Mathematics Online.

 Chapter review

1. State whether these pairs of triangles are congruent. Give reasons for your answers and give the vertices of the triangles in the correct order.

   a 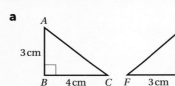   **b**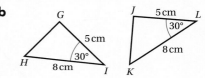

   c 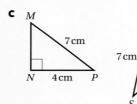   **d**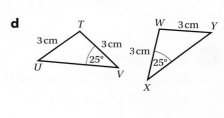

2. Use triangle congruence to prove that triangle EBA is congruent to triangle ECD.

   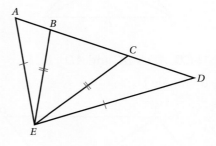

3. *ABCD* is a parallelogram. *CD* is produced to *E* such that when *E* is joined to *B*, it bisects *AD* at *F*. Prove that triangle *ABF* is congruent to triangle *DEF*.

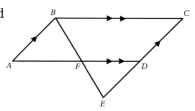

4. Prove that triangle *PQR* is congruent to triangle *RST*.

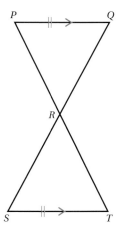

5. Prove that *WX* = *XV* in the diagram.

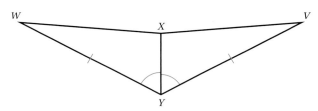

6.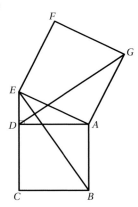

In the diagram,

ADE is a right-angled triangle,

ABCD and *AEFG* are squares.

Prove that triangle *ABE* is congruent to triangle *ADG*.   (3 marks)

© Pearson Education Ltd 2013

30 Pythagoras' theorem

In this chapter you will learn how to …
- develop knowledge and understanding of Pythagoras' theorem.
- apply Pythagoras in 2D problems.
- link the maths to real-life situations.

 For more resources relating to this chapter, visit GCSE Mathematics Online.

Using mathematics: real-life applications

Builders, carpenters, garden designers and navigators all use Pythagoras' theorem in their jobs. It is a method based on right-angled triangles which helps them to work out unknown lengths.

 Calculator tip

Make sure you know how to square a number and calculate a square root using your calculator.

 Tip

When answering questions about Pythagoras' theorem it might be useful to draw a diagram if one isn't provided.

"I use Pythagoras' theorem to help me navigate the ship … I need to know how far away we are from port and I use the theorem to calculate this." *(Navigation officer)*

Before you start …

Chs 4 and 8	You need to calculate confidently with squares and square roots.	**1 a** True or false? 　**i** $\sqrt{100} = \sqrt{10}$　　**ii** $\sqrt{100} = 10$ 　**iii** $\sqrt{100} = \pm 10$　**iv** $10^2 = 100$ **b** True or false? 　**i** $4^2 = 4 \times 2 = 8$　**ii** $3^3 = 3 \times 3 \times 3 = 27$	
Ch 25	You need to recognise different angle types and define them.	**2 a** Which is a right angle? Identify the other angles. 　**i**　　　**ii**　　　**iii** 　**iv**　　　**v** **b** What is the total number of degrees when three right angles are added? *Continues on next page …*	

| Ch 2 | You will need to apply the properties of different types of triangles to solve problems. | 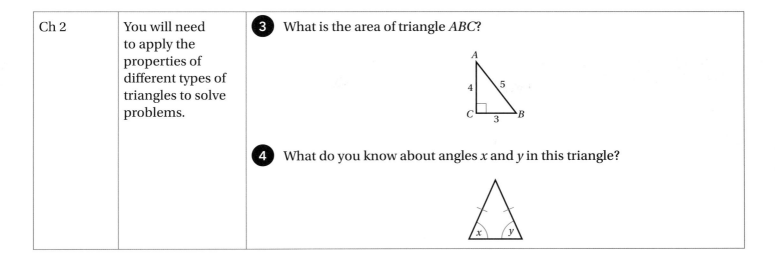 |

3. What is the area of triangle *ABC*?

4. What do you know about angles x and y in this triangle?

Assess your starting point using the Launchpad

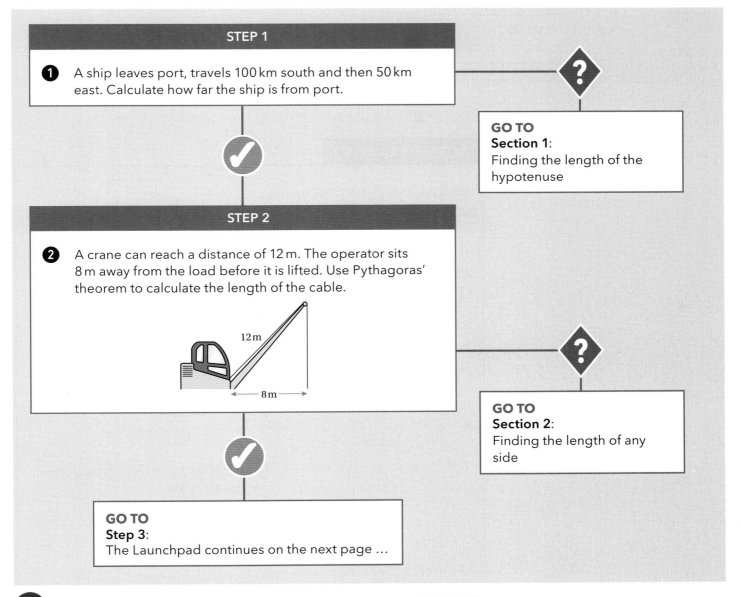

STEP 1

1. A ship leaves port, travels 100 km south and then 50 km east. Calculate how far the ship is from port.

GO TO Section 1: Finding the length of the hypotenuse

STEP 2

2. A crane can reach a distance of 12 m. The operator sits 8 m away from the load before it is lifted. Use Pythagoras' theorem to calculate the length of the cable.

GO TO Section 2: Finding the length of any side

GO TO Step 3: The Launchpad continues on the next page …

Find answers at: cambridge.org/ukschools/gcsemaths-studentbookanswers

Launchpad continued ...

STEP 3

3 A garden designer wants to build some decking that fits against a corner fence. Do the measurements in his design make a right angle?

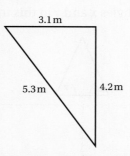

GO TO
Section 3: Proving whether a triangle is right-angled

Section 4: Using Pythagoras' theorem to solve problems

GO TO Chapter review

Section 1: Finding the length of the hypotenuse

WORKED EXAMPLE 1

Find the area of this tilted square.

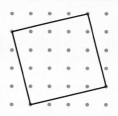

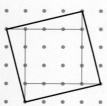

You can divide the square into right-angled triangles and a square.

Area = $\frac{1}{2} \times b \times h$

$= \frac{1}{2} \times 4 \times 1$

$= 2$ squares

Each triangle has a base of four squares and a height of one square. Work out the area.

$4 \times 2 = 8$ squares

Multiply by four to get the total area of the triangles.

$l \times w = 3 \times 3 = 9$ squares

Find the area of the square in the middle.

The area of the tilted square is $8 + 9 = 17$ squares

Add the smaller areas together.

EXERCISE 30A

1 In this diagram, squares have been drawn on the three sides of a right-angled triangle.
Find and record the area of each square in a table like the one below.
Draw some different right-angled triangles and repeat the process.
Make sure that the square labelled C is opposite the right angle.

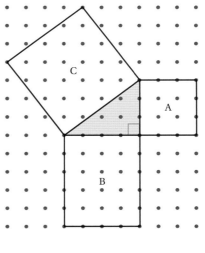

Area of square A	Area of square B	Area of square C

What do you notice about the relationship between the different areas?

What is Pythagoras' theorem?

Pythagoras' theorem describes the relationship between the lengths of the sides of a right-angled triangle.
The **theorem** states:
In a right-angled triangle, the square of the length of the **hypotenuse** is equal to the sum of the squares of the two shorter sides.

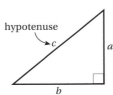

For this triangle, the theorem can be expressed using the formula:
$a^2 + b^2 = c^2$

Key vocabulary

theorem: a statement that can be demonstrated to be true by accepted mathematical operations.

hypotenuse: the longest side of a right-angled triangle; the side opposite the 90° angle.

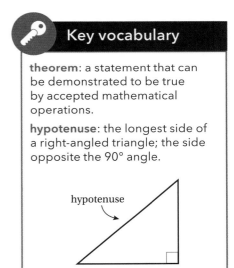

WORK IT OUT 30.1

A right-angled triangle has two shorter sides of 3 cm and 4 cm. Calculate the length of the hypotenuse.

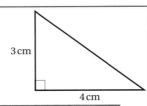

Which of these builders has got the correct answer? Why are the others wrong?

Builder A	Builder B	Builder C
$a^2 + b^2 = c^2$	$a^2 + b^2 = c^2$	$a^2 + b^2 = c^2$
$3 \times 2 = 6$	$3 \times 3 = 9$	$3 \times 3 = 9$
$4 \times 2 = 8$	$4 \times 4 = 16$	$4 \times 4 = 16$
$6 + 8 = c^2$	$9 + 16 = c^2$	$9 + 16 = c^2$
$14 = c^2$	$25 = c^2$	$25 = c^2$
$c = \sqrt{14} = 3.74$ cm	$c = \sqrt{25} = 5$ cm	$c = 25$ cm
(to 2 decimal places)		

Tip

Naming conventions:
always use capital letters for a vertex
hypotenuse is called c
side a is opposite angle A
side b is opposite angle B

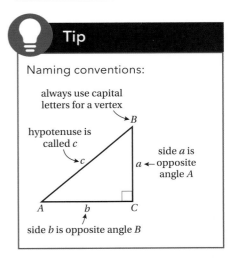

You can use your calculator to find the square of a number by pressing the x^2 button. To find 3^2 press [3] [x^2] [=] and you get 9.

The reverse of squaring is finding the square root. The square root of 9 is 3. This is written as $\sqrt{9} = 3$. You can use the [√▢] button on your calculator to work out a square root.

To find $\sqrt{9}$ press [√▢] [9] [=] and you get 3.

EXERCISE 30B

1 a Calculate:

  i 25^2 **ii** 5.3^2 **iii** 167^2 **iv** 136^2 **v** 14.5^2

  b Calculate, working to 2 decimal places:

  i $\sqrt{3}$ **ii** $\sqrt{7}$ **iii** $\sqrt{4}$ **iv** $\sqrt{17}$ **v** $\sqrt{61}$

2 Find the length of the hypotenuse in each of the following triangles.

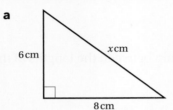
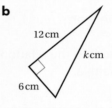
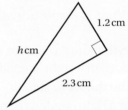

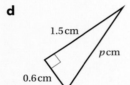
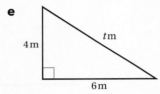

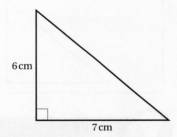

3 a Jamie calculates that the hypotenuse of the triangle on the left is 14 cm. Without calculating the length yourself, just by inspecting the triangle, explain why this must be wrong.

  b He recalculates and gets 7 cm for the length of the hypotenuse. How do you know that this is wrong?

  c What range of answers could Jamie have given where you wouldn't know straight away that his answer was wrong?

Section 2: Finding the length any side

If you know the lengths of any two sides of a right-angled triangle, you can use Pythagoras' theorem to find the length of the third side either by rearrangement or substitution.

In the formula $a^2 + b^2 = c^2$, c is always the hypotenuse and a and b are the two shorter sides.

The formula can be rearranged to make a or b the subject of the formula:

$a^2 = c^2 - b^2$

$b^2 = c^2 - a^2$

WORK IT OUT 30.2

This is the design of an access ramp for the front entrance to a building. What is the vertical height of the step?

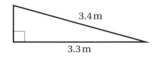

Which of these calculations is correct for this design? Why are the others wrong?

Calculation A	Calculation B	Calculation C
$c^2 - b^2 = a^2$	$c^2 - b^2 = a^2$	$c^2 - b^2 = a^2$
$3.4^2 - 3.3^2 = a^2$	$3.3^2 + 3.4^2 = c^2$	$3.4^2 - 3.3^2 = a^2$
$3.4 \times 2 = 6.8$	$3.3 \times 3.3 = 10.89$	$3.4 \times 3.4 = 11.56$
$3.3 \times 2 = 6.6$	$3.4 \times 3.4 = 11.56$	$3.3 \times 3.3 = 10.89$
$6.8 - 6.6 = a^2$	$10.89 + 11.56 = c^2$	$11.56 - 10.89 = a^2$
$0.2 = a^2$	$22.45 = c^2$	$a^2 = 0.67$
$a = 0.2 \div 2$	$a = \sqrt{22.45}\,\text{m} = 4.74\,\text{m}$	$a = \sqrt{0.67}\,\text{m} = 0.82\,\text{m}$
$a = 0.1\,\text{m}$	(to 2 decimal places)	(to 2 decimal places)

EXERCISE 30C

1 Find the missing length in each of these triangles:

a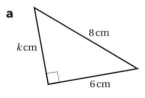

b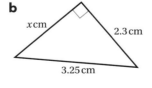

c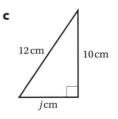

d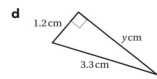

e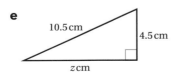

Tip

Always think about whether the answers are realistic and reasonable. The ability to recognise whether values are reasonable will help you to spot mistakes.

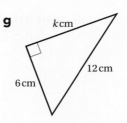
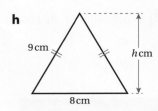

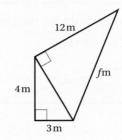

2 Jamil says that the missing side in this triangle is 5.5 cm long.
Without calculating the length, explain why this must be wrong.

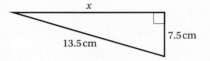

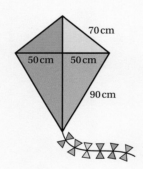

3 In this kite, the dimensions of the two isosceles triangles that make it are given. Using what you know about different types of triangles, work out how long the rod is that holds the kite together from top to bottom.

4 The front of a tent has the dimensions shown in the diagram.
It is a scalene triangle.
The height is 4 m.
What is the length of the base?

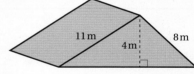

5 Match the missing length to the triangle.

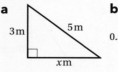
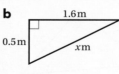
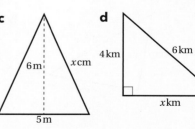

A $x = \sqrt{2.81}$ **B** $x^2 = 6^2 + 2.5^2$ **C** $6^2 - 4^2 = x^2$ **D** $x = 4\,\text{m}$

6 A girl swims across a river. The distance she swims is 20 m from one shore to the other. The distance along the bank she has travelled is 10 m. The banks are parallel. How far apart are they?

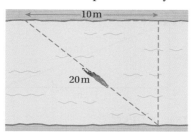

 Tip

Note that there are different types of right-angled triangle:

Scalene right-angled triangle
- one right angle
- two other unequal angles
- no equal sides.

Isosceles right-angled triangle
- one right angle
- two other equal angles, always 45°
- two equal sides.

You can apply Pythagoras' theorem to both types of right-angled triangle.

Section 3: Proving whether a triangle is right-angled

"I use 'Pythagorean triples' in my job – using the 3, 4, 5 rule I know whether the window frame is a true right angle or not. It means the window will fit properly."

(Window fitter)

Carpenters need to make rectangular window frames. If the frames don't have right angles at the corners then the window won't fit. If the sides of the triangle at the corner of the frame have lengths in the ratio of 3, 4 and 5, then the carpenter knows the angle is a right angle because $3^2 + 4^2 = 5^2$. Carpenters know this as the '3, 4, 5 rule'.

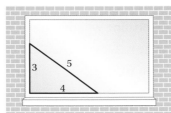

The side lengths 3, 4, 5 are known as a **Pythagorean triple**.

 Tip

There are other Pythagorean triples, such as 5 : 12 : 13.

Key vocabulary

Pythagorean triple: three non-zero integers (a, b, c) for which $a^2 + b^2 = c^2$.

Did you know?

To measure right angles, builders used to carry a coil of rope 12 feet long with knots tied at each 1-foot length. They would put a peg in the ground and then hold the rope taught around it so that there were three knots on one side, four knots on the second side and the remaining five knots on the third side. Putting in pegs at the ends they would then mark out a perfect right angle.

WORK IT OUT 30.3

Is this triangle right-angled?

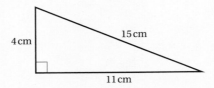

Which of these answers is correct?
Where have the others gone wrong?

Option A	Option B	Option C
$4 + 11 = 15$ So $a^2 + b^2 = c^2$ Yes, the triangle is right-angled!	$a^2 + b^2 = 4^2 + 11^2$ $= 4 \times 2 + 11 \times 2$ $= 8 + 22 = 30$ $c^2 = 15^2 = 15 \times 2 = 30$ $a^2 + b^2 = c^2$ Yes, the triangle is right-angled!	$a^2 + b^2 = 4^2 + 11^2$ $= 16 + 121 = 137$ $c^2 = 15 \times 15 = 225$ $a^2 + b^2 \neq c^2$ No, the triangle is not right-angled!

EXERCISE 30D

1 Use Pythagoras' theorem to help you decide which of the following triangles are right-angled:

a 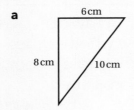   b 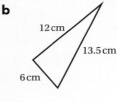   c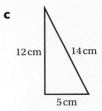

d 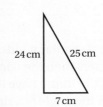   e

2 The lengths of the sides of a number of different triangles are given below. In each case, determine whether the triangles are right-angled.

a 1, 1, 1   b 1, 1, 2   c 5, 7, 9
d 5, 12, 13   e 8, 10, 6   f 1, 2, 2

3 A builder uses 3-4-5 triangles to make right angles.

a Write down the dimensions of three other right-angled triangles where the sides are in the ratio 3 : 4 : 5.

b Some other Pythagorean triples occur regularly. Carry out an investigation to find at least five common triples.

4 A farmer's field has two pairs of straight sides of length 25 m and width 15 m. There is a 35 m long drainage pipe lying diagonally across the field from corner to corner. Could the field be a rectangle with right angles at the corners? Explain your reasoning.

Section 4: Using Pythagoras' theorem to solve problems

WORKED EXAMPLE 2

Find the length of side x and then calculate the perimeter of this composite shape.

All dimensions are in centimetres.

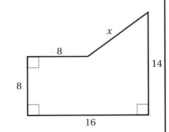

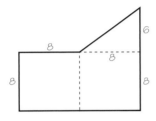

$8^2 + 6^2 = x^2$
$64 + 36 = x^2$
$100 = x^2$
$x = 10\,cm$

Divide the shape to make a right-angled triangle.

Now you can use Pythagoras to find x.

Perimeter $= 16 + 14 + 8 + 8 + 10$
$= 56\,cm$

EXERCISE 30E

In each of the following show your working and give reasons where necessary. Figures are not to scale and all dimensions are in centimetres.

1 Find the length of x in this figure.

2 Determine the length of:
  a BE. **b** BD. **c** BC.

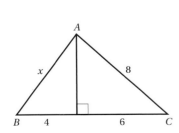

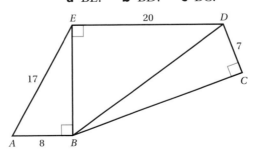

3. Find the length of AB in this figure.

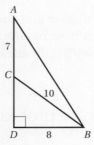

4. Calculate the length of:
   a AC b BC c EC.

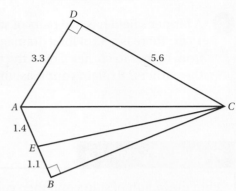

5. For the isosceles trapezium ABCD, calculate the perpendicular height EB and then find the area of the figure.

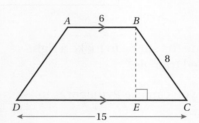

6. Find the length of side AD and calculate the perimeter of this shape.

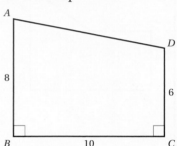

Pythagoras' theorem is useful for solving problems such as finding shortest and longest routes.

Problem-solving framework

You want to cross a 100 metre by 60 metre rectangular football field diagonally from one corner to the other but there is a game playing so you have to go round the edge. Calculate how much further you have to walk.

Steps for approaching a problem-solving question	What you would do for this example
Step 1: If it is useful to have a diagram sketch one and add the information. This may help you visualise the problem.	(diagram of rectangle 100 m by 60 m with diagonal)
Step 2: Identify what you have to do.	Find the difference between the length of the diagonal and the length of the two sides added together.
Step 3: Test the problem with what you know. Can I use a ruler? What type of angle is it?	You could use a ruler, but you would have to draw a very accurate diagram to scale. As you have a right-angled triangle and side lengths you can use Pythagoras' theorem.

Continues on next page …

Step 4: What maths can I do?	Use Pythagoras' theorem to work out the length of the diagonal: $100^2 + 60^2 = c^2$ $10\,000 + 3600 = c^2$ $13\,600 = c^2$ $c = \sqrt{13\,600} = 116.6\,\text{m}$ If you walked straight across the diagonal the distance would be 116.6 m. You have to go round the outside which is 160 m (100 + 60), so you must walk 43.4 m further.
Step 5: Check your workings and that your answer is reasonable.	A diagonal pitch length of 116.6 m seems reasonable given the sides are 60 m and 100 m.
Step 6: Have you answered the question?	You were asked to find how much further you would have to walk. You have found this to be 43.4 m.

This type of problem solving is common in many shape and space questions. It is useful to understand the process that you will go through to begin to solve the problem. These steps can be used to solve lots of problems.

EXERCISE 30F

1 Computer gaming designers use x- and y-coordinates to place characters or objects in a game. They need to know distances between characters or how far players are apart.

If one player is at coordinate (30, 10) and the other at (15, 4), how far apart are they?

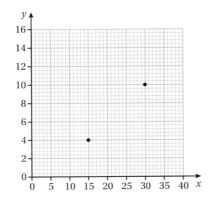

2 In adverts, the size of a screen on a television or computer (for example 15 inches) is actually the length of the diagonal. The diagram on the right shows the length and breadth of a screen. How would this be advertised?

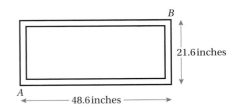

3. You are going to buy a new television. It is an 80-inch television (measured along its diagonal length).

   Your current television is a 52-inch television.

   Both are the same height, 40 inches.

   a How much wider is your new television than your current one?

   b More importantly, will it fit in the 58-inch gap between the chimney and wall?

4. The diagram shows the side view of a shed.
   Calculate the height from the ground to the top of the roof of the shed.

   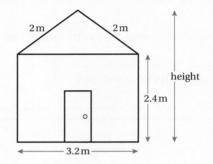

Tip

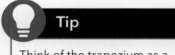

Think of the trapezium as a composite shape. How can you divide it up to solve the problem?

5. Show how you could use Pythagoras' theorem to find the length of side BD in this trapezium.

   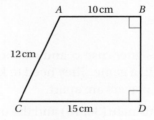

6. The height from a yacht's boom to the top of its mast is 8 m. The longest side of the main sail is 11.4 m.

   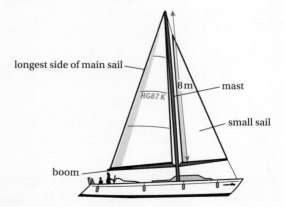

   a How far does the boom of the sail swing out?

   The smaller sail is 1.5 m wide at the bottom.

   b How long is the longest side of the smaller sail?

   State any assumptions you have made about the lengths you used in your calculations.

Checklist of learning and understanding

Pythagoras' theorem

- Pythagoras' theorem only applies to right-angled triangles.
- In a right-angled triangle, the square of the hypotenuse (longest side) equals the sum of the squares of the other two sides. It can be written as:
  $a^2 + b^2 = c^2$
- Rearrange the formula to find either of the other two sides:
  $a^2 = c^2 - b^2$
  $b^2 = c^2 - a^2$
- You can use the theorem to find an unknown length, or to prove there is a right angle within a triangle.
- Always draw a diagram and label it; also write out the formula you are using to fully explain what you have done.
- Pythagorean triples are three whole number lengths that satisfy the formula and therefore prove you have a right angle; learn the common ones, such as 3, 4, 5 and 5, 12, 13.

Chapter review

For additional questions on the topics in this chapter, visit GCSE Mathematics Online.

1 Which of the following statements is true?

A Using Pythagoras' theorem you can find any angle within a right-angled triangle.

B If you know one length in a right-angled triangle you can find the other two.

C You can show that 3, 4, 5 is a Pythagorean triple using Pythagoras' theorem.

D The hypotenuse is always the longest side in a triangle.

2 XYZ is a right-angled triangle.

Diagram **NOT** accurately drawn

Calculate the length of XZ.

Give your answer correct to 3 significant figures. *(3 marks)*

©Pearson Education Ltd 2013

3. What is the perimeter of this kite?

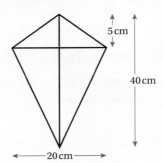

4. A traditional five bar gate measures 4 m wide and 1.5 m high at its highest point where the diagonal pieces of wood meet.

What is the length of each diagonal piece of wood used in the gate?

5. A gardener is laying a rectangular patio from a conservatory that is 9 m wide. To make sure that the angles are right angles and that the edges of the patio are parallel, she measures the diagonal from either end of the conservatory to make sure the length is the same.

If the patio is to be 12 m long, what will the lengths of the diagonals be?

6. On his journey to work, Andrew used to take a shortcut along a farm track.

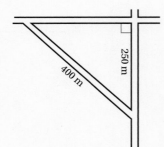

The farm track started 250 m from a crossroads and was 400 m long. The farmer decided to close the track.

a How much further does Andrew now have to travel on his journey (answer to the nearest cm)?

b Andrew does this journey twice each day for five days. How many kilometres extra does he travel in this time (answer to the nearest tenth of a km)?

31 Trigonometry

In this chapter you will learn how to …
- use trigonometric ratios to find lengths and angles in right-angled triangles.
- find and use exact values of trigonometric ratios.
- use trigonometry to solve measurement problems

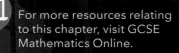

For more resources relating to this chapter, visit GCSE Mathematics Online.

Using mathematics: real-life applications

Trigonometry means 'triangle measurements' and it is very useful for finding the lengths of sides and sizes of angles. Trigonometry is used to determine lengths and angles in navigation, surveying, astronomy, engineering, construction and even in the placement of satellites and satellite receivers.

"I use a theodolite to work out the height of mountains. You basically point it at the top of the mountain. The theodolite uses the principles of trigonometry to measure angles and distances." *(Geologist)*

Before you start …

Ch 30	You should be able to use Pythagoras' theorem to find lengths in triangles.	**1** Find the length of x in each triangle: **a** triangle with legs 7 and 16, hypotenuse x **b** triangle with hypotenuse 7, one leg 5, other leg x	
Ch 9	You must be able to work with approximate values and round to a specified number of places.	**2** What is $\sqrt{53}$ correct to 2 decimal places? **3** If $c^2 = 94.34$, what is c correct to 3 significant figures?	
Ch 19	You need to be able to use ratio and proportion to calculate sides in similar triangles.	**4** Find the length of AC if the ratio of sides $\dfrac{AB}{AC} = \dfrac{5}{3}$ and $AB = 35$ cm.	

Find answers at: cambridge.org/ukschools/gcsemaths-studentbookanswers

Assess your starting point using the Launchpad

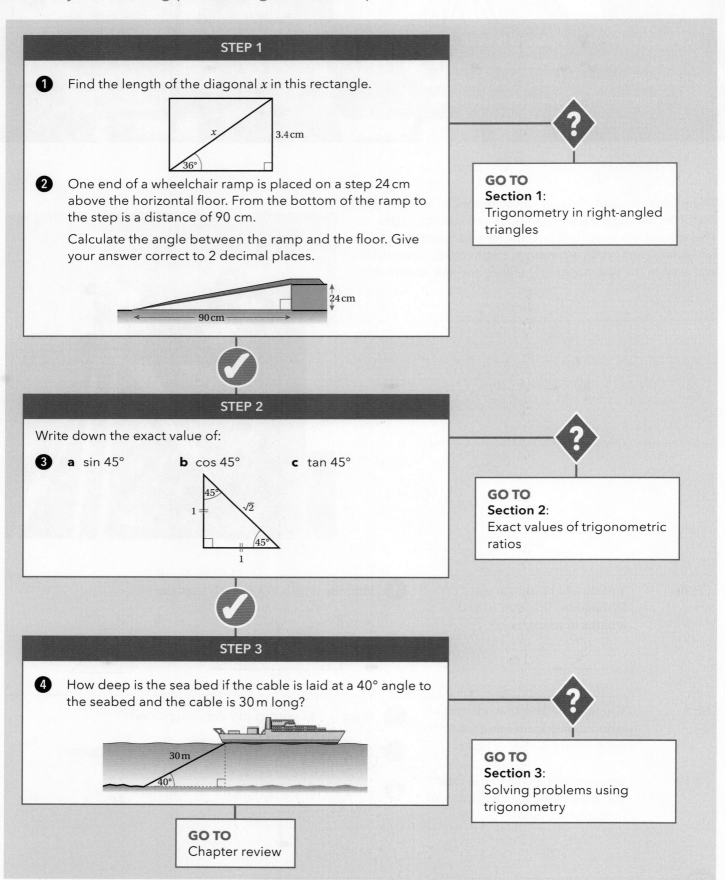

Section 1: Trigonometry in right-angled triangles

There are special relationships between the lengths of the sides of right-angled triangles and you can use Pythagoras' theorem to find missing sides when two sides are known. In similar triangles, the ratio of corresponding pairs of sides is always the same.

These facts are important in understanding and using trigonometry.

Naming the sides of right-angled triangles

The hypotenuse is the longest side of a right-angled triangle, opposite the right angle.

The other two (shorter) sides are named in relation to the acute angles in the triangle.

In this triangle, one of the acute angles is labelled θ (the Greek letter theta).

The sides can then be labelled opposite (that is, opposite angle θ) and adjacent (that is, next to angle θ).

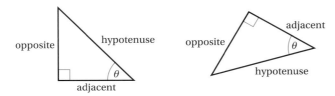

Make sure you understand this system of naming the sides and how it works for triangles in any orientation (drawn any way up).

WORKED EXAMPLE 1

The right-angled triangle XYZ shown here:

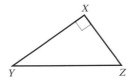

Identify the following sides and angles:

a hypotenuse   b side opposite angle Y c side adjacent to angle Z
d acute angle adjacent to side XY e angle opposite side XZ.

a YZ *The hypotenuse is always opposite the right angle.*
b XZ
c XZ
d Angle Y
e Angle Y

The ratio of sides in similar triangles

This diagram shows three similar right-angled triangles. The green sides are opposite angle θ and the orange sides are adjacent to it. You can see that the ratio of $\frac{\text{opposite}}{\text{adjacent}}$ sides is $\frac{1}{2}$ for all the similar triangles.

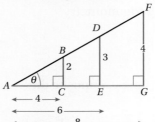

$$\frac{BC}{CA} = \frac{2}{4} = \frac{1}{2} \qquad \frac{DE}{EA} = \frac{3}{6} = \frac{1}{2} \qquad \frac{FG}{GA} = \frac{4}{8} = \frac{1}{2}$$

$$\therefore \frac{\text{opposite}}{\text{adjacent}} = \frac{1}{2}$$

You can extend the diagram forever, but the ratio is always the same for the similar triangles. Depending on which sides you compare, the ratio is given a special name. You will work with three ratios: sine, cosine and tangent.

The trigonometric ratios

The trigonometric ratios (shortened to trig ratios) are named as follows:

- The sine ratio ($\sin \theta$) is the ratio of the side opposite the angle to the hypotenuse.
- The cosine ratio ($\cos \theta$) is the ratio of the side adjacent to the angle to the hypotenuse.
- The tangent ratio ($\tan \theta$) is the ratio of the side opposite the angle to the side adjacent to the angle.

$$\sin \theta = \frac{a}{c} = \frac{\text{opposite}}{\text{hypotenuse}}$$

$$\cos \theta = \frac{b}{c} = \frac{\text{adjacent}}{\text{hypotenuse}}$$

$$\tan \theta = \frac{a}{b} = \frac{\text{opposite}}{\text{adjacent}}$$

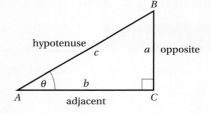

WORKED EXAMPLE 2

1 For this triangle, find the value of:

   a $\sin A$

   b $\cos A$

   c $\tan A$

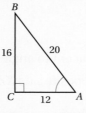

a $\sin A = \dfrac{\text{opposite}}{\text{hypotenuse}} = \dfrac{16}{20} = \dfrac{4}{5}$

                                 You can write this as $\dfrac{\text{opp}}{\text{hyp}}$ for short.

b $\cos A = \dfrac{\text{adjacent}}{\text{hypotenuse}} = \dfrac{12}{20} = \dfrac{3}{5}$

                                 Remember to simplify your answer.

c $\tan A = \dfrac{\text{opposite}}{\text{adjacent}} = \dfrac{16}{12} = \dfrac{4}{3}$

The three ratios have the same value for a particular angle, no matter how long the sides are.

You can find the ratio for any angle using your calculator. Make sure you know how to find and use the trig function keys.

WORKED EXAMPLE 3

Use your calculator to find the value of each of these trig ratios. Give your answers to 3 significant figures where necessary.

a cos 32° b sin 18° c tan 80°

a cos 32° = 0.848 b sin 18° = 0.309 c tan 80° = 5.67

Calculator tip

Most calculators have different angle modes. Make sure yours is set for degrees.

EXERCISE 31A

1. For this triangle, find sin A, cos A and tan A. Give your answers to 3 decimal places where necessary.

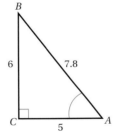

2. For this triangle, find sin B, cos B and tan B. Give your answers to 3 decimal places where necessary.

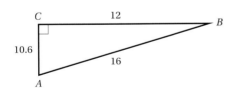

3. For this triangle, find sin A, cos A and tan A. Give your answers to 3 decimal places where necessary.

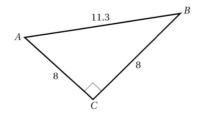

4. Use your calculator to find the values of the following trigonometric ratios to 3 decimal places.

   a sin 40° b cos 65° c tan 30°
   d sin 75° e cos 20° f tan 80°

Find answers at: cambridge.org/ukschools/gcsemaths-studentbookanswers

Tip

If you have two sides of a right-angled triangle you can find the other side using Pythagoras' theorem. If you have one side and at least one of the acute angles, you will need to use the trigonometric ratios.

Tip

Circle the angle you are working with and mark the sides H, A and O to help you see which values you have and which you need.

Solving triangles

Finding unknown sides or angles is called solving the triangle. You can use the three trigonometric ratios to do this.

Finding unknown sides

If you know an angle (other than the right angle) and one side in a right-angled triangle, you can use the ratios to form equations that you solve to find the missing lengths.

You will not be told which ratio to use. You have to pick the right one based on the information that you have about the triangle.

You can remember the ratios using the mnemonic SOH-CAH-TOA and the formula triangles.

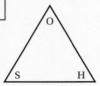 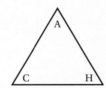 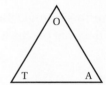

SOH: $\sin = \dfrac{\text{Opposite}}{\text{Hypotenuse}}$ CAH: $\cos = \dfrac{\text{Adjacent}}{\text{Hypotenuse}}$ TOA: $\tan = \dfrac{\text{Opposite}}{\text{Adjacent}}$

WORKED EXAMPLE 4

In triangle ABC, angle $B = 90°$, $AC = 15\,\text{cm}$ and angle $C = 35°$.
Calculate the length of AB correct to 1 decimal place.

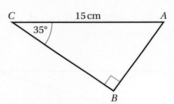

You have angle C.
You want side AB which is opposite to C.
You have CA, which is the hypotenuse.

⮜ Identify which sides you have and want in relation to the acute angle.

Use the ratio with O and H, which is SOH.

⮜ Choose the correct trig ratio.

$\sin 35° = \dfrac{\text{opposite}}{\text{hypotenuse}}$

$\sin 35° = \dfrac{AB}{15}$

$AB = \sin 35° \times 15$

⮜ Multiply both sides by 15 to get AB on its own.

$AB = 8.603646545$

⮜ Use your calculator to find this value.

$AB = 8.6\,\text{cm}$ (to 1 decimal place)

⮜ Round your answer to the required number of decimal places.

WORKED EXAMPLE 5

In triangle XYZ, angle Z is a right angle and angle X = 71°. Side YZ = 7.9 cm.
Calculate the length of XZ correct to 1 decimal place.

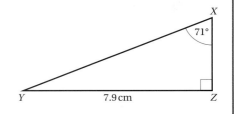

You have angle X.
You want XZ which is adjacent to X.
You have YZ which is opposite to X.
Use the ratio with OA which is TOA.

$$\tan = \frac{\text{opposite}}{\text{adjacent}}$$

$$\tan 71° = \frac{7.9}{XZ}$$

$7.9 = XZ \times \tan 71°$ — Multiply both sides by XZ to get rid of the fraction.

$\frac{7.9}{\tan 71°} = XZ$ — Divide by tan 71° to get XZ on its own.

$XZ = 2.720188145$ — Do the calculation on your calculator.

$XZ = 2.7$ cm (to 1 decimal place) — Round your answer to the required number of decimal places.

EXERCISE 31B

1 For each triangle, choose the appropriate trigonometric ratio and find the length of the side marked with a letter. Give your answers correct to 2 decimal places.

a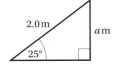
b
c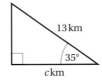
d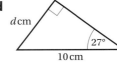

e
f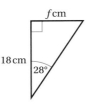
g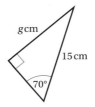
h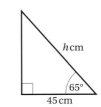

i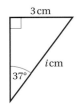
j
k
l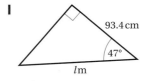

Calculator tip

Your calculator can 'work backwards' to find the size of the unknown angle associated with a particular trigonometric ratio.

To find the angle if you have the ratio, key in the inverse trigonometric function on your calculator, i.e. $\sin^{-1}$, $\cos^{-1}$ or $\tan^{-1}$.

Finding unknown angles

You can use the trig ratios to find the size of missing angles.

To find the size of unknown angles using the trig ratios you need to use the inverse function of each ratio. On most calculators these are the second functions of the sin, cos and tan buttons. They are usually marked $\sin^{-1}$, $\cos^{-1}$ and $\tan^{-1}$.

WORKED EXAMPLE 6

Given that $\tan x$ is 5, what is the size of angle x (to 1 decimal place)?

 78.69006753   Using your calculator.

Angle x is 78.7°

WORKED EXAMPLE 7

Find the size of angle x in each triangle. Give your answers to 1 decimal place.

a 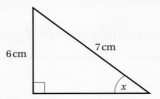   b 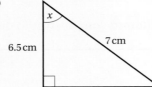   c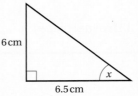

a $\sin x = \dfrac{\text{opposite}}{\text{hypotenuse}} = \dfrac{6}{7} = 0.857...$

$x = \sin^{-1} 0.857... = 59.0°$

Choose the correct trig ratio. You have the side opposite to angle x and also the hypotenuse.

Use the second function button on your calculator.

b $\cos x = \dfrac{\text{adjacent}}{\text{hypotenuse}} = \dfrac{6.5}{7} = 0.928...$

$x = \cos^{-1} 0.928... = 21.8°$

c $\tan x = \dfrac{\text{opposite}}{\text{adjacent}} = \dfrac{6}{6.5} = 0.923...$

$x = \tan^{-1} 0.923... = 42.7°$

WORK IT OUT 31.1

For a ladder to be safe it must be inclined at an angle between 70° and 80° to the ground.

The diagram shows a ladder resting against a wall.

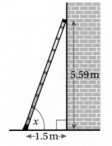

Not to scale

Continues on next page …

Is the ladder positioned safely? Which of these calculations gives you the answer that you need?

Option A	Option B	Option C
$\sin x = \dfrac{1.5}{5.59} = 0.268...$	$\cos x = \dfrac{1.5}{5.59} = 0.268...$	$\tan x = \dfrac{5.59}{1.5} = 3.726...$
$\sin^{-1} 0.268... = 15.57°$	$\cos^{-1} 0.268... = 74.43°$	$\tan^{-1} 3.726... = 74.98°$
Ladder is not safe.	Ladder is safe.	Ladder is safe.

EXERCISE 31C

1 Calculate the value of θ for each ratio.

  a $\sin \theta = 0.682$
  b $\cos \theta = 0.891$
  c $\tan \theta = 2.4751$
  d $\sin \theta = 0.2588$

2 Find the size of each marked angle. Give your answers correct to 1 decimal place.

  a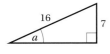
  b
  c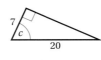

  d
  e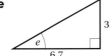
  f

3 PQR (right) is a right-angled triangle; $PQ = 11$ cm and $QR = 24$ cm. Calculate the size of angle PRQ.

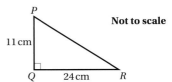

4 What is the size of angle x in the triangle on the right?

5 For each of the triangles described below, sketch the triangle and calculate the missing value.

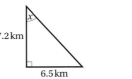

  a In triangle ABC, angle $B = 90°$, $BC = 45$ units and angle $C = 23°$.
  Calculate the length of AB.

  b In triangle PQR, angle $R = 90°$, $PQ = 12.2$ cm and angle $P = 57°$.
  Calculate the length of QR.

  c In triangle EFG, angle $G = 90°$, $EG = 8.7$ cm and angle $E = 49°$.
  Calculate the length of FG.

  d In triangle XYZ, angle $Y = 90°$, $XZ = 36$ units and angle X is $25°$.
  Calculate the length of:
  i XY **ii** YZ.

6. For each triangle described below, draw a sketch and then calculate the required values.

   a In triangle ABC, angle C = 90°, BC = 6.7 units and AB = 9.8 units.
   Calculate angle A.

   b In triangle DEF, angle D = 90°, DF = 13 units and EF = 17 units.
   Calculate angle F.

   c In triangle GHI, angle I = 90°, HI = 8.2 cm and GI = 13.7 cm.
   Calculate the size of:
   i angle G   ii angle H.

   d In triangle JKL, angle J = 90°, JK = 85 mm and KL = 113 mm.
   Calculate the size of:
   i angle K   ii angle L.

   e In triangle MNO, angle M = 90°, NO = 29.8 cm and MN = 20.6 cm.
   Calculate:
   i angle N   ii angle O   iii MO.

   f In triangle PQR, angle Q = 90°, PQ = 57.3 mm and QR = 45.1 mm.
   Calculate:
   i angle P   ii angle R   iii PR.

Section 2: Exact values of trigonometric ratios

When you work out trigonometric ratios on your calculator you often get approximate (or truncated) values.

If left as fractions (or sometimes as square roots) then these ratios are said to be expressed **exactly**.

You need to know the exact values of the sin, cos and tan ratios for the angles 0°, 30°, 60° and 45°, as well as the values of sin and cos for 90°. The tan ratio for 90° is undefined.

Sine, cosine and tangent ratios for 30° and 60°

An equilateral triangle has all angles equal to 60°. Dividing the triangle in half gives two right-angled triangles, each with one 30° angle and one 60° angle.

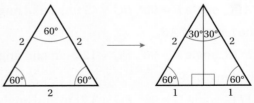

Using Pythagoras' theorem you can find the height h of the right-angled triangle.

$2^2 = 1^2 + h^2$

$4 = 1 + h^2$

$h^2 = 3$ so $h = \sqrt{3}$

You can't write the square root of 3 as an exact decimal number, so leave it in root form. This right-angled triangle with angles of 30° and 60° can be used to work out the values of the trig ratios for these angles.

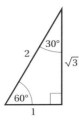

$\sin 30° = \dfrac{\text{opposite}}{\text{hypotenuse}} = \dfrac{1}{2}$ $\qquad$ $\sin 60° = \dfrac{\text{opposite}}{\text{hypotenuse}} = \dfrac{\sqrt{3}}{2}$

$\cos 30° = \dfrac{\text{adjacent}}{\text{hypotenuse}} = \dfrac{\sqrt{3}}{2}$ $\qquad$ $\cos 60° = \dfrac{\text{adjacent}}{\text{hypotenuse}} = \dfrac{1}{2}$

$\tan 30° = \dfrac{\text{opposite}}{\text{adjacent}} = \dfrac{1}{\sqrt{3}}$ $\qquad$ $\tan 60° = \dfrac{\text{opposite}}{\text{adjacent}} = \dfrac{\sqrt{3}}{1} = \sqrt{3}$

Sine, cosine and tangent ratios for 45°

The diagram below shows a right-angled isosceles triangle with the equal sides, 1 cm long.

The hypotenuse is $\sqrt{2}$ cm long.

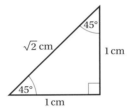

The acute angles are both 45°.

$\sin 45° = \dfrac{\text{opposite}}{\text{hypotenuse}} = \dfrac{1}{\sqrt{2}}$

$\cos 45° = \dfrac{\text{adjacent}}{\text{hypotenuse}} = \dfrac{1}{\sqrt{2}}$

$\tan 45° = \dfrac{\text{opposite}}{\text{adjacent}} = \dfrac{1}{1} = 1$

You will find the exact values of the ratios for 0° and 90° as part of the next exercise.

EXERCISE 31D

1 Copy this table and complete it. Use it to help you memorise the exact values of the trigonometric ratios for the different angles.

Angle θ	sin θ	cos θ	tan θ
0°			
30°			
45°			
60°			
90°			tan 90° is undefined

2 Without using your calculator, find:

    a sin 30° + cos 60°   **b** sin 45° + cos 45°   **c** cos 30° + sin 60°.

Section 3: Solving problems using trigonometry

The trigonometric ratios can be applied to many different types of measurement problems. If the question does not include a sketch, it is useful to draw one. Make it large and clear and mark what you know on it. This will help you to identify the correct ratio to use to solve the problem.

Angles of elevation and depression

Many trigonometry problems involve lines of sight.

Problems that involve looking up to an object can be described in terms of an **angle of elevation**. This is the angle between an observer's line of sight and a horizontal line.

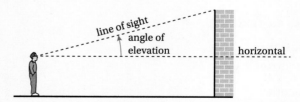

When an observer is looking down, the **angle of depression** is the angle between the observer's line of sight and a horizontal line.

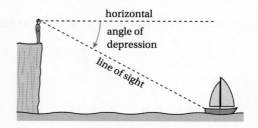

> **Key vocabulary**
>
> **angle of elevation:** when looking up, the angle between the line of sight and the horizontal.
>
> **angle of depression:** when looking down, the angle between the line of sight and the horizontal.

WORKED EXAMPLE 8

From the top of a lighthouse 105 m above sea level, the angle of depression of a boat is 5°.

How far is the boat from the shore? Give the answer to the nearest metre.

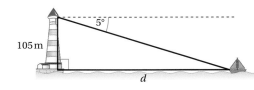

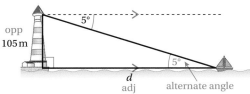

Mark on the diagram what you know.

The angle of depression is alternate to the angle of elevation from the boat.

d is the side adjacent to 5° and the height to the lighthouse is opposite.

You need the ratio with OA, which is TOA.

Choose the correct trigonometric ratio.

$$\tan 5° = \frac{105}{d}$$

$\tan 5° \times d = 105$

$$d = \frac{105}{\tan 5°} = 1200.155492$$

$d = 1200\,m$

Problem-solving framework

A ship is laying cable along the sea bed.

The angle of the cable to the sea bed is 40° and the length of cable to the sea bed is 40 metres.

The ship is 50 miles offshore and travelling northwest.

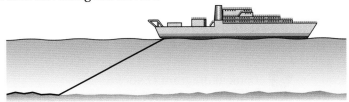

What is the depth of the seabed?

Steps for solving problems	What you would do for this example
Step 1: What have you got to do?	Use Pythagoras' theorem or trigonometry to find the depth of water.
Step 2: What information do you need?	Angle 40° and length of cable as the hypotenuse is 40 m.
Step 3: What information don't you need?	50 miles offshore and the direction of travel.

Continues on next page …

Find answers at: cambridge.org/ukschools/gcsemaths-studentbookanswers

Step 4: What maths can you do?	Use a right-angled triangle.
	Cannot use Pythagoras as you do not know two lengths, so use trig.
	Decide which trigonometry function – sin, cos or tan?
	One angle and the hypotenuse are known – need to find the opposite length.
	$\sin 40° = \dfrac{\text{opp}}{\text{hyp}} = \dfrac{\text{opp}}{40}$
	$\sin 40° \times 40 = $ opposite length $= 25.7$ m
Step 5: Have you done it all?	Yes
Step 6: Is it correct?	Yes – double-checked and estimated. The length must be less than 40 m.
	$\sin 30° = 0.5$ so half of 40 m would be 20 m.

EXERCISE 31E

1 Say whether the marked angles in each of these diagrams is an angle of depression or an angle of elevation.

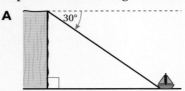
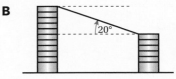

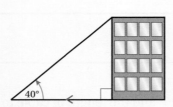

2 Look at diagram A above.

If the boat is 80 m from the base of the cliff, work out the height of the cliff.

3 Look at diagram D above.

If the distance from the top of the mast to the point where the cable is attached to the ground is 35 m, how high is the mast?

4 Two trees stand opposite one another, at points X and Y, on opposite banks of a river. The distance YZ along one bank is perpendicular to XY, and is measured to be 28 m. Angle YZX is measured to be 59°.

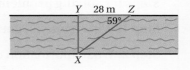

How far apart are the trees?

5 A child's playground slide is 4.2 m long and makes an angle of 33° with the horizontal.

Calculate the height of the slide.

6 Carol is in a hot air balloon at point C in the sky. CG is the vertical height of the hot air balloon above the ground. David is standing on the ground at point D. The distance between points D and G is 27 m. The angle of elevation from David to the hot air balloon is 53°.

Calculate the height, CG.

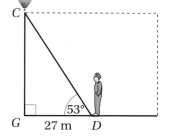

7 Mike is standing 15 m away from a flagpole. His eye level is 1.6 m above the ground. The angle of elevation from his line of sight to the top of the flagpole is 60°.

  a How tall is the flagpole?

  b If he moves another 10 m further away from the flagpole, how will the angle of elevation to the top of the flagpole change?

8 Two observers in different positions at A and B are watching a rare bird on a tree at C. The angle of elevation from A to C is 56° and the angle of elevation from B to C is 25°.

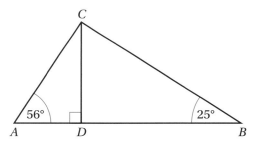

  a If person B is standing 15 m from D (the base of the tree), calculate the height of the bird above the ground (the length of CD).

  b Calculate the distance of person A from D.

9 A person is standing at a point P, 30 m away from a signal tower for a mobile phone operator. The angle of depression from the top of the tower to P is 56°.

Calculate the height of the tower.

10 A tree surgeon uses an instrument to measure that the angle of elevation from her line of sight to the top of a tree is 20°. She is standing 10 m away from the tree. Her eye level is 1.5 m above the ground.

  a If she assumes that the tree is perfectly perpendicular to the ground, how would she calculate the height of the tree?

  b To check her calculation, she moves another 10 m away from the tree in a straight line along flat ground, and measures the angle of elevation again. What should the angle measurement be now if her first measurement and calculation were correct?

Checklist of learning and understanding

Trigonometric ratios

- In a right-angled triangle, the longest side is the hypotenuse. For a given angle θ, the other two sides can be labelled opposite (to angle θ) and adjacent (to angle θ).
- The sine, cosine and tangent ratios can be used to find unknown sides and angles in right-angled triangles.
  - $\sin \theta = \dfrac{\text{opposite}}{\text{hypotenuse}}$
  - $\cos \theta = \dfrac{\text{adjacent}}{\text{hypotenuse}}$
  - $\tan \theta = \dfrac{\text{opposite}}{\text{adjacent}}$
- You can find the value of a ratio using the sin, cos and tan buttons on your calculator. To find the size of an angle, use the inverse functions.

Exact values

- You can find the exact values of sin, cos and tan for special angles. Some exact values contain square roots.

 For additional questions on the topics in this chapter, visit GCSE Mathematics Online.

 Chapter review

1 The diagram shows a triangle ABC. Angle $A = 20°$ and angle $C = 90°$; $AB = 32$ m.

Calculate the height BC.

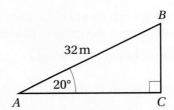

 2 LMN is a right-angled triangle.

 $MN = 9.6$ cm.

$LM = 6.4$ cm.

Calculate the size of the angle marked $x°$.

Give your answer correct to 1 decimal place. *(3 marks)*

©*Pearson Education Ltd 2012*

3) A ladder leans against the side of a house. The ladder is 4.5 m in length, and makes an angle of 74° with the ground.

How high up the wall will it reach? (This length is marked x in the diagram.)

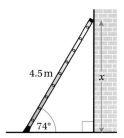

Not drawn accurately

4) The same ladder is now placed 0.9 m away from the side of the house.

What angle does the ladder now make with the ground? (This angle is marked y in the diagram.)

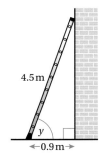

Not drawn accurately

5) Triangles ABC and PQR are similar. $AC = 3.2$ cm, $AB = 4$ cm and $PR = 4.8$ cm.

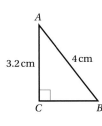 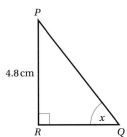

Explain why $\sin x = 0.8$

32 Growth and decay

In this chapter you will learn how to …
- solve problems involving simple and compound interest.
- solve problems involving depreciation and decay.

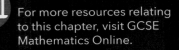

For more resources relating to this chapter, visit GCSE Mathematics Online.

Using mathematics: real-life applications

Many real-life situations involve growth (increase) or decay (decrease) as time passes. Population numbers, growth of bacteria, disease infection rates, world temperature patterns and the value of money or possessions may all increase or decrease over time.

> "My computer program calculates interest on a daily basis. This means whatever is in the account gains interest, not just the initial investment."
> *(Investment broker)*

Before you start …

Chs 7 and 18	You must be able to convert percentages to decimals.	**1**	Write each of the following as a decimal: **a** 5% **b** 190% **c** 0.4% **d** 12.5%
Ch 18	You must be able to find a percentage of a quantity using multiplication.	**2**	Find, using multiplication only: **a** 68% of £300 **b** 2% of $80 **c** 4.5% of £56 **d** 114% of $650 **e** 99.5% of £1540
Ch 18	You must be able to increase or decrease a quantity by a given percentage by multiplying by a suitable decimal.	**3**	Carry out the following increases and decreases using only multiplication: **a** increase $44 by 22% **b** increase £35 by 5.5% **c** decrease £13 by 44% **d** decrease $170 by 8%

32 Growth and decay

Assess your starting point using the Launchpad

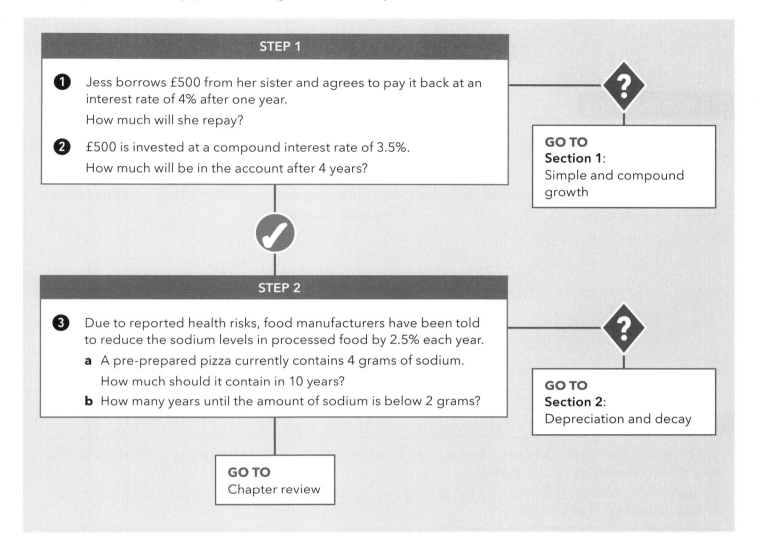

STEP 1

① Jess borrows £500 from her sister and agrees to pay it back at an interest rate of 4% after one year.
How much will she repay?

② £500 is invested at a compound interest rate of 3.5%.
How much will be in the account after 4 years?

GO TO Section 1: Simple and compound growth

STEP 2

③ Due to reported health risks, food manufacturers have been told to reduce the sodium levels in processed food by 2.5% each year.
  a A pre-prepared pizza currently contains 4 grams of sodium. How much should it contain in 10 years?
  b How many years until the amount of sodium is below 2 grams?

GO TO Section 2: Depreciation and decay

GO TO Chapter review

Section 1: Simple and compound growth

Simple interest

When you borrow money, or buy things on credit, you are normally charged interest for the use of the money. When you invest or save money, you are paid interest by the bank or other institution in return for leaving your money with them.

The amount of money you invest or borrow is called the principal (P).

Simple interest (I) is interest paid on the original principal. The same interest is paid for each time period (T). The rate of interest (R) is given as a percentage.

Simple interest can be worked out using a formula:

$I = PRT$

> **Tip**
>
> The total simple interest could be found using the formula:
> $I = PRT$
> $= 500 \times 0.05 \times 6 = 150$
> So, final value = 500 + 150 = £650

Compound interest

When using compound interest, the interest earned in the first time period is added to the principal so in the next time period you calculate the interest on the principal amount plus any interest that has been added to it.

WORKED EXAMPLE 1

Fatima deposits £500 into a savings account for 6 years. The bank pays interest on the savings at 5% per year. Compare the amount each year for 6 years when the interest is calculated using simple interest to the amount when the interest is compounded annually.

The table shows the value of the investment over the 6 years when using both simple interest and compound interest.

Year	Simple interest	Compound interest
1	£500 + 5% = £525	£500 + 5% = £525
2	£525 + (5% of £500) = £550	£525 + (5% of £525) = £551.25
3	£550 + (5% of £500) = £575	£551.25 + (5% of £551.25) = £578.81
4	£575 + (5% of £500) = £600	£578.81 + (5% of £578.81) = £607.75…
5	£600 + (5% of £500) = £625	£607.75… + (5% of £607.75…) = £638.14…
6	£625 + (5% of £500) = £650	£638.14… + (5% of £638.14…) = £670.04…

WORK IT OUT 32.1

Three students attempt the question below. In pairs, decide who has got not only the correct answer but also the most efficient method to find the answer.

The population of Europe is growing at a rate of 0.2% per year. The current population is 739 million. What will the population be in 3 years' time?

Kayleigh	Tom	Zac
Find 0.2%: 0.2% of 739 000 000 = 0.002 × 739 000 000 = 1 478 000 The same growth for 3 years: 3 × 1 478 000 = 4 434 000 Add it on: 739 000 000 + 4 434 000 = 743 434 000	Year 1: Find 0.2% of 739 000 000 = 0.002 × 739 000 000 = 1 478 000 Add it on: 740 478 000 Year 2: Find 0.2% of 740 478 000 = 0.002 × 740 478 000 = 1 480 956 Add it on: 741 958 956 Year 3: Find 0.2% and add it on 1 483 918 + 741 958 956 = 743 442 874	Increase by 0.2% means there is 100.2%, do this three times in a row. 739 000 000 × 1.002 × 1.002 × 1.002 = 739 000 000 × 1.002^3 = 743 442 874

Working with compound interest and growth rates is very much like working with function machines. Each time an output is produced it goes back to becoming an input and the process is repeated. This keeps going for the allotted period of time. This kind of process is called iterative – it iterates or repeats.

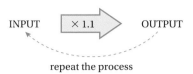

repeat the process

Tip

Always show your working for these questions. Write down what you type into the calculator so you can check it several times and so that your teacher can see how you have calculated your answer.

WORKED EXAMPLE 2

The population of starlings in a park increases at a rate of 10% per year. If the population started with 80 starlings, how many would there be in 5 years' time?

$80 \times 1.1 = 88$

The first input would be 80.

To increase a quantity by 10%, you multiply by 1.1

Tip

100% + 10% = 110%, or 1.1

$88 \times 1.1 = 96.8$

Use the output from year 1 as the input for year 2.

(Notice that if you stop here, you would round sensibly. However, you use the unrounded value in the next stage and only round at the end.)

$96.8 \times 1.1 = 106.48$
$106.48 \times 1.1 = 117.128$
$117.128 \times 1.1 = 128.8408$
≈ 129 starlings in 5 years' time.

Instead of rewriting the output each time, this could be written as:

$80 \times 1.1 \times 1.1 \times 1.1 \times 1.1 \times 1.1 = 128.8408$

or $80 \times 1.1^5 = 128.8408$

EXERCISE 32A

1 Copy and complete this table for simple interest at the given rate.

Investment	Interest rate	1 year	2 years	6 years	n years
£250	2%				
£1500	4.5%				
	3%	£51.50			

2 £300 is invested in an account with a compound interest rate of 2%.

How much is in the account after:

   a 1 year   **b** 3 years   **c** 8 years

3 £1000 is invested with a compound interest rate of 3%.

Plot a graph showing how much money is in the account over the first 10 years of the investment.

4 A colony of bacteria grows by 4% every hour. At first the colony has 100 bacteria.

How many will exist after 24 hours?

5 The population of Ireland is growing at an annual rate of 1.7%. In 2014 the population was 4.6 million.

   a If this growth rate remains constant how many people will be living in Ireland in 2024?

   b How many new inhabitants are there?

   c Use this model to show how many people there were in Ireland in 2012. Comment on the validity of your answers.

6 The Bank of England's target inflation rate is 2%. This tells you how much the cost of living, food, fuel and rent is likely to go up each year. In 2015 a month's rent is given as £450.

Assuming that the Bank targets are correct, how much is this likely to be 20 years later?

7 Gavin is saving for a new bike. The model he wants costs £255. So far he has saved £200. Gavin's dad has offered to pay him 8% interest on the total of his saving for every month that Gavin does his homework, cleans his room and loads the dishwasher.

How long is he going to have to wait for the bike? Show clear working to explain your answer.

8 Population growth models help predict the spread of invasive species. Zebra mussels are one such species. Their population can increase by 1900% each year. Two mussels are found in a freshwater lake.

Should biologists be worried that this will have a significant impact over the next 10 years? Give details to explain your response.

9 Two investors are having an argument. They want to maximise their profit. They are investing for 5 years and have a choice. They can either have 6% simple interest or 5.5% compound.

Which should they choose? Would the answer change if they were investing for 4 years?

10 Jenny is saving for her first car. She needs a deposit of £2775. Each month she saves £200 in an account offering 1% (compound) interest a month.

Will Jenny have enough money after a year to buy a car?

11 Between 1980 and 2010 the price of a chocolate bar went from 25p to 65p.

  a By what percentage did the cost rise overall?

  b What annual (compound) percentage increase is this?

  c If the cost of the bar keeps increasing at the same rate how much will it cost in 2040?

  d When will the chocolate bar first cost more than £1?

12 £100 000 is invested at a rate of 5% for 10 years.

  a How much more money is earned using compound interest compared to simple interest?

  b What simple interest rate would be needed to achieve the same earnings?

13 House prices are rising. A two-bedroomed house cost £195 000 last year and now costs £216 450.

If the price keeps rising at the same rate how much will this house cost in 3 years' time?

14 Which of the following investment models gives the highest earnings?

Model 1	Model 2	Model 3
Year 1: 5% interest	Years 1–3	Years 1–3
Year 2: 4% interest	4% compound interest	3.9% simple interest
Year 3: 3% interest		

15 A colony of bacteria grows by 5% every hour.

How long does it take for the colony to double in size?

Section 2: Depreciation and decay

When the value of something goes down, you say it has depreciated. For example, a new car will show a **depreciation** in value of about 30% in the first year of ownership alone.

Key vocabulary

depreciation: the loss in value of an object over a period of time.

decay: the reduction in a quantity over time or some other measure such as distance. The opposite of growth.

When the number of items in a population declines over time or a physical measure such as atmospheric pressure declines over a distance, it is called **decay** rather than depreciation.

For example, if the population of squirrels is in decay, it means that each year there are fewer and fewer animals in the population. If the rate of decline is 10%, each year 10% of the squirrels disappear, leaving 90%. So from one year to the next the number of animals is $n \times 0.9$.

WORKED EXAMPLE 3

A new computer depreciates by 30% per year. If it cost £1200 new, what will it be worth in 2 years' time?

Method 1	Method 2
Value after 1 year = £1200 − (30% of £1200) = £1200 − £360 = £840	Value after 1 year = 70% of £1200 = £840 Value after 2 years = 70% of £840 = £588
Value after 2 years = £840 − (30% of £840) = £840 − £252 = £588	

WORK IT OUT 32.2

Three students attempt the question below. In pairs, decide who has got not only the correct answer but also the most efficient method to find the answer.

For every 1000 metres you climb, the atmospheric pressure decreases by 12% of the sea-level value. If the sea-level atmospheric pressure is 100 300 pascal (Pa), what would the pressure be for a skydiver at an altitude of 4000 metres?

Belle	Jordan	Ethan
12% of 100 300 = 0.12 × 100 300 = 12 036 4 × 12 036 = 48 144 100 300 − 48 144 = 52 156 Pa	Decrease by 12% leaves 88% 88% of 100 300 = 0.88 × 100 300 = 88 264 88% of 88 264 = 77 672.32 88% of 77 672.32 = 68 351.6416 88% of 68 351.6416 = 60 149.444 608 Pa	Decrease by 12% leaves 88%, do this four times in a row. 100 300 × 0.88 × 0.88 × 0.88 × 0.88 = 100 300 × 0.88^4 = 60 149.444 608 Pa = 60 149.4 Pa (1 d.p.)

EXERCISE 32B

1 A car depreciates in value each year by 8%. A new Compact car costs £11 000. How much will this be worth in:

  a 1 year

  b 3 years

  c 8 years?

2 Copy and complete this table:

Initial cost	Depreciation rate	1 year	2 years	6 years
£400	2%			
£2500	15%			
£50 000	3.5%			

3 The pesticide DDT is banned in many countries, including the UK, because it builds up in animal tissue over time to dangerous levels. However, DDT in the soil is absorbed by the soil (i.e. it decays) at a rate of 7% a year.

If a farmer used 2 kg of DDT on a field in 2000, how much would remain in the field in 2014?

4 The rate at which water flows out of a tank with an opening at the bottom depends on the amount of water left in the tank. For one particular tank, every 5 minutes the height of the water left reduces by 15% of its value at the start of the 5-minute period. Which of the following graphs depicts this?

A

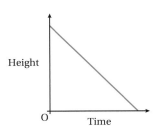

B

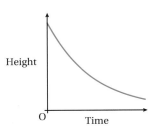

C

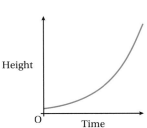

5 At the start of an experiment there are 8000 bacteria. A lethal pathogen is introduced to the population causing a reduction of 1600 in the population in an hour.

  a What percentage decrease in population is this?

  b Assuming the same rate of decrease, how many bacteria would you expect to be alive after 8 hours?

  c How long until fewer than 100 bacteria are alive?

6. For every 1000 m that you climb, the atmospheric pressure decreases by 12% from its sea-level value. If the sea-level atmospheric pressure is 100 300 pascal (Pa), what was the pressure for world record holder Felix Baumgartner, who jumped off a balloon flying 39 km above sea level?

7. The population of Bulgaria is decreasing at a rate of 0.6% per year. In 2014 the population was 7.4 million people.

    a How many people are expected to be living in Bulgaria in 2020?

    b How many years until the population dips below 7 million?

8. The cost of mobile phones has been falling. Three years ago the latest model cost £400 with no contract; now the latest model costs £342.95.

    If the price keeps falling at the same rate, how long until the current model costs less than two-thirds of today's price?

Checklist of learning and understanding

Simple and compound growth

- Simple growth, such as interest, is a fixed rate of growth, calculated on the original amount.
- The formula $I = PRT$ can be used to calculate simple interest.
- Compound growth, such as compound interest, is calculated on the principal for the first period and then compounded by calculating it on the principal plus any interest paid or due for each previous period.
- You can work out compound growth using a multiplier for each period.

Depreciation and decay

- A drop in value of an object over time is called depreciation.
- A decline in a population or physical quantity is called decay.

For additional questions on the topics in this chapter, visit GCSE Mathematics Online.

Chapter review

1. A camera has a cash price of £850. Nasief buys it on credit and pays a 10% deposit, with the balance to be paid over 2 years at a simple interest rate of 10%. Calculate:

    a the amount of his deposit

    b the balance owing after deducting the deposit

    c the amount of interest paid in total over 2 years

    d the monthly payment amount for 24 equal monthly instalments

    e the difference between the cash price and what Nasief actually paid in the end.

2 Salma invests her money in an account that pays 6% interest compounded half-yearly.

If she puts £2300 in the account and leaves it there for 2 years, how much money will she have at the end of the period?

3 A car valued at £8500 depreciates by 30% in the first year, 20% in the second year and a further 12% in the third year.

How much is it worth after 3 years?

4 Each year the education department arranges a quiz. There are 128 students in the competition to start with. During each round, half of the quiz contestants are eliminated.

How many students will still be participating after round 4?

5 Due to pollution in an estuary, the number of freshwater shrimp is declining. In 2012, it was estimated that there were 15 000 shrimp in the estuary, and in 2013, 14 100.

   a What is the percentage depreciation?

   b At the same rate, what will the population be in 2020?

   c During which year will the population have halved?

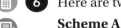

6 Here are two schemes for investing £2500 for 2 years.

Scheme A

gives 4% **simple** interest each year.

Scheme B

gives 3.9% **compound** interest each year.

Which scheme gives the most total interest over 2 years?

You must show all your working.   *(4 marks)*

©*Pearson Education Ltd 2013*

33 Proportion

In this chapter you will learn how to ...
- understand proportion and the equality of ratios.
- solve problems involving direct and inverse proportion, including graphical and algebraic representations.
- understand that x is inversely proportional to y is equivalent to x is proportional to $\frac{1}{y}$.
- interpret equations that describe direct and inverse proportion.

For more resources relating to this chapter, visit GCSE Mathematics Online.

Using mathematics: real-life applications

Proportional reasoning is very common in daily life. You use proportional reasoning when you mix ingredients for a recipe, convert between units of measurement or work out costs per unit. It is an area of maths where you can use many different methods to solve particular problems.

> **Tip**
>
> Review the sections in Chapter 19 on equivalent ratios and fractions to prepare for this chapter.

"I test out new dishes on my family. Then I have to scale up the recipes in proportion so that they taste just as good. Sometimes it may be for just a few people at one table in my restaurant; at other times it may be for a whole room of wedding guests."

(Chef and restaurant owner)

Before you start ...

Chs 6 and 10	You need to know how many minutes there are in fractions of an hour.	**1** How many minutes in 　a half an hour? 　b a quarter of an hour? 　c a third of an hour? 　d a fifth of an hour?
Chs 6 and 10	You need to be able to find what fraction of an hour a given time is.	**2** What fraction of an hour is 　a 5 minutes? 　b 24 minutes? 　c 54 minutes?
Chs 5 and 16	You should know how to substitute values into formulae.	**3** $g = 3b$ 　a What is the value of g when $b = 7$? 　b What is the value of b when $g = 72$? 　c What is b when $g = 1.2$?

Assess your starting point using the Launchpad

STEP 1

1
a. A recipe for blueberry muffins makes 12 muffins. It uses 180 g of blueberries. What weight of blueberries is needed to make 30 muffins?
b. A car is travelling at 80 km per hour. How far would it travel in 75 minutes?
c. €1 = $1.40
   How many euros is a t-shirt that costs $24.50?

GO TO Section 1: Direct proportion

STEP 2

2 The cost of carpeting a hallway is proportional to the area of the hall. One hallway measuring 15 m² costs £97.50.
a. Find a formula for the cost, c, of carpeting a hallway with area, a.
b. How much would it cost to carpet an area of 32 m²?
c. What area can be carpeted for £328.90?

GO TO Section 2: Algebraic and graphical representations

STEP 3

3 Ten people have enough food for a six-day camping trip.
a. How long would the food last if there were only five people?
b. Two more people join the group unexpectedly. How long would the food last if there were 12 people?

GO TO Section 3: Inverse proportion

GO TO Chapter review

Find answers at: cambridge.org/ukschools/gcsemaths-studentbookanswers

Section 1: Direct proportion

Key vocabulary

ratio: the relationship between two or more groups or amounts, explaining how much bigger one is than another.

direct proportion: two values that both increase in the same ratio.

When two quantities vary but remain in the same **ratio** they are said to be in **direct proportion**. A simple example would be the quantity and price of petrol. The more petrol a driver puts into the car, the more it costs.

Scaling up recipe ingredients also involves direct proportion. If you want to make double the amount of food, you need to use double the amount of ingredients. Other examples are the number of hours someone works and the amount they get paid at an hourly rate, and exchange rates between two currencies.

In direct proportion problems, you might be given a rate such as price per litre. If not, it might be helpful to find this rate.

WORKED EXAMPLE 1

A car travels 12 miles in 15 minutes at a constant speed.

a At what speed is the car travelling?

b How far would the car go in 75 minutes?

c How long would it take the car to travel 80 miles?

a $12 \times 4 = 48$ miles per hour

> 15 minutes is $\frac{1}{4}$ of an hour, so multiply 12 by 4 to get the speed in mph.

b $\frac{75}{15} \times 12 = 60$ miles

> How many 15s in 75 minutes? Use this to multiply 12.

c $80 \div 48 = 1.66...$ hours
This is 100 minutes or 1 hour 40 minutes

> Time = distance ÷ speed.
> Here it is sensible to give the final answer in hours and minutes.

Tip

Often it helps to write down a proportion fact you know and consider what would happen if one side is halved, doubled, multiplied by 10 and so on. For example, if you know that six eggs make two cakes then half the number of eggs, three, will make just one cake.

EXERCISE 33A

1. A bluefin tuna fish can travel 3 km in just 20 minutes.

   List some other distance–time facts about the fish assuming that it always travels at a constant rate.

2. Each day a cat eats 40 grams of dried cat food.

   How many grams would it eat in a fortnight (two weeks)?

3. Patrick works for 4 hours and gets paid £22.

   What is his rate of pay per hour?

4   Jelly beans cost £1.20 for 100 g.

   a   How much would 50 g cost?

   b   How much would 300 g cost?

   c   How much would 1 kg cost?

   d   What weight of jelly beans could you buy with £4.20?

5   On holiday Ben uses his mobile to call home. A 12-minute call costs £4.20.

   a   How much would it cost to ring home for 18 minutes?

   b   Danny calls home for 20 minutes and it costs him £6.40.

   Whose phone is better value, Ben's or Danny's? Why?

6   Look at this pancake recipe. It serves eight people.

   100 g plain flour

   2 eggs

   300 ml semi-skimmed milk

   a   Copy and complete this table.

Ingredients	8 people	4 people	16 people	12 people	20 people
plain flour	100 g				
eggs	2				
semi-skimmed milk	300 ml				

   b   If you have 2 litres of milk, 500 g of plain flour and 9 eggs and make as much pancake mixture as possible, how many will it serve?

7   The fastest train in Europe is the French TGV from Paris to Le Mans. It travels at 320 kilometres per hour. Assume that the train is going at its full speed.

   a   How far does it travel in 2 hours?

   b   How far does it travel in 30 minutes?

   c   How far does it travel in 15 minutes?

   d   How far does it travel in 1 minute?

   e   How far does it travel in 10 seconds?

   f   The Equator is approximately 40 000 km long.

   If it were possible, how long would it take to travel around the Equator in a TGV train?

8 The fastest animal on land is the cheetah. It can reach speeds of up to 120 kilometres per hour but for only a short burst of time.

How far would it travel at this speed in 15 seconds?

Unitary method

Sometimes, scaling up or down by an easy figure, halving, trebling, and so on is not possible. In these cases, you use the given relationships to find the value of one item.

WORKED EXAMPLE 2

While in Florida, Danny bought a t-shirt for $18. He paid on a debit card. When Danny returned home the charge on his debit card statement was £10.71. He also bought a pair of jeans for $32. Assuming the bank uses the same exchange rate, what would this charge appear as on his debit card statement?

$18 = £10.71   Divide both sides by 18
$1 = £0.595

Using the information for $18, you need to find the equivalent information for $1.

$32 = £0.595 × 32 = £19.04

The charge on Danny's debit card statement should say £19.04

$1 = £0.595 is the exchange rate of dollars to pounds. You can use this to convert any number of dollars to pounds.

EXERCISE 33B

1 £1 = $1.68. Copy and complete this table:

Pounds £	£1	£2	£5	£15				£124.53
Dollars $					$26.88	$71.40	$80.50	

2 Before going to Australia, Finley exchanges £175 into Australian dollars. He gets an exchange rate of £1 = AU$1.81.

How many dollars does he get?

3 When returning from her holiday, Amber exchanges her $44 back into pounds. The exchange rate is £1 = $1.68.

How many pounds does she receive?

4 Paint is sold in a variety of tins. However, the price per litre remains unchanged. Find the cost of each of these tins of paint:

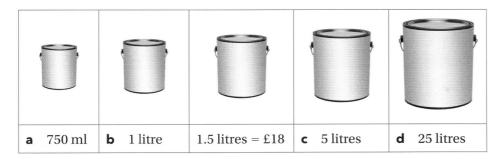

| a | 750 ml | b | 1 litre | 1.5 litres = £18 | c | 5 litres | d | 25 litres |

5 Lucy exchanges £50 for €60.50.

  a What is the exchange rate from pounds to euros?

  b What is the exchange rate from euros to pounds?

6 When planning a skiing holiday in Switzerland, Ethan compares two resorts. The exchange rate from pounds to Swiss francs is £1 = CHF1.48

Which resort is a better deal? How many pounds cheaper is it?

Resort	Accommodation	Food	Ski rental	Flights
Bun di Scuol	£340	£65	£300	£69
Flims-Laax-Falera	CHF444	CHF148	CHF164.28	CHF213.12

7 The graph shows how to convert between pounds and Bulgarian lev (ЛВ).

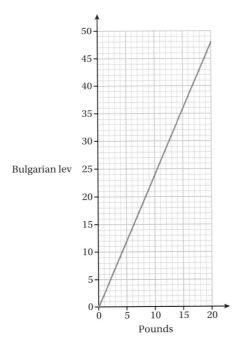

  a What is the exchange rate from pounds to lev?

  b What is the exchange rate from lev to pounds?

8 During a year abroad Aaron had to change currencies on a regular basis. Use the following exchange information to answer the questions below.

UK pound	Euro	Kenyan shilling	Indian rupee	Mongolian tughrik	New Zealand dollar	Brazilian real
£1	€1.21	KSh145	₹102	₮3000	$1.95	R$3.77

a Aaron exchanged £350 into euros.

How many euros did he get?

b Aaron had 40 600 Kenyan shillings.

How many pounds is this?

c When in Kenya, Aaron went on a safari drive. He had booked this before going out at a cost of £185. He paid in Kenyan shillings.

How many Kenyan shillings did it cost?

d When he left India, Aaron exchanged 5202 Indian rupees into Mongolian tughrik.

How many tughrik did he receive?

e Aaron paid €11 for a hostel in France, 1305 shillings in Kenya, 500 rupees in India, 6500 tughrik in Mongolia, 15 dollars in New Zealand and 20 real in Brazil.

Put these prices in order of expense, cheapest first.

9 What information would you need to collect to compare the 'crowdedness' of two school playing fields? How would you carry out the comparison?

Section 2: Algebraic and graphical representations

Direct proportion problems can also be represented graphically or generalised through the use of algebra. This allows you to solve problems concerning the same relationship, either by reading information off a graph or by using an algebraic formula.

WORK IT OUT 33.1

Which of these graphs show a pair of variables that are directly proportional to each other?

Explain your choice. How can you tell that the variables in the other graphs are not directly proportional to each other?

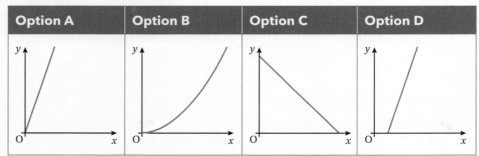

The graph of a directly proportional relationship has a fixed gradient and goes through the origin.

The mathematical symbol $\propto$ is used to indicate that two values are proportional.

If you pay per minute to use your mobile phone, the cost of the call, c, is proportional to the time you speak, t. This can be written algebraically as: $c \propto t$.

This means that for some fixed value k (called the constant of proportionality), you can write a formula:

$c = kt$

Tip

Be careful with units in questions. In this example the cost is in pence and time in minutes. To use the formula you'd need to make sure all quantities were in pence and minutes and convert any that were not.

WORKED EXAMPLE 3

If you pay 75p for a 15 minute call, find:

a the value of the constant of proportionality

b a formula connecting cost and time of call

c the cost of a call taking 13 minutes.

a $c \propto t$

The cost (c) is proportional to the time (t) of the call.

$c = kt$

Rewrite as a formula, replacing the $\propto$ symbol with = and introducing the constant of proportionality.

$75 = k \times 15$
$k = 5$

Substitute the values you know and solve for k.

b $c = 5t$

Replace k with 5 in the formula.

c $c = 5 \times 13 = 65p$

Substitute $t = 13$ into the formula.

WORK IT OUT 33.2

Which of these formulae represent variables that are directly proportional to each other? Why or why not?

Option A	Option B	Option C	Option D
$y = 3x + 5$	$10w = h$	$\dfrac{s}{t} = 7$	$d^2 = 4f$

Find answers at: cambridge.org/ukschools/gcsemaths-studentbookanswers

EXERCISE 33C

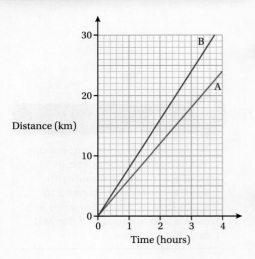

1. On the left is a time–distance graph for two runners.
   a How far had runner A travelled after 30 minutes?
   b How long did it take runner B to travel 18 km?
   c Which runner is going faster?
   d What is the speed of each runner?
   e What assumptions have been made when drawing this graph?

2. The graph on the right shows the cost of telephone cable.
   a What is the cost per metre?
   b Complete this formula: cost = _____ × length.
   c Write a formula linking the cost, in pounds (c), and length, in metres, of wire (l).

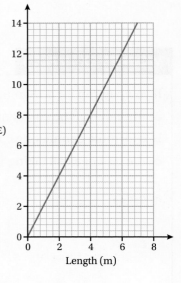

3. The graph below shows the number of new sports cars a factory can make. The number of cars being produced is directly proportional to the number of days that the factory stays open.

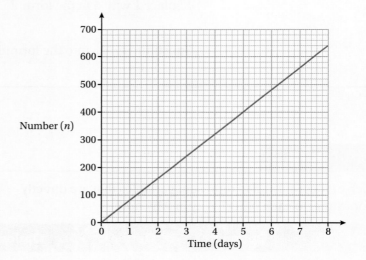

Find a formula for the number of cars (n) being produced in terms of the number of days (d) the factory remains open.

What assumptions have been made?

4 The length of an object's shadow is directly proportional to the object's height. Use this diagram to help explain why this is true.

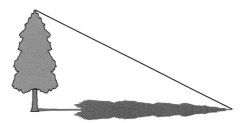

At one specific time in the day a man of height 1.8 metres has a shadow of 1.35 m.

a Find a formula for the length of an object's shadow (s) in terms of its height (h).

b The Angel of the North in Gateshead is 20 m tall. How long would its shadow be at the same specific time?

c The shadow of the tallest upright stone at Stonehenge at this specific time is 502.5 cm. How tall is the stone?

d Measure your height and use the method of comparing shadows to find the height of your school building.

5 Two variables p and q are directly proportional. When p = 6.5, q is 52.

a Find a formula for q in terms of p.

b Find the value of q when p is 3.8

c Find the value of p when q is 14.8

6 Wheelchair ramps have to be designed with a specific steepness allowing their safe use. One such, well-designed, ramp has a horizontal distance of 4 metres and a height gain of 60 cm.

Find a formula for the horizontal distance (d) in terms of the height gain (h).

Section 3: Inverse proportion

If you increase your speed, the time it takes to travel a fixed distance is reduced. If you add more workers, the time it takes to complete the job goes down. In these examples, one quantity decreases as the other one increases. These types of relationship are inversely proportional.

The graph of a pair of inversely proportional variables never quite touches the x- or y-axis, but comes closer and closer to them.

WORK IT OUT 33.3

A rectangle has a fixed area of 24 cm². Its length, x, and height, y, can vary. Which of the graphs below represents this situation? How did you come to your decision?

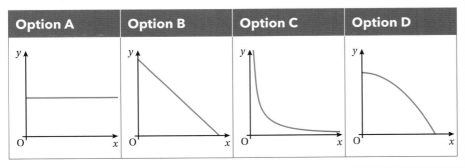

Key vocabulary

inverse proportion: a relation between two quantities such that one increases at a rate that is equal to the rate that the other decreases.

In the example of a rectangle with a fixed area, its length and height are in **inverse proportion**. When one of these dimensions increases, the other must decrease for the area to remain the same.

In a similar way to direct proportion, inverse proportion can be written as:

a proportionality: $y \propto \dfrac{1}{x}$

a formula: $y = \dfrac{k}{x}$

Substituting related values for x and y allows you to find the value of the constant of proportionality.

WORKED EXAMPLE 4

The length (x) and height (y) of a rectangle are inversely proportional. If $x = 2$cm when $y = 12$cm, find the height when the length is 3cm.

$y \propto \dfrac{1}{x}$ — Write as an inverse proportionality.

$y = \dfrac{k}{x}$ — Introduce the constant of proportionality to produce a formula.

$12 = \dfrac{k}{2}$

$k = 12 \times 2 = 24$ — Use known, related values to find the value of k.

$y = \dfrac{24}{x}$

$= \dfrac{24}{3}$

$= 8$cm — Use your formula to find the height.

EXERCISE 33D

 It takes four people three days to paint the school hall.

  a How many person-days is this?

  b How long would it take two people?

  c How long would it take six people?

  d The job needs to be completed in a day.
  How many people are needed?

  e What assumptions are being made?

 While on holiday you budget to buy five souvenirs at $2.40 each.

  a How much money do you intend to spend?

  b How many souvenirs costing $0.80 each could you buy with your budget?

  c If you need eight souvenirs of equal value, how much should you pay for each souvenir to keep within your budget?

33 Proportion

3 Speed (s miles per hour) and travel time (t hours) are inversely proportional. The faster you travel the less time a journey takes. For a journey between Cambridge and Manchester this can be represented by $s = \dfrac{180}{t}$

  a It takes 4 hours to make the journey.
  What speed is this?

  b How long will it take to do the journey at 60 miles per hour?

  c It takes 2 hours 15 minutes to make the journey.
  What speed is this?

4 A water tap is running at a constant rate (r litres per minute) filling a pond (in m minutes). The greater the flow of water the less time it takes to fill the pond up. This can be represented as $r = \dfrac{600}{m}$.
Copy and complete the table and then draw a graph to represent this situation.

m minutes	10	20	30	40	50	60	70	80	90	100
r litres per minute										

5 A group of friends play a lucky draw game. The more friends they get the more tickets they can buy but the more people they have to share the prize with. The graph shows the relationship between the number of people in the group and their share of the winnings.

  a If 10 people buy tickets how much do they each win?

  b How much is in the prize fund?

  c Find a formula for the winnings, w, in terms of the number of people buying the winning ticket, n.

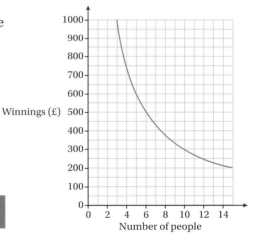

 Checklist of learning and understanding

Direct proportion

- If two quantities are directly proportional to each other they increase and decrease at the same rate. For example, if one is tripled so is the other, if one is halved so is the other.

- A formula for a direct proportion relationship between two variables x and y is $y = kx$, where k is the constant of proportionality and can be found by substituting in known values.

- The graph of two directly proportional variables is a straight-line graph of the form $y = mx$, where m is positive.

Inverse proportion

- If two quantities are inversely proportional to each other, when one increases the other decreases. For example, if one is tripled the other is divided by three, if one is halved the other is doubled.

Find answers at: cambridge.org/ukschools/gcsemaths-studentbookanswers

- A formula for an inverse proportion relationship between two variables x and y is $y = \dfrac{k}{x}$, where k is the constant of proportionality and can be found by substituting in known values.
- The graph of two inversely proportional variables is of the form $y = \dfrac{m}{x}$, where m is positive. It looks like:

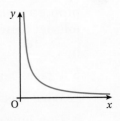

For additional questions on the topics in this chapter, visit GCSE Mathematics Online.

Chapter review

1 Wine gums cost 90p for 200 grams.

    a How much would 800 grams cost?

    b What weight of wine gums would you get for £2.70?

2 Look at this stir-fry recipe. It serves six people.

    120 g chicken

    300 g vegetables

    15 tbsp of soy sauce

If you have 300 g of chicken, 500 g of vegetables, and 60 tbsp of soy sauce and make the most stir-fry possible, how many will it serve?

3 Here are the ingredients needed to make 16 chocolate biscuits.

Chocolate biscuits

Makes 16 chocolate biscuits

100 g of butter

50 g of caster sugar

120 g of flour

15 g of cocoa

Sabrina has 250 g of butter

             300 g of caster sugar

             600 g of flour

       and 60 g of cocoa

Work out the greatest number of chocolate biscuits Sabrina can make.

You must show your working. *(3 marks)*

©Pearson Education Ltd 2013

4. The graph shows the cost of buying electrical wire.

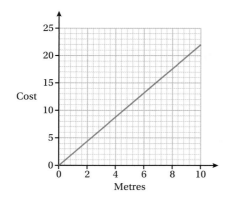

Write a formula for the cost, c, in terms of the number of metres bought, m.

5. It takes three hairdressers an hour to style the hair of models for a fashion show.

How long would it take nine hairdressers?

6. The rate at which water flows into a pond is inversely proportional to the time it takes to fill up. It takes 2 hours for the pond to fill when water flows in at a rate of 50 litres an hour.

Find a formula for the time in hours, t, in terms of the flow rate of the water, w.

34 Algebraic inequalities

In this chapter you will learn to ...
- use the correct symbols to express inequalities.
- understand and interpret inequalities.
- solve linear inequalities in one variable and represent the solution set on a number line.

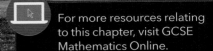

For more resources relating to this chapter, visit GCSE Mathematics Online.

Using mathematics: real-life applications

Inequalities are one way of showing the ranges of values that have to be met and considered in running a successful business. For example, a business might want wastage to be less than a certain figure, or profit to be greater or equal to a particular amount.

"I work in quality control in food production. One of my jobs is to check that the quality and size of the ingredients we use are within an acceptable range for a product, for example greater than 10 cm, but smaller than 11 cm."   *(Quality controller)*

Before you start ...

Ch 14	You must be able to solve linear equations.	1	a If $3x + 2 = 2x + 5$, then $x = ?$ b If $4(n + 3) = 6(n - 1)$, then $n = ?$
Chs 1 and 4	You should be confident with ranking numbers in ascending or descending order	2	Write this set of numbers in numerical ascending order: -2 50 -27 $\frac{1}{3}$ 1.25 2%
Ch 1	You need to remember the rules for operations with negative integers.	3	Evaluate the following: a $5 \times (-2)$ b $-5 \times (-2)$ c $12 \div 6$ d $12 \div (-6)$ e $7 - (-1)$

Assess your starting point using the Launchpad

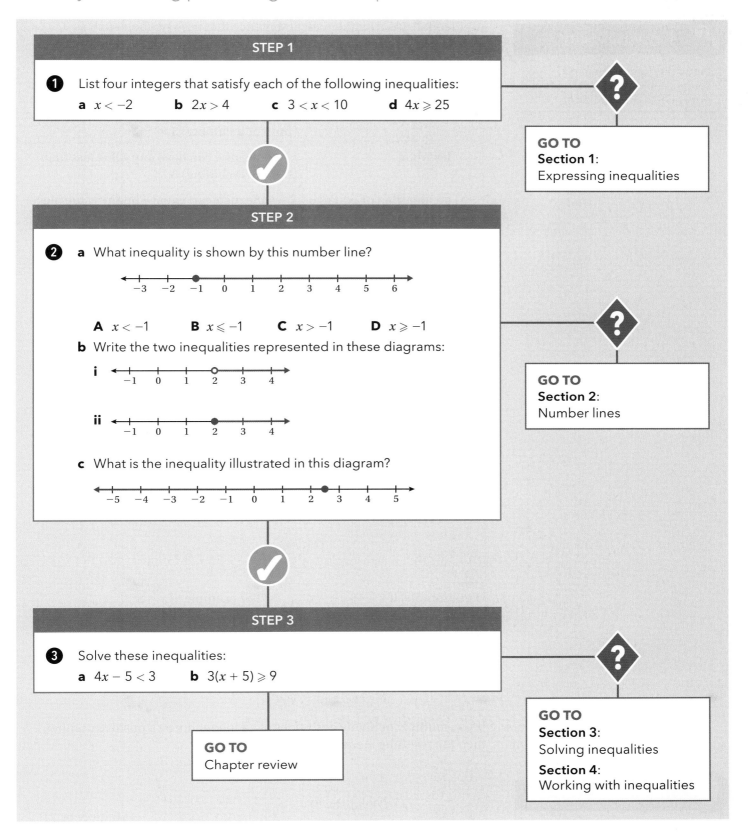

STEP 1

① List four integers that satisfy each of the following inequalities:
  a $x < -2$
  b $2x > 4$
  c $3 < x < 10$
  d $4x \geqslant 25$

GO TO Section 1: Expressing inequalities

STEP 2

② a What inequality is shown by this number line?

  A $x < -1$ **B** $x \leqslant -1$ **C** $x > -1$ **D** $x \geqslant -1$

  b Write the two inequalities represented in these diagrams:
    i
    ii

  c What is the inequality illustrated in this diagram?

GO TO Section 2: Number lines

STEP 3

③ Solve these inequalities:
  a $4x - 5 < 3$
  b $3(x + 5) \geqslant 9$

GO TO Chapter review

GO TO Section 3: Solving inequalities
Section 4: Working with inequalities

Find answers at: cambridge.org/ukschools/gcsemaths-studentbookanswers

Section 1: Expressing inequalities

An **inequality** is a mathematical sentence that uses symbols such as $<$, $\leqslant$, $\neq$, $>$ or $\geqslant$ in place of an equals sign. The expressions on either side of the symbol are not equal.

The most common inequality symbols are:

$>$	greater than	$x > 5$ means x can have any value bigger than (but not including) 5.
$<$	less than	$x < 8$ means x can have any value less than (but not including) 8.
$\geqslant$	greater than or equal to	$x \geqslant 4$ means x can equal 4 or any value bigger than 4.
$\leqslant$	less than or equal to	$x \leqslant 7$ means x can equal 7 or any value smaller than 7.

Two inequality symbols can be used to give a limited range of values. For example,

$2 < x < 6$

means 2 is *less than* x and x is *less than* 6.

Another way to read this statement is to say x lies between 2 and 6.

Sometimes, whole number (integer) values of x are asked for. In the above example, the integer values that satisfy the expression are 3, 4 and 5.

Addition and subtraction

If you add or subtract the same number to both sides of an inequality, then the resulting inequality is true.

$x > 3$ $x + 4 > 3 + 4$ $x + 4 > 7$	$x > 7$ $x - 6 > 7 - 6$ $x - 6 > 1$
For example, if $x = 10$ $\quad 10 > 3$ $10 + 4 > 3 + 4$ $\quad 14 > 7$	For example, if $x = 8$ $\quad 8 > 7$ $8 - 6 > 1$ $\quad 2 > 1$

Multiplication and division

If you multiply or divide both sides of an inequality by a positive number, then the resulting inequality is true.

$3x \leqslant 15$ $\dfrac{3x}{3} \leqslant \dfrac{15}{3}$ (Divide by 3) $x \leqslant 5$	$\dfrac{x}{2} > 5$ $\dfrac{x}{2} \times 2 > 5 \times 2$ (Multiply by 2) $x > 10$

Continues on next page ...

Key vocabulary

inequality: a mathematical sentence in which the left side is not equal to the right side.

Tip

You will use these skills in Section 3 to solve inequalities.

For example, if $x = 4$, then $3x = 12$	For example, if $x = 18$, then $\dfrac{x}{2} = 9$
$12 \leqslant 15$ (Divide by 3)	$9 > 5$ (Multiply by 2)
$4 \leqslant 5$	$18 > 10$

If you multiply or divide both sides of an inequality by a negative number, then you must **reverse** the inequality sign to make the resulting inequality true.

$7 > 3$ (Multiply both sides by -2)
$-14 < -6$ (Reverse the inequality sign)

This result is useful if you need x and not $-x$. For example,

$-x < 3$ (Multiply or divide both sides by -1)
$x > -3$ (Remember to reverse the inequality sign)

EXERCISE 34A

1. Complete the statements with the correct inequality symbol:
   a $7 > 3$, then $4 + 7 \square 4 + 3$
   b $8 < 13$, then $8 - 5 \square 13 - 5$
   c $-5 < -1$, then $-5 + 3 \square -1 + 3$
   d $-4 > -11$, then $-4 - 6 \square -11 - 6$

2. Complete the statements with the correct inequality symbol:
   a $7 > 3$, then $2 \times 7 \square 2 \times 3$
   b $8 < 13$, then $2 \times 8 \square 2 \times 13$
   c $7 > 3$, then $7 \div 2 \square 3 \div 2$
   d $8 < 13$, then $8 \div 2 \square 13 \div 2$

3. Complete the statements with the correct inequality symbol:
   a $7 > 3$, then $-2 \times 7 \square -2 \times 3$
   b $8 < 13$, then $-2 \times 8 \square -2 \times 13$
   c $7 > 3$, then $7 \div -2 \square 3 \div -2$
   d $8 < 13$, then $8 \div -2 \square 13 \div -2$

4. List four whole numbers that satisfy the following inequalities:
   a $x > 14$
   b $x \geqslant 6$
   c $x \leqslant -2$
   d $x + 3 \geqslant 7$
   e $x - 4 \leqslant 5$

5. If $x > 6$, how many values can x take?

6. If $3 < x < 8$, how many whole number (integer) values can x take? How many values can x take if we include decimal values or fractions?

7. List the whole number values given by $6 > x > 2$.

> **Tip**
>
> Remember that we read inequalities from left to right.

Section 2: Number lines

You can use a number line to illustrate an inequality.

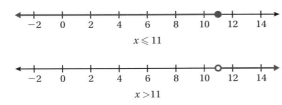

Tip

When drawing inequalities on a number line, the convention is to use an open dot (small circle) if the starting value is not included and a solid dot if the starting point is included.

The expression $x \leqslant 11$ means numbers less than 11 and including 11. So a number line representing $x \leqslant 11$ shows values starting from and including 11 with a **solid** dot at 11.

The expression $x > 11$ means numbers greater than 11. So a number line representing $x > 11$ starts at 11 but with an **open** dot to show that 11 is not included.

EXERCISE 34B

1 Draw number lines to indicate the following inequalities:

   a $x > 4$ **b** $x \leqslant -1$ **c** $x \geqslant -5$ **d** $3 \leqslant x \leqslant 10$

   e $-3 \leqslant x \leqslant 10$ **f** $-10 \leqslant x \leqslant -3$

2 Write the inequalities that are shown in the number line diagrams:

   a, b, c, d, e (number line diagrams)

3 Write an inequality to describe each of these diagrams:

   a, b, c, d (number line diagrams)

Section 3: Solving inequalities

Solving the linear inequality $4x - 5 < 3$ means finding all of the values for x that satisfy that inequality.

You can solve inequalities using the same methods that you used for linear equations.

However, you must also apply the rules that you learned in Section 1.

WORKED EXAMPLE 1

Solve for x. Show your solutions on a number line.

a $4x - 5 < 3$ **b** $\dfrac{5x-3}{2} \geq 11$ **c** $-5 \leq 3x + 4 \leq 13$ **d** $8 - 3x > 14$

a $4x - 5 < 3$
$4x < 8$ — Add 5 to both sides.
$x < 2$ — Divide both sides by 4.

b $\dfrac{5x - 3}{2} \geq 11$
$5x - 3 \geq 11 \times 2$ — Multiply both sides by 2.
$5x \geq 22 + 3$
$5x \geq 25$ — Add 3 to both sides.
$x \geq 5$ — Divide both sides by 5.

c $-5 \leq 3x + 4 \leq 13$
$-5 - 4 \leq 3x \leq 13 - 4$ — Subtract 4 from **each** expression.
$-9 \leq 3x \leq 9$
$-3 \leq x \leq 3$ — Divide all the terms by 3.

d $8 - 3x > 14$
$-3x > 14 - 8$ — Subtract 8 from both sides.
$-3x > 6$
$x < \dfrac{6}{-3}$ — Divide both sides by -3.
$x < -2$ — Don't forget to reverse the direction of the inequality.

Check: substitute $x = -3$ ($x < -2$)
$8 - 3(-3) > 14$
$8 + 9 > 14$
$17 > 14$

You can check your answer by choosing a suitable value and substituting in the original inequality.

EXERCISE 34C

1 Solve these inequalities:

a $x + 3 \geq 7$ **b** $x - 7 \leq 4$ **c** $x + 12 > 9$
d $x - 2 < 3$ **e** $x + 4 \geq -8$ **f** $x - 10 > -6$
g $x - 5 > -12$ **h** $2x \leq 6$ **i** $3x > -15$
j $3(x + 5) \leq 9$ **k** $2(5x - 2) > 5$ **l** $2(x - 3) \leq 5$
m $\dfrac{x + 3}{2} \leq \dfrac{3 - x}{2}$ **n** $-5x + 3 \leq 78$

Tip

Remember, if you multiply or divide both sides of an inequality by a negative number, then you must reverse the inequality sign to make the resulting inequality true.

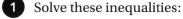
Find answers at: cambridge.org/ukschools/gcsemaths-studentbookanswers

Tip

If you have $2 < x$ and you want x on the left-hand side, you can swap the sides of the inequality but you must also reverse the signs so, you get $x > 2$. If 2 is less than x then x must be greater than 2. (Think of this as reading the inequality from right to left.)

2 Solve each inequality. Leave fractional answers as fractions in simplest form.

a $3(h - 4) > 5(h - 10)$
b $\dfrac{y + 6}{4} \leqslant 9$
c $\dfrac{1}{2}(x + 50) \leqslant 2$
d $3 - 7h \leqslant 6 - 5h$
e $2(y - 7) + 6 \leqslant 5(y + 3) + 21$
f $6(n - 4) - 2(n + 1) < 3(n + 7) + 1$
g $5(2v - 3) - 2(4v - 5) \geqslant 8(v + 1)$
h $\dfrac{z - 2}{3} - 7 > 13$
i $\dfrac{3k - 1}{7} - 7 > 7$
j $\dfrac{2e + 1}{9} > 7 - 6e$

3 Solve these inequalities:

a $5(x - 2) - 2(3x + 1) > 0$
b $5(2x - 3) < 4(x + 3)$
c $3(x + 4) - 4(x + 2) > 0$
d $2(3x - 7) - 5(2x + 3) \leqslant 0$

Section 4: Working with inequalities

WORK IT OUT 34.1

Problem: $2x - 5 < 1$

Which is the correct solution? Identify the errors made in the incorrect solutions.

Option A	Option B	Option C
$2x - 5 + 5 < 1 + 5$	$2x - 5 + 5 > 1 + 5$	$2x - 5 - 5 < 1 - 5$
$2x < 6$	$2x > 6$	$2x < -4$
$\dfrac{2x}{2} < \dfrac{6}{2}$	$\dfrac{2x}{2} > \dfrac{6}{2}$	$\dfrac{2x}{2} < -4$
$x < 3$	$x > 3$	$x < -2$

EXERCISE 34D

1 Solve each inequality and draw a number line to show the solution set.

a $x + 1 > 5$
b $2x - 1 < 6$
c $\dfrac{x + 1}{2} \geqslant -4$
d $-2x + 1 \leqslant 6$
e $1 - 5x \geqslant 21$
f $3d + 5 \leqslant 5d + 7$

2 If the values for x are whole numbers and satisfy the inequality $0 < x < 10$, copy and complete the Venn diagram (left) with the correct values for x. Label the circles: prime numbers, square numbers, even numbers.

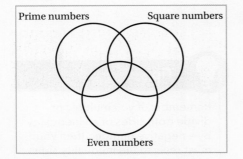

3 To solve the following problems first change the statements given into inequalities and then solve for the variable.

a When 5 is added to twice p, the result is greater than 17. What values can p take?

b When 16 is subtracted from half of q, the result is less than 18. What values can q take?

c The sum of 4d and 6 is greater than the sum of 2d and 18. What values can d take?

d A number a is increased by 3 and this amount is then doubled. If the result of this is greater than a, what values can a take?

4 a Suppose you work part-time for £12 an hour and you save 25% of what you earn.

How many hours will you need to work in order to save at least £75 in a week?

b Suppose you work for £p an hour and save 25% of what you earn.

Write an inequality that expresses the number of hours you will need to work in order to save at least £75 in a week.

Checklist of learning and understanding

Describing inequalities

- Inequalities use the symbols >, <, ⩾, ⩽ and indicate a range of values.
- Inequalities can be illustrated on a number line:

    A closed dot (●) indicates that the starting value is included.

    An open dot (○) indicates that the starting value is not included.

Addition and subtraction rules

- If $a > b$ then $a \pm c > b \pm c$; likewise, if $a < b$ then $a \pm c < b \pm c$.

Multiplication and division rules

- Multiplying or dividing by a positive number: if $a > b$ and $c > 0$ then $ac > bc$ and $\frac{a}{c} > \frac{b}{c}$.
- Multiplying or dividing by a negative number: $10 > 8$, but if you multiply by -2 then $-20 < -16$; if you divide by -2 then $-5 < -4$. You must reverse the sign in each case.
- Multiplying or dividing by a negative number: if $a > b$ and $c < 0$ then $ac < bc$ and $\frac{a}{c} < \frac{b}{c}$.

Solving inequalities

- Linear inequalities in one unknown can be solved as equations but the answer indicates a range of values for the variable.

Chapter review

For additional questions on the topics in this chapter, visit GCSE Mathematics Online.

1 Which of the following statements are true?

a $x + 11 > x - 11$

b $x \geqslant 12$ means that a number x is greater than 12.

c This number line shows the inequality $x \leqslant -1$:

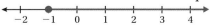

Find answers at: cambridge.org/ukschools/gcsemaths-studentbookanswers

2 x is a whole number such that $-3 \leqslant x < 5$ and y is a whole number such that $-4 \leqslant y \leqslant 2$.

Write down the greatest possible value of:

  a $x + y$ **b** $x - y$ **c** xy

3 I am thinking of an integer. I double it and add 1. The result is less than -7.

What is the largest integer I can be thinking of?

4 x is an integer.

List all the values of x such that $-1 < 2x \leqslant 8$.

5 Frozen chickens will be sold by a major chain of supermarkets only if they weigh at least 1.2 kg and not more than 3.4 kg.

  a Represent this region on a number line.

  b Write an inequality to represent this region.

6 m is an integer such that $-2 < m \leqslant 3$

  a Write down all the possible values of m.   *(2 marks)*

  b Solve $7x - 9 < 3x + 4$ *(2 marks)*

©*Pearson Education Ltd 2012*

35 Sampling and representing data

In this chapter you will learn how to ...

- infer properties of populations or distributions from a sample, recognising the limitations of sampling.
- interpret and construct appropriate tables, charts and graphs.
- choose the best form of representation for data and understand the appropriate use of different graphs.

For more resources relating to this chapter, visit GCSE Mathematics Online.

Using mathematics: real-life applications

We live in a very information-rich world. Knowing how to construct accurate graphs and how to interpret the graphs we see is important. Many graphs in print and other media are carefully designed to influence what we think by displaying the data in particular ways.

"When we have data, we need to display it so that our message has the maximum impact."

(Newspaper editor)

> **Tip**
> The key to displaying data is to choose the graph or chart that clearly shows what the data tells us without the reader having to work too hard.

> **Tip**
> Getting the scale and labelling right makes a big difference when creating graphs and charts.

Before you start...

KS3	You need to be able to sort and categorise data.	1	What would be suitable categories for a set of adult heights ranging from 1.39 m to 1.85 m?
Chs 9 and 10	You need to be able to use scales properly.	2	**a** What is each division on this scale? **b** A scale between 0 and 100 has five divisions. Which numbers should go alongside each division?
Chs 21 and 25	You need to be able to measure and draw angles to create pie charts.	3	**a** Measure these angles: **b** Draw an angle of 72° accurately.

Find answers at: cambridge.org/ukschools/gcsemaths-studentbookanswers

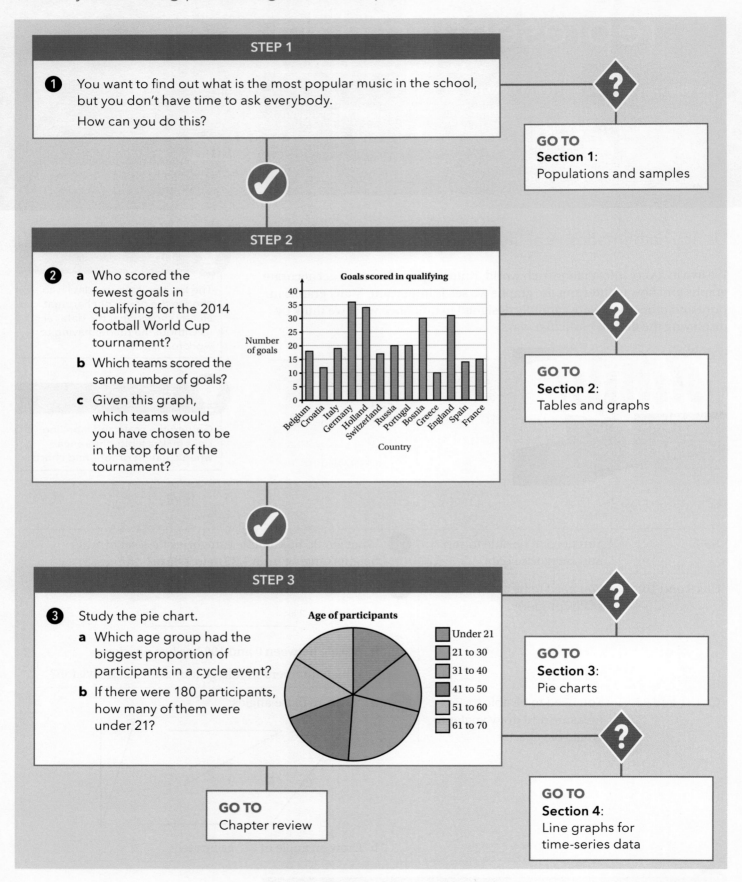

35 Sampling and representing data

Section 1: Populations and samples

A statistical **population** is a set of individuals or objects of interest.

A school might want to find the mean height of students to decide what size of equipment to buy for the gymnasium. In this example, the population would be all the students in the school.

In a large school it would be impractical to measure each student's height. Instead, a **sample** of the students (population) would be used. The sample needs to be a **representative sample** to provide useful data.

A representative sample would come from a mix of male and female students from different years. It would not be a good idea to measure just the Year 7 students.

A representative sample can be created by taking a factor that is unrelated to age or gender, for example, all the students whose first name begins with a letter drawn at random.

As the starting letter of your name has no effect on your height, this should give you a representative sample.

Although the sample cannot guarantee you will get the heights of the biggest and smallest students, it should make sure that you get a good idea of the spread of the data.

> **Key vocabulary**
>
> **population**: the name given to a data set.
>
> **sample**: a set of data collected from a population.
>
> **representative sample**: a smaller quantity of data that represents the characteristics of a larger population.

"I collect data on behalf of my company so that they can find out how likely people are to buy new products. We use quota sampling in our work. This involves choosing people with particular characteristics. For example, I may only be interested in collecting data on teenagers who play video games."

(Market researcher)

Random sampling

In a random sample, each member of the population is equally likely to be chosen. If you wanted a 10% sample and had a list of all the students in the school, you could use a computer program to pick them at random from the list.

You could number all the items in the population and then choose numbers at random using a random number generation application (or just draw names/numbers out of a hat).

In reality, it is often difficult to choose a genuinely random sample. If you are doing a survey for a school project, you are likely to use convenience sampling because you would probably survey friends and family members. This method could result in a biased sample.

Find answers at: cambridge.org/ukschools/gcsemaths-studentbookanswers

In statistics, the population may be divided into groups using a particular system. Stratified sampling is quite common. This involves dividing the sample into groups (strata) and then choosing a random sample from each group. The size of the sample chosen from each group should be in proportion to the size of the group within the population.

For example, in a school population you could use the year groups as strata. If the Year 9 students make up 30% of the school population, then 30% of the sample should come from that group.

WORK IT OUT 35.1

A market researcher has been asked to survey a random sample of shoppers at a shopping centre. She suggests the following four options.

a Which is the only option that would produce a random sample?

b Explain why each of the other three options does not produce a random sample.

Option A	Option B	Option C	Option D
Ask all the women with children.	Ask people between 8 a.m. and 8.30 a.m.	Stand outside a book shop and ask everyone who comes out.	Stop and ask every 10th person who walks by the researcher.

EXERCISE 35A

1 Which of these methods are likely to give a random sample?

  A Selecting all the odd numbered houses in a street.

  B Calling people on their home telephones during the day.

  C Selecting everybody who is wearing trainers.

  D Calling the person whose name is at the top of each page of the phone book.

  E Drawing a series of names from a hat.

  Give reasons for your answers.

2 A market research company wants to find out how many people are likely to buy a new baby food.

  a Suggest a good place to conduct a survey of young parents.

  b 35 of the 50 parents asked said they would be interested.

  How many parents would you expect to be interested in a population of 1000 parents?

3 A gym owner wants to know how many running machines to buy. She asks every member whose surname begins with an 'S' whether they will use a running machine.

  a If 15 of the 28 in her sample say 'yes', what would be a sensible number of machines to buy if there are 300 members overall?

  b Does she really need this many machines?

  c Is there a better way of sampling her members to make sure she gets a realistic number of machines?

4 At the end of 2012 there were 28.7 million cars on the roads of Great Britain.

Surjay and his friends conduct a random survey of the cars passing the school and discover that of the 50 cars recorded, three had no sun roof, one had a faulty exhaust and four had chips on the windscreen.

Use this information to estimate how many cars in Great Britain have:

 a no sun roof   b faulty exhausts   c chips in the windscreen.

5 A building firm employs joiners (30), electricians (40), plumbers (20) and bricklayers (10). For a survey, they want a 10% representative sample in proportion to the different trades. How many of each trade will be in the survey?

Section 2: Tables and graphs

Using tables to organise data

When you have many pieces of data you can use a table to organise them and make them simpler to work with.

A frequency table is used to collect or record data that shows the 'frequency' of an event (how often it happens).

These are the number of goals scored by each of the 20 premiership teams one weekend:

| 5 | 1 | 3 | 0 | 1 | 2 | 4 | 1 | 1 | 2 |
| 0 | 3 | 1 | 0 | 0 | 4 | 0 | 1 | 3 | 0 |

In a frequency table these results would look like this:

Number of goals scored	Tally	Frequency
0								6
1								6
2				2
3					3
4				2
5			1

> **Tip**
> You worked with frequency tables in Chapter 20 and Chapter 26 when you dealt with probability.

> **Tip**
> Tally marks are used as you work systematically through the data. Adding the tallies gives you the frequency.

Using bar charts to display data

The data in the frequency table above can be shown on a bar chart.

Find answers at: cambridge.org/ukschools/gcsemaths-studentbookanswers

Key vocabulary

discrete data: data that can be counted and can only take certain values.

categorical data: data that has been arranged in categories.

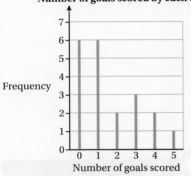

The chart has a title, a scale on the left and accurately drawn bars (of the same width).

Note that there is an equal gap between each bar and each one is labelled.

Bar charts are used to display **discrete data**. The number of goals scored by each team is discrete data because it can only have certain values. It must be a whole number – you can't score $\frac{1}{2}$ a goal or 2.34 goals.

Sometimes it is helpful to sort data into categories. Pairs of shoes in a cupboard could be categorised into 'brown shoes', 'black shoes', and so on. Each piece of data can only be in one category. This is known as **categorical data**.

A vertical line graph (left) is very similar to a bar graph but the number of items in each category is represented by a line rather than a bar, as shown on the left.

This data could also be shown using a pictogram.

In a pictogram, a symbol is used to represent either each team or a number of teams.

In this example each ball represents two teams:

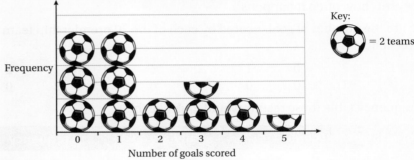

Note that a pictogram should also have a key to indicate what the symbol represents.

WORK IT OUT 35.2

Ramiz records the number of mistakes he makes in a series of maths tests.

2 3 1 3 4 2 0 3 2 6 1 1 3 2 4 2

Which graph or chart best shows this data? What is wrong with the other two?

Option A

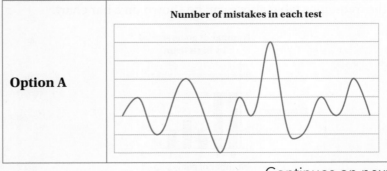

Continues on next page ...

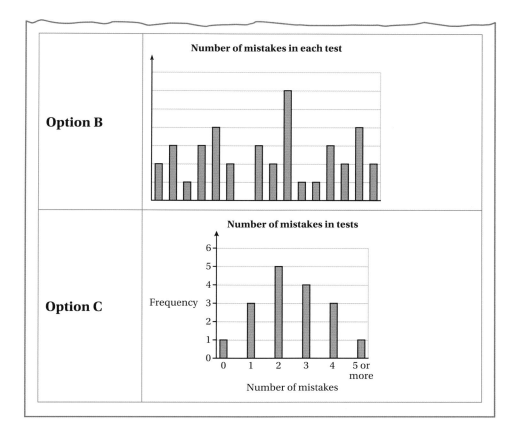

It is not always possible to say that one type of graph is better than another.

These guidelines can help you choose an appropriate graph for different kinds of data:

- Use bar graphs or vertical line charts for discrete data that can be categorised.
- Use a pie graph or a composite bar graph if you want to compare different parts of the whole or show proportions in the data.
- Use a line graph for numerical data when you want to show trends (changes over time).
- Use scatter diagrams when you want to show relationships between different sets of data (you will deal with these in Chapter 36 page 548).

> **Tip**
> You will see examples of these graphs later in this chapter.

EXERCISE 35B

1. In an extended family of 30 members, 10 have blond hair, nine have black hair, six have brown hair and five have grey hair.

   Draw a vertical line graph to show this information.

2. The table (right) shows the percentages of people who use a particular mode of transport to get to work in a factory.

   Show this information in a bar chart.

Mode of transport	Percentage
car	36
bus	27
cycle	19
walk	18

3) 30 students were asked how many times in the last week they had visited the snack shop. Their responses were:

1 2 1 2 1 5 1 3 2 1 2 1 3 2 1 2 0 2 3 2 0 2 0 1 2 0 0 3 1 2

   a Draw a frequency table for the data.
   b Present this information on a suitable graph.

4) Construct a bar graph for the data in the table below.

Favourite holiday destination	UK	Spain	France	USA	Greece
Frequency	9	15	17	12	8

5) A group of students were asked to choose their favourite snacks. The results are in the table below.

Favourite snack	Number of students
fruit	6
crisps	8
chocolate bar	9
pizza slice	12
cookie	7

Draw a pictogram to show these results.

6) The graph below shows the monthly rainfall in Lowestoft in 2012.

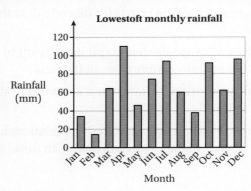

   a In which month was the rainfall heaviest?
   b Estimate the amount of rain that fell in April.
   c Which was the driest month?
   d Spring is March, April and May. Estimate how much rain fell in the spring.
   e The average annual rainfall for Lowestoft is approximately 575 mm. Was 2012 a wetter or drier year than average?

7 Jenny is carrying out a survey of what sort of snacks are bought from a shop outside her school.

She writes down the items that people buy:

Chocobar	Apple	NRG drink	Juicebar	Crisps	NRG drink	Chocobar	NRG drink	Juicebar
Juicebar	Crisps	Cheese puffs	Gum	Cheese puffs	Fruit chews	NRG drink	NRG drink	Chocobar
Chocobar	Juicebar	Chocobar	Crisps	Chocobar	Gum	Chocobar	Cheese puffs	Crisps
Cheese puffs	Crisps	NRG drink	Fruit chews	NRG drink	Cheese puffs	NRG drink	Juicebar	Gum
NRG drink	Chocobar	Apple	NRG drink	Chocobar	Juicebar	Crisps	Chocobar	Cheese puffs
Gum	Fruit chews	Gum	Crisps	Apple	Crisps	Fruit chews	Fruit chews	Fruit chews
Juicebar	Crisps	Cheese puffs	Fruit chews	Gum	Cheese puffs	Fruit chews	Crisps	Cheese puffs

a How could Jenny have been better organised before she started her survey?

b Use Jenny's data to create a table to show what was bought in the shop.

c Jenny will get extra credit if she can categorise her data.

Adjust your table so that the data is classified in an appropriate way.

Multiple and composite bar charts

A multiple bar chart is useful when you want to compare data for two or more groups. For example, to compare shoe sizes of male and female students in Year 9 you would show the data for male and female students in matching pairs of bars like this:

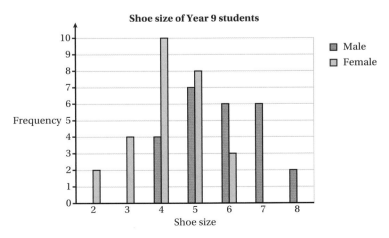

Notice that:

- The graph has a key to show what each colour bar represents.
- The two bars for male and female students who wear each size touch each other, but there is an equal space between each pair of bars.

Composite bar graphs are used to show parts of a whole. The total height of each bar represents a total amount. The length of the bar is divided into parts that show each category's share of the total amount.

To interpret a composite bar chart you need to work out what each bar represents and then do a calculation to find the fraction or percentage of the total that each part represents.

WORKED EXAMPLE 1

This composite bar chart shows the amount of water used by three different households over a 4-month period.

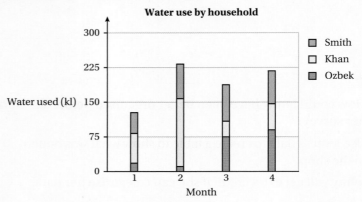

a What does each bar show?
  The top of each bar shows the total water used by three households in a month. Each coloured segment shows the fraction of the total used by each household.

b Which household used the greatest amount of water in month 1?
  The Khan household. The yellow section is bigger than the other two.

c Describe the trend in water use for the Ozbek household over this period.
  In months 1 and 2 they used very little water. In month 3 the amount of water used increased quite dramatically and in month 4 it went up a little more.

d One household had a leaking pipe in this period.
  Can you work out from the graph who this was and when it happened?
  It is most likely the Khan household as they had a massive jump in consumption in month 2. However, it could be the Ozbeks as well. If their water pipe started leaking in month 3 and wasn't fixed, it could account for the big increase in their consumption.

EXERCISE 35C

1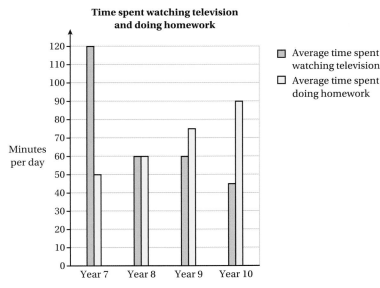

a What two sets of data are shown on the graph above?

b Describe the trend in the amount of time spent watching TV as students move into higher years.

c What happens to the amount of time spent on homework as TV watching time decreases?

d How much time do Year 10 students spend on average each day on

   i homework   ii watching TV?

2 Naresh runs a computer company. He keeps a record of his costs and his income for four large projects in a year.

He drew this graph (right) to compare his costs and his income for each project.

a Which project brought in most money?

b Which project brought in least money?

c Which project had the highest costs?

d Which project had the lowest costs?

e Which project gave Naresh the biggest profit? (Remember, profit = income − cost.)

f On which project did Naresh lose money? How can you tell?

g How much profit did Naresh make altogether?

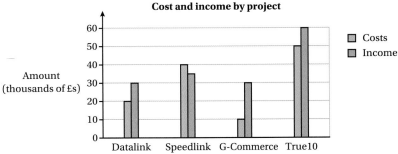

3 This composite bar graph shows the proportion of total sales and how they are made for four different companies.

a Can you work out the value of each company's total sales from this graph? Explain your answer.

b Which company does most of its sales direct from the shop?

c Which company makes the least of its sales through agents?

d Which company makes 40% of its sales through agents?

e What percentage of Company A's sales are by catalogue mail order?

f Describe the breakdown of sales by type for Company D.

4 The World Health Organization (WHO) reported the following malaria data for different regions of the world in 2010.

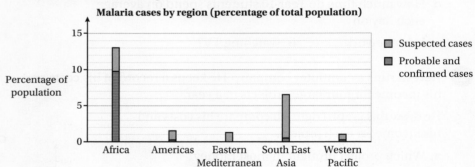

a What percentage of the population of Africa had or were suspected of having malaria?

b In South East Asia what percentage of the population were suspected of having malaria?

c Were any of the suspected cases in the Eastern Mediterranean confirmed? How can you tell?

d Which region of the world has the biggest problem with malaria?

Section 3: Pie charts

A pie chart is useful for displaying data when you are interested in the relative sizes or the proportions of the data.

Pie charts are always circular, and so the sum of the angles at the centre must always be 360°.

When drawing pie charts that have data as percentages, each 1% will be represented by 3.6° because 360 ÷ 100 = 3.6.

This data shows the area in which students in a class live:

Area	Frequency	Percentage	Angle
Reepham	12	40.0%	144°
Whitwell	6	20.0%	72°
Booton	3	10.0%	36°
Cawston	2	6.7%	24°
Salle	7	23.3%	84°

There are 30 students altogether.

The pie chart below shows this data:

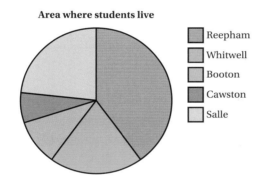

Tip

Each percentage is worked out by dividing the number of students by the total number of students and multiplying by 100; for example, $\frac{12}{30} \times 100 = 40\%$

WORK IT OUT 35.3

A survey of how late 20 trains are in minutes gives the following results:

1 0 2 0 3 1 5 4 1 3 6 4 3 5 2 4 3 2 2 4

Which of the pie charts best shows this information? Explain what is wrong with the other two pie charts.

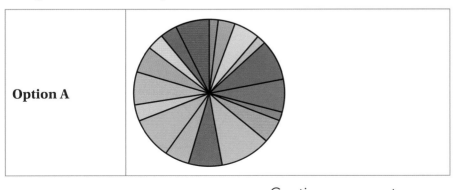

Continues on next page …

Find answers at: cambridge.org/ukschools/gcsemaths-studentbookanswers

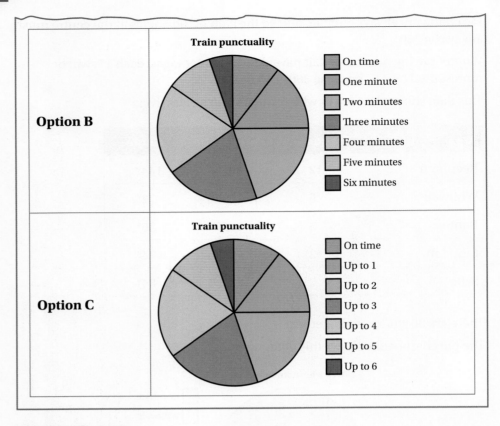

EXERCISE 35D

1 Create a pie chart to represent this data:

Electricity generation	Proportion used
Gas	28%
Other fuels	2.6%
Coal	39%
Nuclear	19%
Renewables	11.4%

2 The pie charts below show the population of two different countries by age.

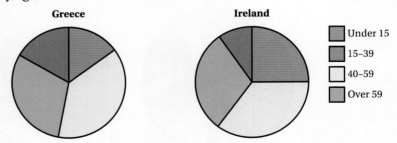

a Write down two differences between Greece and Ireland.

b Which country has the bigger proportion of over 59s?

c There are more under 15s in Ireland than Greece. Is this statement true?

3. This pie chart (right) shows the favourite leisure activity of 72 students.

   a Use a protractor to measure the sector for music. Use this measurement to work out how many students prefer music.

   b Which is the most popular activity?

   c How many students prefer reading?

4. The department of transport maintains data for the different types of vehicles on the road in the UK. Data for vehicles other than cars is shown for 1994 and 2013 (below).

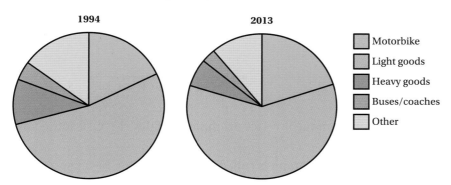

   a Give two differences between the proportions for 1994 and 2013.

   b If there were 6.075 million vehicles other than cars on the road in 2013, calculate the number of light goods vehicles there were.

   c What percentage of vehicles other than cars were motorbikes in 2013?

5. This data shows the destinations of students leaving a sixth form college:

Destination	College A	College B
Higher education	32	46
Further education	45	72
Employment	28	31
Gap year	12	24
Unemployment	15	22

Create two pie charts and use them to argue that one college is more successful than the other.

Section 4: Line graphs for time-series data

Some data that you collect change with time. For example, the average temperature each month for a year, the number of cars that pass through a junction each hour, or the amount of credit you have left on your mobile each week.

Line graphs are useful for showing how data changes over time.

When time is one of the variables it is always plotted on the horizontal axis of the graph.

The maximum temperature at a weather station is recorded at noon every day. Data which shows change over time like this is called time-series data.

Monday	Tuesday	Wednesday	Thursday	Friday	Saturday	Sunday
15 °C	17 °C	18 °C	21 °C	16 °C	20 °C	14 °C

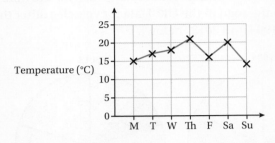

Each of the points on the line (for example, Monday, 15 °C) is joined by a straight line.

The type of data in this graph is called **continuous data** because it can take any numerical value (within a range) and can be measured. The height of students in your class would also be continuous data.

Key vocabulary

continuous data: data that can have any value.

Problem-solving framework

The average temperature each month in Alicante is as follows:

Jan 17 °C   Feb 18 °C   Mar 20 °C   Apr 21 °C   May 24 °C   Jun 28 °C
Jul 30 °C   Aug 31 °C   Sep 29 °C   Oct 25 °C   Nov 20 °C   Dec 18 °C

How can we display this data to show how the temperature changes?

Steps for approaching a problem-solving question	What you would do for this example
Step 1: If it is useful to have a table draw one.	<table><tr><td>J</td><td>F</td><td>M</td><td>A</td><td>M</td><td>J</td><td>J</td><td>A</td><td>S</td><td>O</td><td>N</td><td>D</td></tr><tr><td>17 °C</td><td>18 °C</td><td>20 °C</td><td>21 °C</td><td>24 °C</td><td>28 °C</td><td>30 °C</td><td>31 °C</td><td>29 °C</td><td>25 °C</td><td>20 °C</td><td>18 °C</td></tr></table>
Step 2: Identify what you have to do.	You need to choose a suitable means of displaying the data and then draw it.

Continues on next page …

Steps for approaching a problem-solving question	What you would do for this example
Step 3: Start working on the problem using what you know.	You know that a time-series graph shows changes over time, so it will be a good visual way of showing how the temperature varies. Choose a suitable scale and plot each point on the axes using the table. Join the points to make a line. 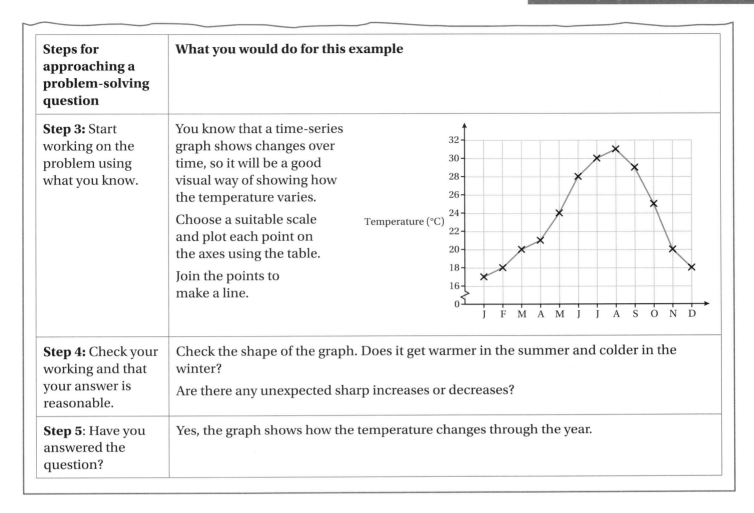
Step 4: Check your working and that your answer is reasonable.	Check the shape of the graph. Does it get warmer in the summer and colder in the winter? Are there any unexpected sharp increases or decreases?
Step 5: Have you answered the question?	Yes, the graph shows how the temperature changes through the year.

EXERCISE 35E

1 a Construct a time-series graph for the average temperature (in °C) in a particular city, which is given in the table below.

Month	Jan	Feb	Mar	Apr	May	Jun	Jul	Aug	Sep	Oct	Nov	Dec
Average Temp (°C)	15.2	16.5	17.2	19.1	19.6	20.1	22.2	24.1	21.3	19.3	17.6	16.6

b Use the time-series line graph to write a brief description of how the average temperature varies in this particular city.

2 The table below gives the annual profit (in £million) of a company over a 10-year period. Construct a time-series graph of the information.

Year	Year 1	Year 2	Year 3	Year 4	Year 5	Year 6	Year 7	Year 8	Year 9	Year 10
Profit (£million)	2.2	1.8	2.3	1.2	0.6	1.1	2.2	3.1	3.7	4.2

3 The table below gives the number of teeth extracted at a dentist's surgery each month for a year.

Month	Jan	Feb	Mar	Apr	May	Jun	Jul	Aug	Sep	Oct	Nov	Dec
Number of teeth	54	47	49	60	41	45	36	11	38	42	32	22

  a Represent this information on a time-series graph.

  b Briefly describe how the number of teeth extracted each month changed over the year.

  c Why might the number of teeth extracted fall during August?

4 The table below gives the position of a particular five-a-side football team in a league of 10 teams at the completion of each week throughout the season.

Round	1	2	3	4	5	6	7	8	9
Position	2	3	5	7	6	5	6	7	5
Round	10	11	12	13	14	15	16	17	18
Position	5	4	5	3	4	3	3	4	3

  a Represent this information on a time-series graph.

  b Describe the progress of the team throughout the season.

5 The data below shows the value of sales at a service station on a main road over a period of 3 years. Each quarter represents three months (a quarter) of the year. The quarters are labelled 1–12 in the corresponding time-series graph.

Sales quarter	Sales £thousand
Quarter 1	64
Quarter 2	82
Quarter 3	83
Quarter 4	65
Quarter 5	77
Quarter 6	89
Quarter 7	96
Quarter 8	58
Quarter 9	79
Quarter 10	92
Quarter 11	101
Quarter 12	66

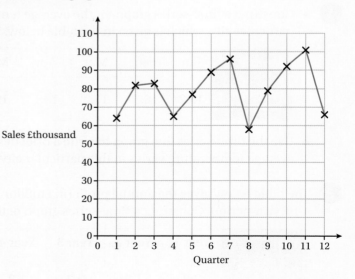

a In which quarter of each year is the value of sales highest?

b In which quarter of each year is the value of sales lowest?

c Compare the sales figures for the first quarter of each year. Are the sales figures improving from one year to the next?

6 The table below gives the numbers of garden sheds sold each quarter during 2012–2014.

Number of sales	Q1	Q2	Q3	Q4
2012	27	32	56	41
2013	33	35	65	45
2014	38	41	72	51

a Represent this information on a time-series graph.

b Describe how the shed sales have altered over the given time period.

c Does it appear that shed sales are seasonal?

7 Study the following graph.

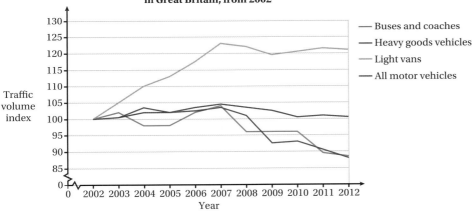

a Describe the trend in numbers of light vans.

b What has happened to the number of all motor vehicles?

c Suggest why the number of heavy goods vehicles might have decreased. Can you answer this by just using the graph?

8 This graph (right) shows how the water level in a pond varies from month to month.

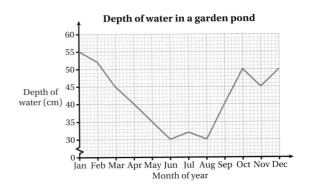

a When is the depth of water lowest?

b What do you think might have happened in July?

c When does the water level drop most rapidly?

d How much water is in the pond in May?

e What is the difference in depth between August and September?

Find answers at: cambridge.org/ukschools/gcsemaths-studentbookanswers

Checklist of learning and understanding

Sampling

- A sample is a representative group chosen from a population.
- In a random sample, each member of the population is equally likely to be chosen for the sample.

Tables and graphs

- A frequency table is a method of organising data by showing how often a result appears in the data.
- Data can be displayed using a number of different graphs and charts.
- All graphs should be clearly labelled and scaled, and include a title.
- Vertical line graphs and bar charts are good ways of showing discrete data, where the height of each bar, or line, determines the frequency.
- Pictograms are an interesting visual way of displaying discrete data.
- Pie charts are used to compare categories of the same data set.

Line graphs

- Line graphs for time-series data show trends and changes over time.

For additional questions on the topics in this chapter, visit GCSE Mathematics Online.

Chapter review

1 Bonita needs to find out how students travel to school, so she decides to take a representative sample.

  a Suggest two ways in which she could do this.

  b Bonita asks a representative sample of 50 students and gets the results shown in the table:

  If there are 600 students in the school, what is a sensible estimate of the number of students who walk to school?

Car	15
Walk	17
Bus	6
Taxi	7
Bike	5

2 A business employs 18 office workers and 63 factory floor workers. The manager wants nine workers for a survey.

  a What would be wrong with a simple random survey?

  b How many should be chosen from each group of workers?

  c How could the workers be chosen from each group?

3 Kimberley surveys her classmates to find their favourite pizza. The results are shown in the table.

Choose a suitable scale and draw a pictogram to represent this data.

Cheese and tomato	5
Seafood	8
Meat	2
Roast vegetable	6
Pepperoni	7

4. Two adults are comparing how much money they spend each month.

Expenditure per month (£)	Josh	Ben
Rent	840	450
Food	250	300
Transport	350	160
Savings	250	40
Entertainment	110	250

a Draw suitable graphs to enable you to compare the proportions of money they spend.

b Write two sentences comparing their spending habits.

5. The profits for two companies are given for each quarter of a 2-year period below:

Company profits (£)	Company A	Company B
1st quarter 2013	134 820	125 912
2nd quarter 2013	138 429	189 355
3rd quarter 2013	140 721	130 969
4th quarter 2013	131 717	156 548
1st quarter 2014	103 746	219 357
2nd quarter 2014	197 028	151 296
3rd quarter 2014	187 883	249 216
4th quarter 2014	168 414	102 158

Use the data to plot a suitable graph to compare how profits change.

a Which company is the more successful?

b Which is the biggest change between quarters?

6. The table shows some information about the minimum and maximum temperatures in Paris each month from January to May.

The temperatures are in °C.

	Jan	Feb	Mar	Apr	May
Minimum temperature	2	3	5	7	10
Maximum temperature	7	8	12	15	19

Show this information in a suitable diagram.

©Pearson Education Ltd 2013

36 Data analysis

In this chapter you will learn to ...

- calculate and compare summary statistics for ungrouped and grouped data.
- recognise when data is being misrepresented.
- plot and interpret scatter diagrams and lines of best fit and use them to describe correlation and predict results.
- identify outliers and understand how they can indicate errors in data.

For more resources relating to this chapter, visit GCSE Mathematics Online.

Using mathematics: real-life applications

Analysing large sets of data enables financial and insurance companies to make predictions about what might happen in the future. Car insurance premiums are worked out according to typical or 'average' behaviour of large groups of people.

Tip

Knowing how to calculate averages and measures of spread gives you tools to compare different sets of data. Make sure you know what these are and when to use the different measures.

"We group drivers together by age and gender and use statistics to find typical driving behaviour for each group. Young drivers have more accidents, so their insurance costs more." *(Insurance broker)*

Before you start ...

KS3	You should remember how to find the mean, median, mode and range of a set of data.	**1** Find the mean, median, mode and range of the following sets of data. Give your answers correct to 1 decimal place. **a** 2 4 2 7 3 5 4 2 3 1 **b** 40 20 30 60 50 10
Ch 23	You should be able to plot coordinates on a set of axes.	**2** Write down the coordinates of points A, B and C on the line.
Ch 23	You should be able to recognise whether a gradient is positive or negative.	**3** Use the graph above. **a** What is the gradient of the graph? **b** What is the equation of the line?

Assess your starting point using the Launchpad

STEP 1

1 Answer the questions about these three sets of data:

A	10	5	9	10	8	12	7
B	3	4	5	6	6	10	12
C	7	10	11	14	18		

   a What is the median of set C?
   b Which data set has a mode of 10?
   c Which set of data has the smallest mean?
   d Which data set does not have a mode?
   e How would the mode, median and mean of set B change if we added the value 14 to the set?

2 The frequency table shows the ages of a group of students visiting the Natural History Museum.

Age (years)	14	15	16	17
Frequency	23	17	13	9

   a What is the modal age?
   b What is the mean age of the students in this group?
   c What is the range of ages in the group?

GO TO
Section 1: Summary statistics

STEP 2

3 This graph appeared in a newspaper article. Explain how this graph could be misleading.

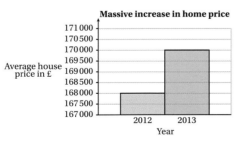

GO TO
Section 2: Misleading graphs

GO TO
Step 3:
The Launchpad continues on the next page …

Find answers at: cambridge.org/ukschools/gcsemaths-studentbookanswers

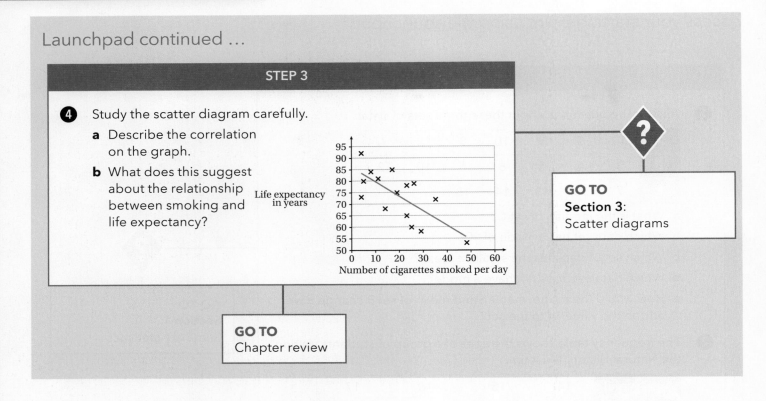

Section 1: Summary statistics

The mean, median and mode are all types of averages, or measures of central tendency.

The range is a measure of spread or dispersion. The range is useful for determining whether the mean is distorted or not.

- the mean: $\dfrac{\text{sum of values}}{\text{number of values}}$
- the mode: the value with the highest frequency
- the median: the middle value when the values are arranged in size order
- the range: the difference between the highest value and the lowest value.

To describe and compare two sets of data, calculate the averages and range and write sentences to summarise what you notice.

Choosing the correct average

The type of average that you choose depends on the situation and what you want to know.

The mean is the average that is used most often. The mean is useful when you want to know a typical value. If the data is very spread out (it has a big range) then the mean will not be typical.

For example, the boss in a company earns £20 000 per month. Her nine employees earn £2000 each. This gives a mean salary of £3800, which is not typical.

In this situation, the median salary is £2000 and the modal salary is £2000. Both are more representative than the mean.

When the data is not numerical, you have to use the mode as the average.

The mode is most useful when you need to know which item is most common or most popular.

You would use the mode when you wanted to show:

- which clothing size was bought most often
- which shoe size is most common
- which brand of mobile phone is the most popular.

The average you choose can affect how you see the data.

For example, Jeanne asks her friends how many takeaways they have had in the past year. These are their answers:

1   1   2   1   3   40   1

She works out the mean number of takeaways:

$$\frac{(1 + 1 + 2 + 1 + 3 + 40 + 1)}{7} = \frac{49}{7} = 7$$

The mean suggests that Jeanne's friends have had an average of seven takeaways each in the past year.

But that is not a very representative average and it does not give the typical number of takeaways. The mean number of takeaways is high because one friend had many more takeaways than the others.

Jeanne doesn't think the mean is a good average, so she finds the median:

1   1   1   1   2   3   40

This gives her an average of one takeaway. That seems more typical. The mode is also 1 because most people have only had one takeaway. The median and the mode are more representative and they are not affected by outliers in the data, so they are better averages in this case.

Tip

A very high or a very low value in a set of data is called an outlier. If there is an outlier then the mean will not be a typical value.

WORKED EXAMPLE 1

Josh has developed a website and is monitoring how many hits it receives per hour.

In the first two days (48 hours) it receives the following numbers of hits per hour (arranged in numerical order):

100	105	106	106	107	107	108	110	117	118
135	137	145	148	148	148	153	155	157	159
162	171	171	179	183	183	185	185	189	199
201	203	204	209	216	220	223	224	224	227
229	230	231	233	234	235	237	238		

Continues on next page …

a Find the following summary statistics:

   i the mean, median and mode

   ii the range.

   i Mean of the data: 8394 ÷ 48 = 174.88 hits.
   Median of the data is halfway between the 24th and the 25th data values
   $= \frac{179 + 183}{2} = 181$ hits.
   The value 148 is the mode because it occurs three times.

   ii The range = 238 − 100 = 138.

b Josh is trying to sell advertising on his website.

   Write a mathematical sentence he could use about the number of hits his site is receiving.
   There is a consistent hit rate of over 100 hits per hour with 175 hits per hour on average.

c Josh compares his data to a similar website run by Delia. Delia's data set has the following data values:

   mean = 180 hits

   median = 140 hits

   mode = 135 hits

   range = 200

   What can Josh say to compare the two sets?
   Although the mean of the hits is a bit higher for Delia's set, the median is much lower. This indicates that the mean of Delia's hits is influenced by a few high values, but usually the number of hits is lower. Delia's data shows a wider range, showing that the data is more spread out and therefore less consistent.

Analysing grouped data

Data is sometimes grouped together before it is analysed. The groups are known as **class intervals**. Note that they do not overlap.

For example:

Marks scored	Frequency
0–9	6
10–19	6
20–29	4
30–39	5
40–50	9
Total	30

When you only have the grouped data in a frequency table, it is not possible to calculate precise values for the mean, median, mode and range because you don't know the individual values.

You can identify the modal and median classes from the previous table.

- The modal class is the class interval that has the most elements, not the individual value that appears the most. In this table the modal class is 40–50 marks.
- To estimate the median of grouped data, find the class interval in which the middle value occurs. It is only possible to say that the median is within that group. In the table there are 30 values, the middle value is between the 15th and 16th values, so it must fall into the class 20–29 marks.

Estimating the mean of a frequency distribution

To estimate the mean of grouped data find the midpoint of each class interval. The midpoint is found by adding the lowest and highest possible values for each class interval and dividing by 2. Multiply each midpoint by the frequency for each class interval. Find the total of these values, and divide by the total number of values you have.

WORKED EXAMPLE 2

Ben goes fishing and records the masses of the fish he catches in the table below:

Mass (m) in kg	Frequency	Midpoint	Midpoint × frequency
$2 \leqslant m < 4$	5		
$4 \leqslant m < 6$	8		
$6 \leqslant m < 8$	4		
$8 \leqslant m < 10$	9		
$10 \leqslant m < 12$	3		

a Copy and complete the table.

b Find the modal class.

c Estimate the mean, median and range.

a *Completed table:*

Mass (m) in kg	Frequency	Midpoint	Midpoint × frequency
$2 \leqslant m < 4$	5	3	15
$4 \leqslant m < 6$	8	5	40
$6 \leqslant m < 8$	4	7	28
$8 \leqslant m < 10$	9	9	81
$10 \leqslant m < 12$	3	11	33

Continues on next page …

Find answers at: cambridge.org/ukschools/gcsemaths-studentbookanswers

b The modal class is $8 \leqslant m < 10$ kg.

c Estimating the mean:
Total of the midpoint × frequency values is 197.
Estimated mean is 197 ÷ 29 = 6.79 kg
The median class:
There are 29 values, so the middle value is value number 15. This occurs in the interval $6 \leqslant m < 8$ kg.
The range is 12 − 2 = 10 kg.

EXERCISE 36A

1 A large company has kept a record of how many days each employee is absent from work each year.

The results are in the table below:

Days absent (d)	Frequency	Midpoint	Midpoint × frequency
$0 \leqslant d < 5$	15		
$5 \leqslant d < 10$	23		
$10 \leqslant d < 15$	19		
$15 \leqslant d < 20$	12		
$20 \leqslant d < 25$	6		
Total			

a Copy and complete the table.

b Use the information to find the modal class.

c Estimate the mean, median and range.

2 For a charity event, several students are throwing darts at a dart board while blindfolded.

The scores they achieve are given on the right:

a Choose suitable class intervals and group the data.

b Estimate the mean, the median and the range of the scores.

c Is it sensible to estimate the range?

d What is the modal group?

89	11	57	25	55	78
28	35	15	90	83	38
57	37	28	14	36	40
74	59	57	9	18	70
25	18	22	2	37	53
74	61	79	53	87	46
30	29	4	90	83	77

3. A health club has measured its members' heights (in metres) before buying some new gym equipment. The data is given on the right:

1.68	1.68	1.58	1.72	1.58	1.75	1.89
1.84	1.55	1.65	1.66	1.84	1.55	1.81
1.47	1.55	1.58	1.66	1.55	1.61	1.68
1.57	1.57	1.69	1.65	1.75	1.55	1.73
1.64	1.85	1.53	1.65	1.77	1.66	1.75
1.75	1.59	1.88	1.82	1.62	1.69	1.67
1.63	1.66	1.84	1.77	1.52	1.84	1.53

a Use group intervals of every 5 cm, starting with the group $1.45 \leq h < 1.50$

Estimate the mean and the median.

b What is the modal class?

c Use class intervals of every 10 cm, starting with the class $1.40 \leq h < 1.50$

What difference does this make to your estimates of the mean and median?

4. 30 runners complete a marathon race. Their times are given below (to the nearest minute):

2 h 45 min, 3 h 25 min, 3 h 46 min, 4 h 15 min, 5 h 8 min, 4 h 49 min,
4 h 18 min, 3 h 38 min, 3 h 43 min, 3 h 5 min, 2 h 55 min, 4 h 23 min,
4 h 25 min, 3 h 39 min, 3 h 20 min, 4 h 1 min, 3 h 33 min, 4 h 6 min,
5 h 11 min, 2 h 51 min, 4 h 35 min, 3 h 19 min, 4 h 47 min, 4 h 28 min,
5 h 5 min, 4 h 19 min, 2 h 46 min, 3 h 18 min, 3 h 53 min, 4 h 35 min.

a Group the data into suitable class intervals.

b Find the modal class.

c Estimate the mean, median and range.

5. The mass of fruit harvested in a week from an orchard is recorded in the table below.

Mass (m) of fruit in kg	Frequency	Midpoint of class interval	Midpoint × frequency
$2000 \leq m < 2500$	15		
$2500 \leq m < 3000$	11		
$3000 \leq m < 3500$	13		
$3500 \leq m < 4000$	7		
$4000 \leq m < 4500$	2		
$4500 \leq m < 5000$	2		
$5000 \leq m < 5500$	2		
Total			

a Calculate an estimate of the mean mass of the fruit harvested in a week.

b In which interval does the median lie?

Comparing two or more sets of data

The summary statistics allow you to compare sets of data and make decisions about them. One summary statistic on its own does not give enough information about the whole set. Think about the following:

Set A might have a similar mean value to Set B, but the median is lower than the median of Set B. This shows that there are a few higher values in the set that have made the mean higher, but that more of the values are low.

 Find answers at: cambridge.org/ukschools/gcsemaths-studentbookanswers

One set could have a higher mean than the other, but the range of the data might be much wider, which shows that many of the values are very different from the mean or median.

EXERCISE 36B

1 The results from two maths tests are given.

A	35	68	55	52	49	63	61	69	35	53
B	47	34	71	41	60	44	57	74	67	64

Describe and compare the results.

2 Two cricketers are having an argument about who has had the better season.

They have both batted 16 times, and the number of runs they have scored in each innings is given below.

Ahmed   27 16 36 27 55 35 51 38 44 17 41 53 7 43 48 49

Bill   2 30 44 11 26 32 13 46 40 44 0 45 15 34 14 24

a Compare and describe their records.

b Who do you think has had the better season?

3 Yusuf has recorded the time in minutes it takes him to get home on two different buses.

Which bus route should he use?

Bus 127   17 17 21 23 19 20 19 18 21 22 19 22 21 20

Bus 362   23 26 20 15 15 20 26 19 18 15 16

Explain your answer. Does it matter that he has more data about the 127 bus?

4 A factory needs to choose between two machines that are both capable of bottling soft drinks.

Both manufacturers have provided data about how many bottles each machine fills per hour.

Machine A	
Bottles (b)	Frequency
$200 < b \leqslant 250$	36
$250 < b \leqslant 300$	48
$300 < b \leqslant 350$	59
$350 < b \leqslant 400$	61
$400 < b \leqslant 450$	21

Machine B	
Bottles (b)	Frequency
$200 < b \leqslant 250$	16
$250 < b \leqslant 300$	58
$300 < b \leqslant 350$	63
$350 < b \leqslant 400$	78
$400 < b \leqslant 450$	15

Use estimates of the mean, median and the range along with the modal group to decide which machine to choose.

5 The following statement in a newspaper seems to be incorrect.

According to latest figures, half the population weighs more than 70 kg. The "average" person weighs 80 kg.

Can you give an example of a sample of five people with mean weight of 70 kg and median weight of 80 kg?

6 a According to the Office of National Statistics, the 'average' price of a house in the UK in July 2014 was £272 000.

What would be the best type of average to measure house prices? Explain why you think so.

b Give an example of when using the mean as a measure of central tendency would be the most useful.

c Give two examples when using the mode as a measure of central tendency is most useful.

Section 2: Misleading graphs

One of the advantages of using graphs is that they show information at a glance. But graphs can also be misleading because most people do not look at them very closely.

When you look carefully at a graph you might find that it has been drawn in a way that gives a misleading impression. Sometimes this is intentional, sometimes it is not. You need to be able to look at graphs and know if they are misleading or wrong.

When you look at a graph, you have to think about:

- the scale and whether or not it has been exaggerated in any way to give a particular impression
- whether or not the scale starts at 0 – this can affect the information shown and give a misleading impression
- whether bars or pie diagrams have sections that look three-dimensional, which make some parts look much bigger than others
- whether the scales are labelled and whether or not the graph has a title
- whether the source of the data is given.

Here are some examples of misleading graphs.

This graph seems to suggest that the price of solar energy is dropping quickly while the cost of nuclear power is increasing.

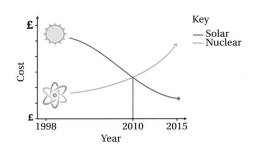

There are no values on the scale, so it is not possible to decide whether that is really true and no source is given for the information. The scale could be £5 million at the bottom and £10 million at the top, in which case the graph would be very misleading.

You also don't know which costs are being compared. The graph could be comparing the cost of building a nuclear power station (very expensive) and the cost of installing 25 solar panels (much less expensive).

The graph below is suggesting that recycling has increased dramatically from 1975. The graph uses proportion to mislead. If you look at the scale, you will see that the amount of recycled material has increased from 100 kg to 250 kg, so 2.5 times more material is recycled. The bin, however, is about six times bigger, so it looks like much more is recycled.

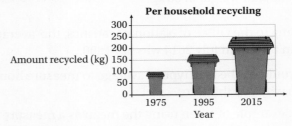

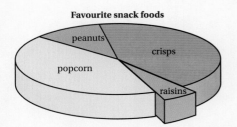

By drawing the pie chart (on the left) in this orientation and making the sectors appear 3D, it looks as if raisins are just as popular as peanuts and that popcorn is more popular than crisps. The real figures show that only 5% chose raisins and 11% chose peanuts, so the ▨ sector that sticks out represents less than half of the ▨ sector. The other two sectors each represent 42% but they do not look the same size in this graph.

WORKED EXAMPLE 3

What is wrong with this graph?

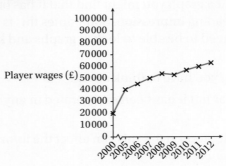

The scale of the vertical axis goes as high as £100 000, which makes the increases in wages from year to year appear to be less significant. The small scale makes the slope appear flatter. The years 2001–2004 are omitted, making the increase between 2000 and 2005 seem more significant.

EXERCISE 36C

1 Identify the error in this graph.

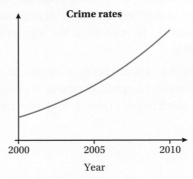

2 How is this graph misleading?

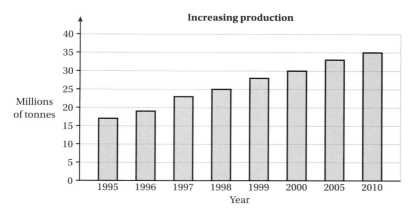

Why might someone have drawn the graph like this?

3 Look carefully at the graph below.

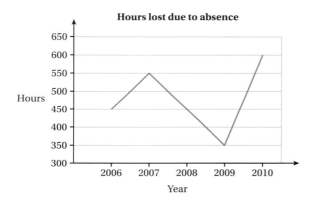

a What is misleading about this graph?

b Why do you think it has been drawn this way?

4 The same data as in question **3** has been presented in this 3D graph.

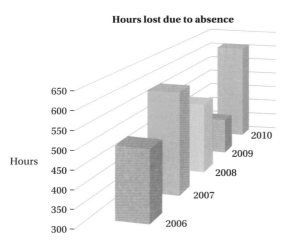

a Which year has the most hours lost, 2006 or 2008?

b Why is it hard to tell?

5. You are given the following data showing viewing figures for various TV programmes (in millions).

Week beginning	Britain's Got Talent	The Crimson Field	Gogglebox
7/4/14	10.03	6.89	2.75
14/4/14	8.45	6.31	3.37
21/4/14	8.63	6.25	3.48
28/4/14	8.45	6.01	3.47
5/5/14	8.58	6.33	3.54

Choose one of the TV programmes and create a graph which shows how well it has performed. You can use any type of graph, but you must not change the numbers.

Section 3: Scatter diagrams

A scatter diagram is used to show whether or not there is a relationship between two sets of data collected in pairs. Data that is collected in pairs is called **bivariate data**.

For example, you could record the number of hours different students spend studying and the results they get in a test. This would give two pieces of data for each learner: time spent studying and results. You can think of these as a pair of number coordinates (x, y).

In bivariate data, both sets of data are numerical, so each pair of data can be plotted as a point using coordinates on a pair of axes.

Once you have plotted the data, you can look for a pattern to see whether there is a relationship or **correlation** between the two variables or not. The diagrams below show the typical patterns of correlation and what they mean.

Key vocabulary

bivariate data: data that is collected in pairs.

correlation: relationship or connection between data items.

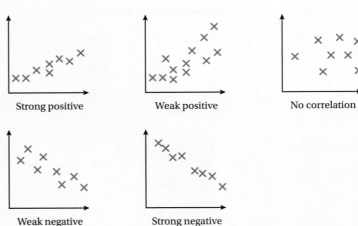

WORKED EXAMPLE 4

Nick says people who are good at maths are also good at science.

Use this data to draw a scatter diagram and comment on whether Nick is correct or not.

	Maths average (%)	Science average (%)
Student A	20	22
Student B	32	30
Student C	45	39
Student D	38	40
Student E	60	60
Student F	80	70
Student G	80	72
Student H	90	90
Student I	80	25
Student J	60	65

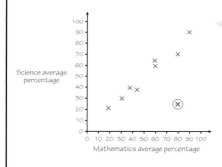

The points slope up towards the right, so it seems there is a positive correlation between maths achievement and science achievement. Nick seems to be correct.

Look at the graph in the Worked example 4.

Note that maths is on the horizontal axis and science on the vertical axis.

Science is the **dependent variable** in this case. Nick's statement is that science achievement is dependent on whether or not you are good at maths. Maths is the independent variable so it goes on the horizontal axis.

The pattern of points is used to decide whether there is a relationship. In this case, the points are grouped fairly closely and they form a thick line that slopes up to the right, so you can say there is a positive correlation between the scores.

One point is far away from the others and doesn't seem to fit the pattern (shown circled). It shows a student with a high mark for maths but a low mark for science. This point is an **outlier** in this set of data.

It is important to note that *correlation is not causation*. This means that although there might be a relationship between two variables, the change in one cannot be said to be the definite reason for the change in the other; one does not necessarily cause the other.

> **Key vocabulary**
>
> **dependent variable**: data that is measured in an experiment.
>
> **outlier**: data value that is much larger or smaller than others in the same data set.

Lines of best fit

A line of best fit is used to show a general trend on a scatter diagram.

This is a line drawn on the graph passing as close to as many points as possible.

This is a line of best fit for the scatter diagram in Worked example 4.

You can use a line of best fit to make predictions based on the collected data.

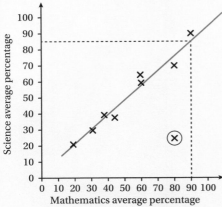

For example, if you wanted to predict the science results for a student who got 90% (or any other value) for maths, you could find this is 85% using the line. This is shown by the dotted line on the diagram (left).

Estimating a value in this way is known as **interpolation**, but can only be used as a safe estimate if the values lie within the range of the original data. **Extrapolation** is a similar process but here the line of best fit is used to predict a result whose values lie outside the range of the original data. Results from extrapolation should be treated with care as they can be unreliable.

EXERCISE 36D

1 Draw a scatter diagram for the following data and draw a line of best fit:

Homework (minutes)	10	25	38	65	84	105	135	158
TV viewing (minutes)	60	55	50	20	30	15	10	8

What kind of correlation is this?

Key vocabulary

interpolation: using two known values to estimate an unknown value.

extrapolation: using values within a known range to estimate an unknown value that lies outside of that range.

2

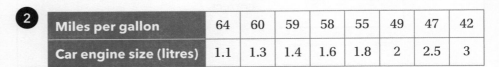

a Draw a scatter diagram to show the relationship between car engine size and fuel economy (miles per gallon).

b Draw a line of best fit for this data.

c Use the line of best fit to estimate the fuel economy of a car with a 1.5-litre engine.

d Why would it not be a good idea to use the line of best fit to estimate the economy of a car with a 6-litre engine?

3 Mike has an ice cream stall in the local park.

He writes down the number of ice creams he sells and the maximum temperature each day for a week.

Ice creams sold	86	89	45	69	84	25	78
Maximum temperature	25 °C	26 °C	19 °C	23 °C	25 °C	15 °C	21 °C

a Draw a scatter diagram to show the correlation between ice cream sales and the temperature.

b Can you suggest any other factors that might affect sales?

4 During a census the number of people living in each house is recorded.

House number	1	3	5	7	9	11	13	15	17	19	21
Number of residents	1	5	1	4	2	5	6	3	5	3	6

a Draw a scatter diagram to show this data.

b What kind of correlation is this?

5 The table below shows the athlete's height and the height jumped by the last 10 men's high jump world record holders.

Athlete	Athlete height (m)	Height jumped (m)
Sotomayor	1.95	2.45
Sjoberg	2.00	2.42
Paklin	1.91	2.41
Povarnitsyn	2.01	2.40
Jianhua	1.93	2.39
Wessig	2.00	2.36
Mogenburg	2.01	2.35
Wszola	1.90	2.35
Yashchenko	1.93	2.34
Stones	1.96	2.32

a Draw a scatter diagram showing this data.

b Is there a correlation between the height of the jumper and the height he jumped?

Outliers

Outliers are data values that lie outside the normal range for a set of data. In science experiments they might be 'freak' results or the result of inaccurate measurements. It can be difficult to decide when it is reasonable to disregard an outlier, but if it is an obvious error then the value is usually just ignored.

Outliers can't be ignored just because they spoil a pattern. Outliers will have an impact on calculating the mean and the range of a set of data, but less so when finding the median and the mode.

On a scatter diagram an outlier will be a point that is away from the main scatter of points, or might fit the line of best fit but be at an extreme value.

Find answers at: cambridge.org/ukschools/gcsemaths-studentbookanswers

WORKED EXAMPLE 5

A coach records the 100 m times of her 10 athletes at the start and the end of a week of intense training.

Nine of the athletes improve by a mean of 0.2 seconds, but one athlete is 2 seconds slower.

Can the coach claim to be making an impact on her athletes?
Yes, the coach is making an impact on the athletes, although the mean would show a reduced performance, because the single athlete's performance has reduced by much more than the others have improved. The athlete with reduced performance is an outlier, and her performance may be affected by ill health.

EXERCISE 36E

1 Several students sit a maths test and their scores are given below:

| 54 | 50 | 47 | 42 | 54 | 44 | 36 | 37 | 45 | 36 | 55 | 55 | 52 | 85 | 39 |

a What is the mean score in the test?
b What is the range?
c What is the median score?
d What is the median without the outlier?
e What is the mean without the outlier?

2 Several students sit a maths and an English exam. Their scores are given below:

| English | 49 | 42 | 46 | 44 | 53 | 41 | 64 | 14 | 44 | 53 | 55 | 42 |
| Maths | 46 | 47 | 43 | 45 | 49 | 48 | 69 | 39 | 33 | 46 | 53 | 44 |

a Plot their scores on a scatter diagram.
b Draw a line of best fit on your scatter diagram.
c Are any of the points outliers?

3 The time taken to travel by train from Norwich to London in minutes is recorded for 20 journeys:

| 109 | 129 | 98 | 106 | 109 | 156 | 128 | 98 | 99 | 113 |
| 126 | 99 | 105 | 110 | 126 | 98 | 106 | 114 | 122 | 107 |

On a normal day the journey should take between 95 and 115 minutes, depending on the number of stops at stations.

a What is the mean journey time?
b The train company claim that the mean journey time is 111 minutes on a normal day.

Is this right?

36 Data analysis

Checklist of learning and understanding

Summary statistics

- The mean, median and mode are all measures of central tendency, which are often called 'averages'. They can be found precisely for populations that are ungrouped, and estimated for grouped data.
- The range is a measure of spread. It can be found precisely for ungrouped data and estimated for grouped data.

Misleading graphs

- The way that data is presented in graphs can be misleading. Watch out for uneven scales on the axes and graphs that show comparisons using areas, which exaggerate increases.

Scatter diagrams and correlation

- Scatter diagrams can be used to look for correlations in bivariate data. A correlation is a relationship, such as one quantity increasing as another decreases. Some bivariate data has no correlation.
- Correlation does not mean causation; in other words, identifying a relationship does not necessarily mean that a change in one data set is causing the change in the other.
- Outliers are pieces of data that sit outside the pattern or expected result. They can be ignored if an obvious error, but otherwise should be considered and explained.

Chapter review

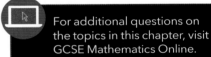

For additional questions on the topics in this chapter, visit GCSE Mathematics Online.

1 Faisel weighed 50 pumpkins.

The grouped frequency table gives some information about the weights of the pumpkins.

Weight (w kilograms)	Frequency
$0 < w \leq 4$	11
$4 < w \leq 8$	23
$8 < w \leq 12$	14
$12 < w \leq 16$	2

Work out an estimate for the mean weight. *(4 marks)*

©*Pearson Education Ltd 2012*

Find answers at: cambridge.org/ukschools/gcsemaths-studentbookanswers

2. Windsurfers need a certain amount of wind to surf, but too much can be dangerous.

A learner would typically surf in a speed of 7–18 knots, but an expert would prefer to surf at above 30 knots.

Use the data on wind speed in knots measured at the same time each day for the two lakes below to decide which lake is better for beginners and which for experts. Use measures of central tendency and spread to support your argument.

First lake (knots)	0	21	33	13	20	11	35	3	5	3	31
	28	19	19	22	26	40	40	4	25	21	26
Second lake (knots)	15	11	19	11	10	19	23	25	10	18	10
	16	23	15	15	20	22	10	13	11	18	18

3. Study the two line graphs.

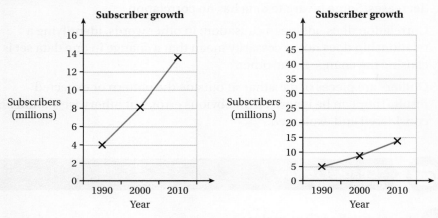

a These two graphs show the same data. Explain why they look different.

b Which graph would you use if you were a mobile phone service provider who wanted to suggest that there had been a huge increase in subscribers over this period? Why?

c Who might find the other graph useful? Why?

4. Data for the price of chocolate bars and their mass is given:

Price	45p	80p	£1.50	£3.00	£5.00	£10
Mass	35 g	80 g	175 g	320 g	540 g	1 kg

a Plot the data on a scatter diagram and draw a line of best fit.

b State what type of correlation there is.

37 Interpretation of graphs

In this chapter you will learn how to …
- construct and interpret graphs in real-world contexts.
- interpret the gradient of a straight-line graph as a rate of change.

For more resources relating to this chapter, visit GCSE Mathematics Online.

Using mathematics: real-life applications

All sorts of information can be obtained from graphs in real-life contexts. The shape of a graph, its gradient and the area underneath it can tell us about speed, time, acceleration, prices, earnings, break-even points or the values of one currency against another, among other things.

"My car needs to perform at its optimum limits. We generate and analyse diagnostic graphs to calculate the slight changes that would increase power, acceleration and top speed."
(Racing driver)

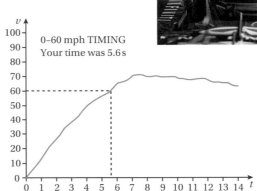

Before you start …

Ch 33	You will need to be able to distinguish between direct and inverse proportion.	**1**	Which of these graphs shows an inverse proportion? How do you know this? 
Ch 23	You'll need to be able to calculate the gradient of a straight line.	**2**	Calculate the gradient of AB.

Find answers at: cambridge.org/ukschools/gcsemaths-studentbookanswers

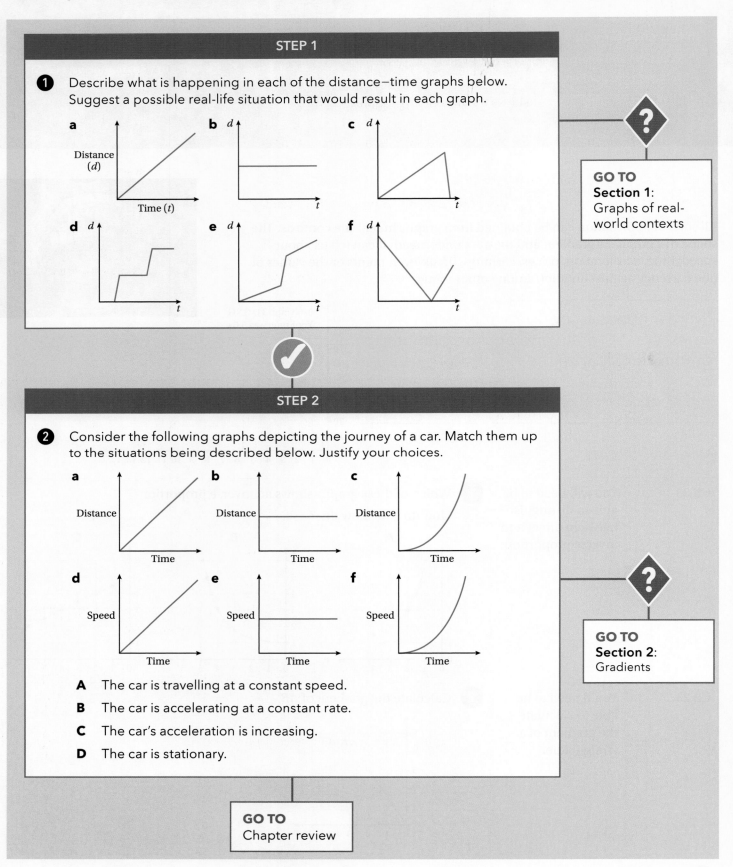

Section 1: Graphs of real-world contexts

Graphs are useful for visually representing the relationships between quantities.

The ticket and transport costs for a group of people to attend a play are shown in the graph (right).

The horizontal axis (or x-axis) shows the number of people attending. The vertical axis (or y-axis) shows the total cost.

The cost depends on the number of people attending (with a fixed minimum charge of £10).

There are six marked points on the graph.

This graph is a linear graph, but it does not show direct proportion because it does not go through the origin.

Read up from 10 people on the x-axis to the straight line. When you reach the line, move across horizontally until you reach the y-axis. The cost is £30. This means that 10 people will need to pay £30 to attend the play.

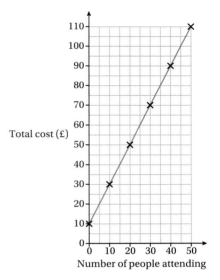

WORKED EXAMPLE 1

This graph shows the relationship between the length and the breadth of a hall with a constant area.

Find the formula for this relationship.

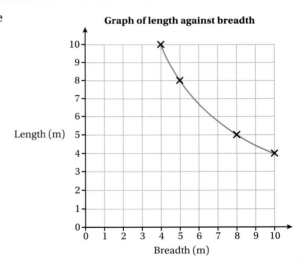

The area of the hall is constant, 40 m²

Reading the points off the graph, you have (4, 10), (5, 8), (8, 5) and (10, 4).
$4 \times 10 = 5 \times 8 = 40$

The formula is $length = \dfrac{40}{breadth}$

This graph shows an inverse proportion.

Because it shows a real-world context, the graph in the example above is only valid for that particular range of values.

Graphs are also useful in the real world for reading off values quickly without having to do the whole calculation. They can serve as conversion charts.

WORKED EXAMPLE 2

This graph shows the number of Indian rupees you would get for different numbers of US dollars at an exchange rate of US$1 : Rs45. This relationship is a direct proportion.

a Use the graph to estimate the dollar value of Rs250.

b Use the graph to estimate how many rupees you could get for US$9.

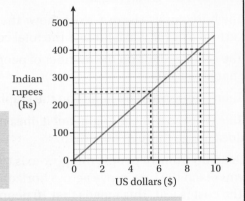

a Rs250 is worth about $5.50.

> Follow across from Rs250 on the vertical axis to reach the graph. Drop down to the dollars axis to find $5.50.

b You could get about Rs400 for $9.

> Go straight up from US$9 on the horizontal axis to reach the graph. Then move left to the rupees axis to find Rs400.

Distance–time graphs

Graphs that show the connection between the distance an object has travelled and the time taken to travel that distance are called distance–time graphs or travel graphs.

Time is always shown along the horizontal axis and distance on the vertical.

The graphs normally start at the origin because at the beginning no time has elapsed (passed) and no distance has been covered.

The graph on the left shows the following journey:

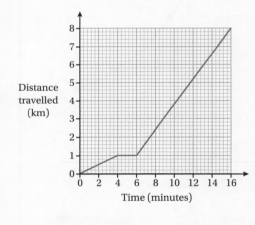

- a cycle ride for 4 minutes from home to a bus stop 1 km away
- a 2 minute wait for the bus
- a 7 km journey on the bus that takes 10 minutes.

The line of the graph remains horizontal while the person is not moving (waiting for the bus) because no distance is being travelled at this time. The steeper the line, the faster the person is travelling.

EXERCISE 37A

1 This graph shows the movement of a taxi in city traffic during a 4-hour period.

a Clearly and concisely describe the taxi's journey.

b For how many minutes was the taxi waiting for passengers in this period?
How can you tell this?

c What was the total distance travelled?

2 This distance–time graph represents Monica's journey from home to a supermarket and back again.

  a How far was Monica from home at 09:06 hours?

  b How many minutes did she spend at the supermarket?

  c At what times was Monica 800 m from home?

  d On which part of the journey did Monica travel faster, going to the supermarket or returning home?

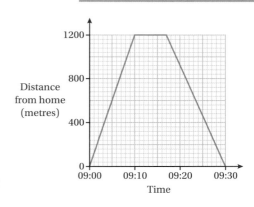

3 A swimming pool is 25 m long. Jasmine swims from one end to the other in 20 seconds.

She rests for 10 seconds and then swims back to the starting point. It takes her 30 seconds to swim the second length.

  a Draw a distance–time graph for Jasmine's swim.

  b How far was Jasmine from her starting point after 12 seconds?

  c How far was Jasmine from her starting point after 54 seconds?

4 A hurricane disaster centre has a certain amount of clean water. The length of time the water will last depends on the number of people who come to the centre.

  a Calculate the missing values in this table.

  b Plot a graph of this relationship.

No. of people	120	150	200	300	400
Days the water will last	40	32			

Section 2: Gradients

Speed in distance–time graphs

The steepness (slope) of a graph gives an indication of speed. A straight-line graph indicates a constant speed.

The steeper the graph is, the greater the speed.

An upward slope and a downward slope represent movement in opposite directions.

The distance–time graph shown on the right is for a person who walks, cycles and then drives for three equal periods of time.

For each period, speed is given by the formula:

$$\text{speed} = \frac{\text{distance travelled}}{\text{time taken}}$$

The average speed for the whole journey $= \dfrac{\text{total distance travelled}}{\text{total time taken}}$

Tip

You worked with kinematic formulae in Chapter 16.

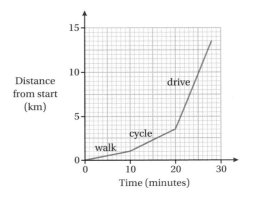

Using gradient triangles to interpret changing gradients

The gradient of a graph, along with the axis labels, gives a large amount of detail – even when, as in the following case, there is no scale given.

Consider these graphs:

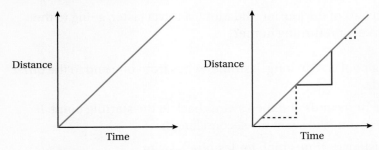

What happens as time moves on? In this case, as time moves on the distance covered increases. So the car is moving.

Look at the gradient triangles drawn on the second copy of the graph. It doesn't matter where these triangles are drawn or how large they are; all the triangles are similar to each other. The ratio of the rise/run is the same for each similar triangle, so the gradient is the same.

This shows that the car is moving at a constant speed.

The graph on the left also shows that as time moves on the distance covered increases. So the car is moving.

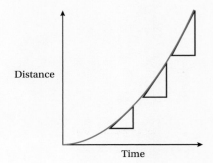

Now consider the gradient triangles drawn on the graph; each has the same base (unit of time).

This time the gradient triangles don't fit the graph as it is not a straight line. Instead, they have been laid against the graph at different places so that the hypotenuse of each forms a tangent to the graph.

You can see by the slope of each triangle's hypotenuse that the speed is changing along the graph. Moving up the slope, the hypotenuse of each triangle is steeper than the one before. The gradient of the graph is increasing. This shows that the car is speeding up, or accelerating.

EXERCISE 37B

1 The following graphs show what is happening to the level of water in a tank.

Describe what is happening in each case. Justify your answers using gradient triangles.

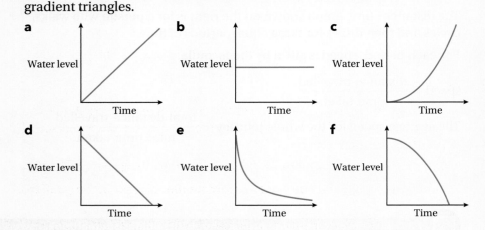

2 The following graphs show what is happening to the price of oil.

Describe what is happening in each case, justifying your answers using gradient triangles.

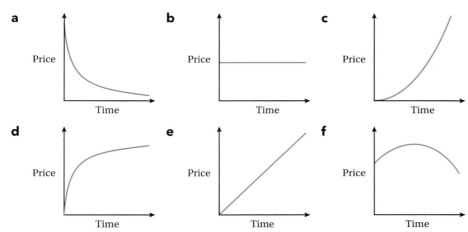

3 The following is a speed–time graph of a parachute jump.

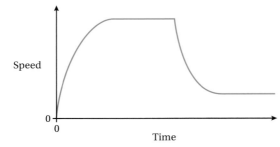

Describe what is happening to the speed and acceleration of the parachutist throughout the jump.

4 This graph (from Exercise 37A) shows the movement of a taxi in city traffic during a 4-hour period.

Calculate the taxi's average speed:

  a during the first 20 minutes
  b during the first hour
  c from 160 to 210 minutes
  d for the full period of the graph.

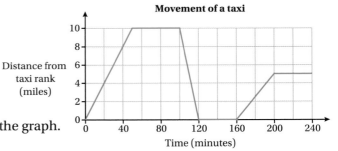

Checklist of learning and understanding

Graphs of real-world contexts

- Real-world graphs show the relationship between variables.

Gradient

- Distance–time graphs show the connection between the distance an object has travelled and the time taken to travel that distance. If speed is constant the gradient is constant.
- Curved graphs have gradients that change along the graph continually.
- Gradient triangles can be used to estimate the changes in the gradient.

Find answers at: cambridge.org/ukschools/gcsemaths-studentbookanswers

Chapter review

1. Debbie drove from Junction 12 to Junction 13 on a motorway.

   The travel graph shows Debbie's journey.

   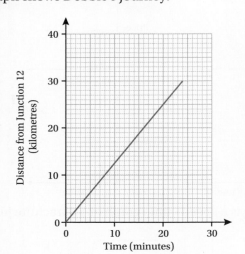

   Ian also drove from Junction 12 to Junction 13 on the same motorway.

   He drove at an average speed of 66 km/hour.

   Who had the faster average speed, Debbie or Ian?

   You must explain your answer. *(4 marks)*

   ©*Pearson Education Ltd 2013*

2. This speed–time graph represents the journey of a train between two stations. The train slowed down and stopped after 15 minutes because of engineering work on the railway line.

   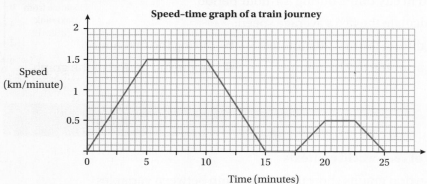

   a Calculate the greatest speed, in km/h, which the train reached.

   b Calculate the deceleration of the train as it approached the place where there was engineering work.

   c Calculate the distance the train travelled in the first 15 minutes.

   d For how long was the train stopped at the place where there was engineering work?

   e What was the speed of the train after 19 minutes?

3) The graph shows how the population of a village has changed since 1930.

a Copy the graph using tracing paper and find the gradient of the graph at the point (1950, 170).

b What does this gradient represent?

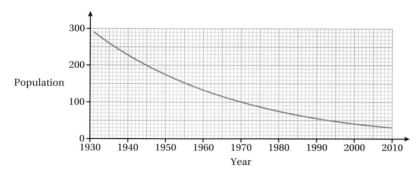

38 Transformations

In this chapter you will learn how to ...
- carry out rotations, reflections and translations.
- identify and describe rotations, reflections and translations.
- describe translations using column vectors.

For more resources relating to this chapter, visit GCSE Mathematics Online.

Using mathematics: real-life applications

You can see examples of reflections, rotations and translations all around you. Patterns in wallpaper and fabric are often translations, images reflected in water are reflections and the blades of a wind turbine are a good example of rotation.

"I use transformations all the time when I program computer graphics. Transformations allow me to position objects, shape them and change the view I have of them. I can even change the type of perspective that is used to show something." *(Computer programmer)*

Before you start ...

Ch 25	You need to know what angles of 90°, 180° and 270° look like and also the directions clockwise and anti-clockwise.	1	How many degrees is each angle? State whether each arrow is showing clockwise or anti-clockwise movement.  a b c
Ch 23	You need to know how to plot straight-line graphs in the form $x = a$, $y = a$ and $y = x$.	2	Find the equation of each of these lines. 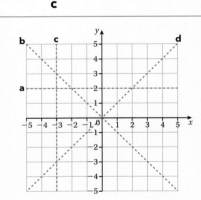
Ch 22	You need to know what a vector is and how they describe movement.	3	a What is the difference between the coordinate (3, 2) and the vector $\begin{pmatrix} 3 \\ 2 \end{pmatrix}$? b What is the difference between the vectors $\begin{pmatrix} -1 \\ 3 \end{pmatrix}$ and $\begin{pmatrix} 3 \\ 1 \end{pmatrix}$?

Assess your starting point using the Launchpad

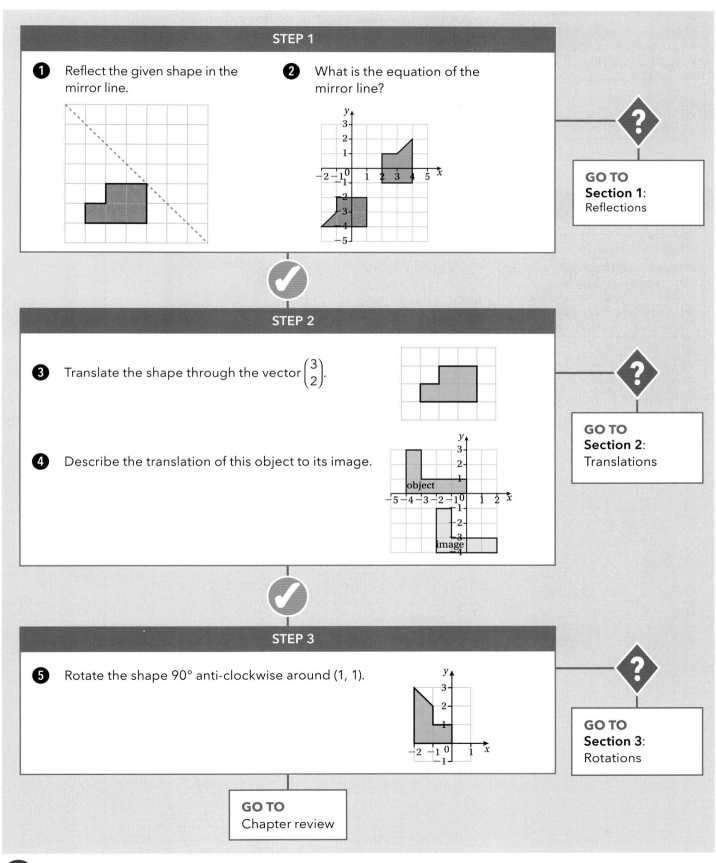

Key vocabulary

object: the original shape (before it has been transformed).

image: the new shape (once the object has been transformed).

mirror line: line equidistant from all corresponding points on a shape and its reflection.

Tip

You worked with congruent triangles in Chapter 29.

Enlargement is also a transformation. Enlargement changes the position of an object and also its size. This was covered in Chapter 28.

Section 1: Reflections

A transformation is a change in the position of a point, line or shape. When you transform a shape you change its position or its size, or both.

The original point, line or shape is called the **object**. For example, a triangle ABC.

The transformation is called the **image**. The symbol ′ is used to label the image. For example, the image of triangle ABC is $A'B'C'$.

Reflection, rotation and translation change the position of an object, but not its size. Under these three transformations an object and its image will be congruent.

Mirrors, windows and water surfaces all reflect objects. You can see the reflection of clouds and trees clearly in the photograph. If you draw a line horizontally across the centre of the image and fold it, the top half will fit exactly onto the bottom half. The fold line is called a **mirror line**.

Mathematically, when a shape is reflected it is 'flipped' over a mirror line to give its image. The object and the image are the same distance from the mirror line.

EXERCISE 38A

1 The following images are painted on three pieces of square paper:

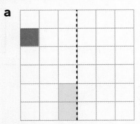
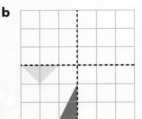
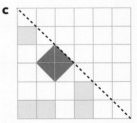

While the paint is still wet, the paper is folded along the dotted lines. What will each image look like? How can you predict this accurately?

2 In this diagram the line of reflection of the shape has been marked. Join corresponding points in the two halves of the diagram. What do you notice?

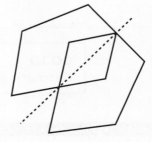

3 Reflect each shape in the given mirror line.

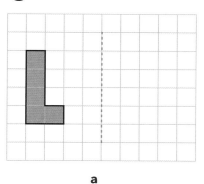
a

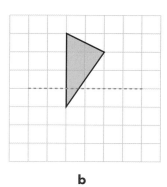
b

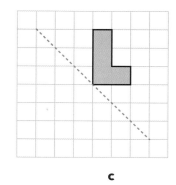
c

Under reflection corresponding points on the object and the image are the same distance from the mirror line. If you join a pair of corresponding points the line formed is cut in half by the mirror line and they meet at 90°. The mirror line is the **perpendicular bisector** of any pair of corresponding points.

> **Tip**
>
> You learned about perpendicular bisectors in Chapter 21.

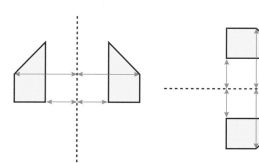

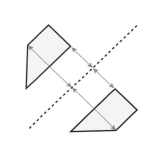

> **Tip**
>
> You can turn your book around so that diagonal mirror lines look vertical or horizontal. Often our brains find this easier than working diagonally.

WORK IT OUT 38.1

This Z shape is reflected in the line $y = -1$. What is its image?

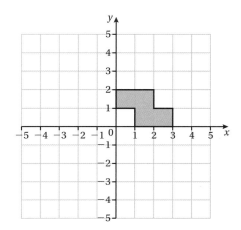

Continues on next page …

Find answers at: cambridge.org/ukschools/gcsemaths-studentbookanswers

Which one of these answers is correct? What has gone wrong in each of the others?

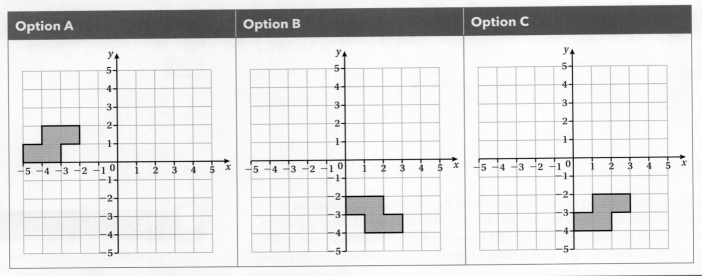

EXERCISE 38B

1 Reflect the triangle in the line $x = 1$, and then reflect the triangle and the resultant image in the line $y = -1$.

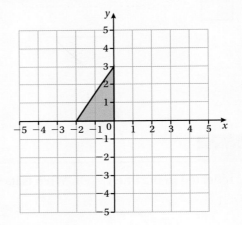

2 Reflect this shape in the line $y = x$, and then reflect the shape and the resultant image in the line $y = -x$.

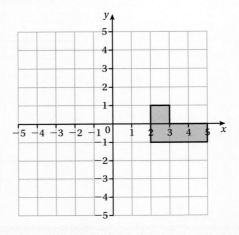

38 Transformations

3 Copy the grid and shapes below. Carry out the 10 reflections on your grid to reveal the picture.

Shape A in the line $y = x$ Shape E in the line $x = 7$ Shape I in the line $y = 6$

Shape B in the x-axis   Shape F in the line $y = -x$ Shape J in the line $y = x$

Shape C in the line $y = -2$ Shape G in the line $x = -4$

Shape D in the line $y = 4$ Shape H in the line $x = 1.5$

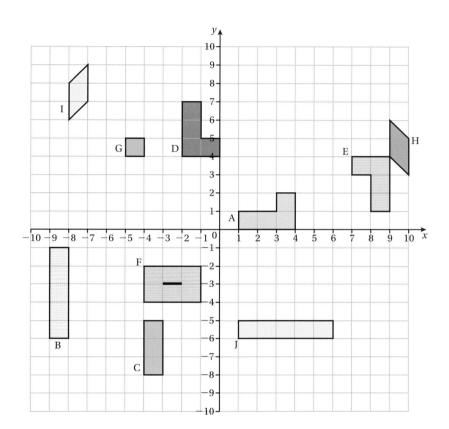

Describing reflections

Since the mirror line is the perpendicular bisector of two corresponding points in a reflection, if you can't 'spot' a mirror line you can join two corresponding points and construct the perpendicular bisector to find it.

You must also be able to give the equation of the mirror line when the reflection is shown on a coordinate grid.

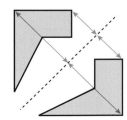

Tip

To check a reflection, trace the **object**, the **image** and the mirror line. Fold the tracing paper along the mirror line; the shapes should match up exactly.

EXERCISE 38C

1 Find the equation of the mirror line in each reflection.

a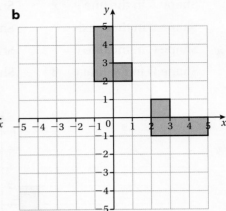

b

c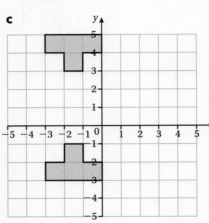

d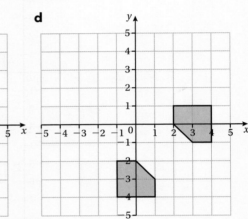

2 a Write the equations of the mirror lines to fully describe each of the following reflections.

  i Shape A to shape E.
  ii Shape C to shape G.
  iii Shape A to shape C.
  iv Shape H to shape D.

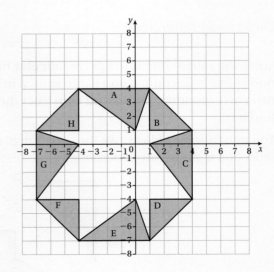

b Challenge another student to describe a reflection of two triangles you choose.

3 Trace each pair of shapes and construct the mirror line for the reflection.

a

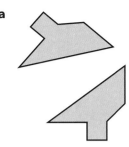

b

c

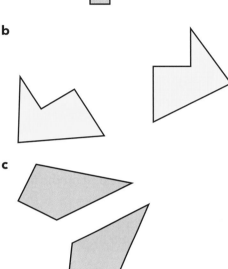

Section 2: Translations

A translation is a 'slide' along a straight line. (Think about pushing a box across a floor.) The translation can be from left to right (horizontal), up or down (vertical) or both (horizontal and vertical), i.e. diagonal.

The image is in the same **orientation** as the object and every point on the shape moves exactly the same distance in exactly the same direction. Translated shapes are congruent to each other.

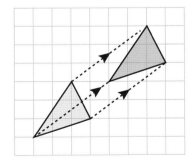

You can describe translations on a coordinate grid using column vectors. Remember, a column vector shows horizontal displacement over vertical displacement.

Key vocabulary

orientation: the position of a shape relative to the grid.

Tip

Remember, shapes that are congruent are exactly the same size and shape.

WORKED EXAMPLE 1

Describe the translation ABC to $A'B'C'$ by means of a column vector.

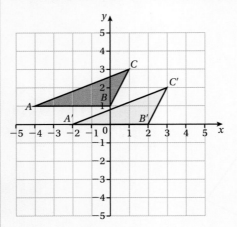

Look at point C and point C'.
→ Take any point on the object and find the corresponding point on the image.

To get from C to C' move:
2 units to the right = +2
1 unit down = −1
→ Work out how the point has been translated horizontally and vertically.

The translation is $\begin{pmatrix} 2 \\ -1 \end{pmatrix}$
→ Write this as a vector.

Tip

Drawing on the grid to show the movements can help you to avoid unnecessary mistakes.

EXERCISE 38D

1 Translate each shape as directed.

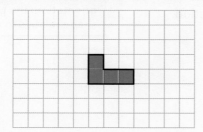 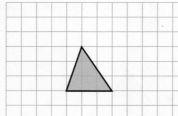 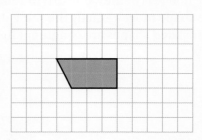

a Translate 6 right and 2 up. **b** Translate 3 left and 1 down. **c** Translate 3 down and 4 right.

2 Write each of the translations in question **1** as a column vector.

WORK IT OUT 38.2

This T shape is translated through a vector of $\begin{pmatrix} -2 \\ 4 \end{pmatrix}$. Draw its image.

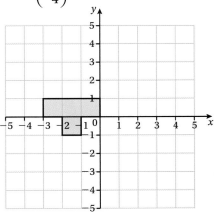

Which one of these answers is correct? What has gone wrong in each of the others?

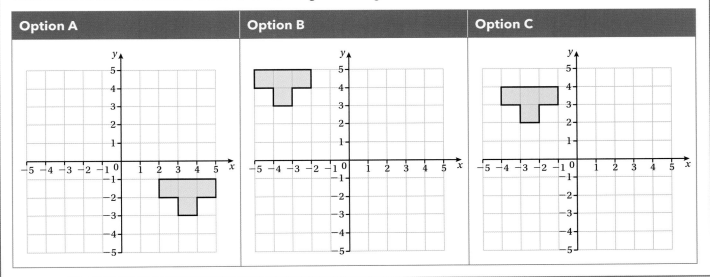

EXERCISE 38E

1. Translate each shape using the given vector.

   a $\begin{pmatrix} 3 \\ -2 \end{pmatrix}$ b $\begin{pmatrix} -1 \\ 2 \end{pmatrix}$ c $\begin{pmatrix} 0 \\ 4 \end{pmatrix}$

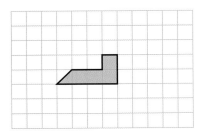 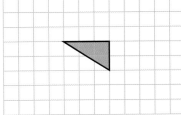 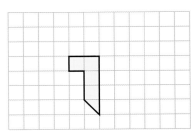

2 Translate each shape by the given vector. Then give the name of the shape you have put together.

a Translate shape A $\begin{pmatrix} -1 \\ -3 \end{pmatrix}$.

b Translate shape B $\begin{pmatrix} 1 \\ 5 \end{pmatrix}$.

c Translate shape C $\begin{pmatrix} 2 \\ -1 \end{pmatrix}$.

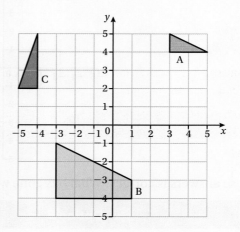

3 Translate each piece of this jigsaw using the vectors on the right.

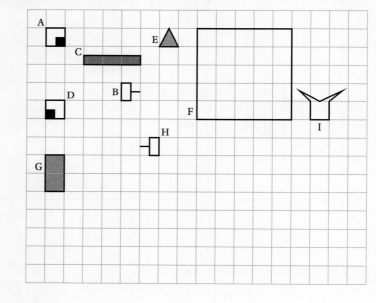

A $\begin{pmatrix} 12 \\ -8 \end{pmatrix}$

B $\begin{pmatrix} 12 \\ -5 \end{pmatrix}$

C $\begin{pmatrix} 10 \\ -9 \end{pmatrix}$

D $\begin{pmatrix} 14 \\ -4 \end{pmatrix}$

E $\begin{pmatrix} 7 \\ -9 \end{pmatrix}$

F $\begin{pmatrix} 3 \\ -7 \end{pmatrix}$

G $\begin{pmatrix} 13 \\ -5 \end{pmatrix}$

H $\begin{pmatrix} 5 \\ -2 \end{pmatrix}$

I $\begin{pmatrix} -1 \\ -2 \end{pmatrix}$

38 Transformations

Describing translations

You should be able to use vectors to describe a translation. Remember to count between corresponding points on the two shapes.

> **Tip**
>
> Make sure you count from the object to the image and write this as a column vector (not a coordinate!).

WORK IT OUT 38.3

Which transformations below are reflections and which are translations? How did you make your decision? Describe the translations as vectors.

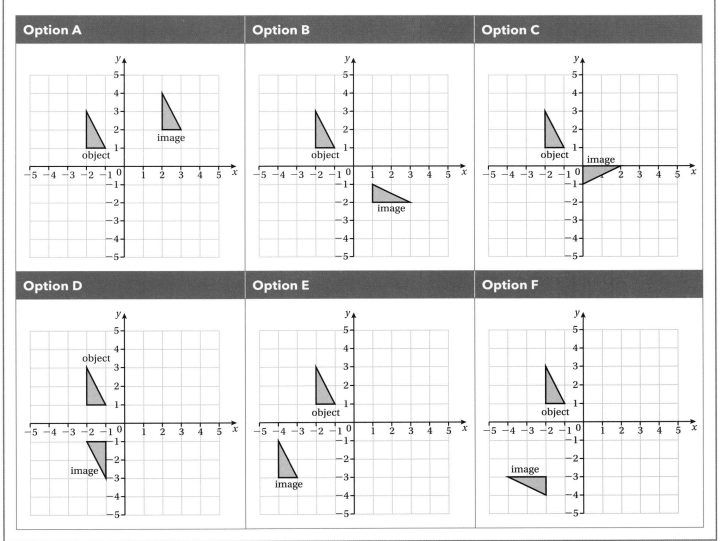

EXERCISE 38F

1 Here are some completed translations. The objects are shown in grey and the images are in colour. Write column vectors to describe the translation from each object to its image.

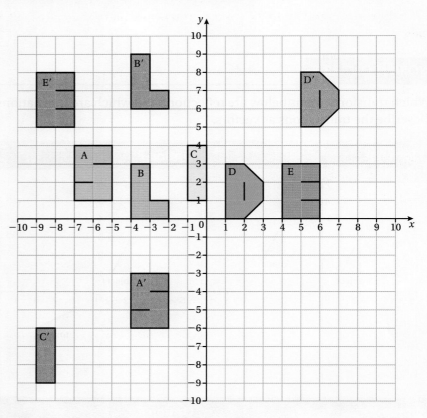

2 Work on a grid. Draw the four objects (A–D) used to make this image in any position on the grid. Make up translation instructions for moving the four objects to form the image. Exchange with a partner and perform the translations to make sure their instructions are correct.

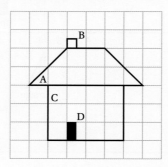

Section 3: Rotations

A rotation is a turn. An object can turn clockwise or anti-clockwise around a fixed point called the centre of rotation. The centre of rotation can be inside, on the edge of or outside the object.

A rotation changes the orientation of a shape, but the object and its image remain congruent.

When you rotate a shape the distance from the centre of rotation to any point on the object remains the same. Each point travels in a circle around the centre. Think about the tip of a blade on a wind turbine or a child sitting on a roundabout. When they rotate, the paths they trace out are circles.

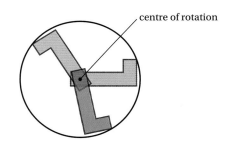
centre of rotation

To carry out a rotation you need to know the centre of rotation as well as the angle and direction of rotation. At this level, all rotations will be in multiples of 90°.

WORK IT OUT 38.4

This L shape is rotated anti-clockwise with centre of rotation (0, 1) through an angle of 90°. What is its image?

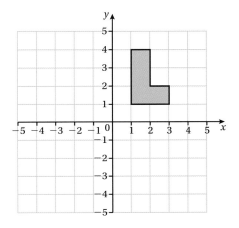

Which one of these answers is correct? What has gone wrong in each of the others?

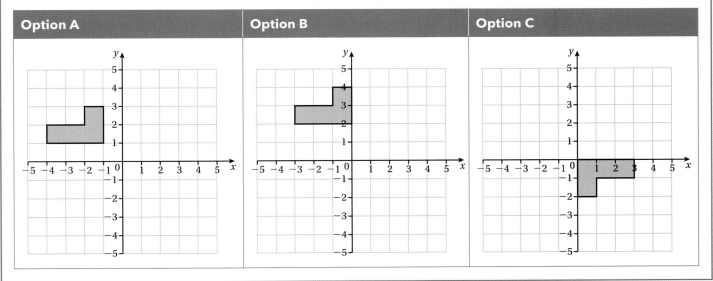

| Option A | Option B | Option C |

EXERCISE 38G

1 Draw the image of the orientation of each shape if the page was rotated as directed.

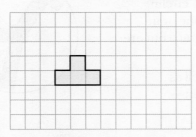 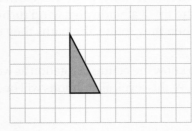 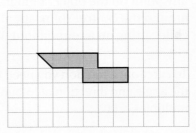

a Rotate 90° clockwise.   b Rotate 180°.   c Rotate 90° anti-clockwise.

2 Rotate each shape as directed about the marked centre

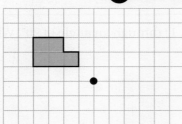 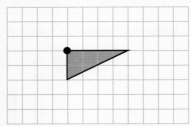 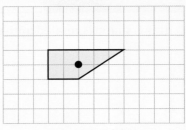

a Rotate 180°.   b Rotate 90° clockwise.   c Rotate 90° anti-clockwise.

Tip

Tracing paper is very useful for work with transformations. Don't be afraid to ask for it in an exam.

3 Rotate the triangle 180° about the origin.

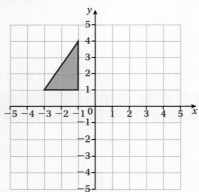

4 Rotate the shape 90° clockwise around the point (1, 1).

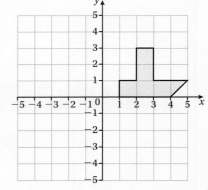

5 Rotate the shape 90° anti-clockwise around the point (−2, 1).

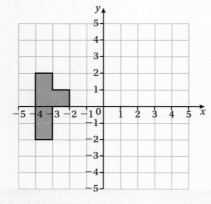

6 Rotate the shape 90° anti-clockwise around the point (2, 1).

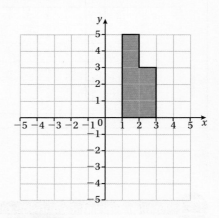

38 Transformations

7 Rotate each shape as directed.

Shape A: 90° anti-clockwise around the point (−1, −1).
Shape B: 180° around the point (2, 3).
Shape C: 90° clockwise around the point (1, 0). Label this D.
Shape D: 180° around the point (−3.5, 2).
Shape E: 180° around the point (3, 1).

8 The image below was designed by drawing a triangle and rotating this around the origin in multiples of 90°. What do you notice about the coordinates of the vertices of the triangle? Would this work if you rotated an image around a different point? Why?

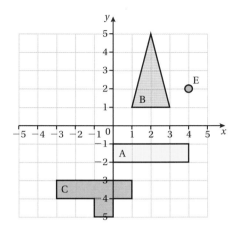

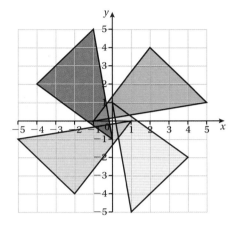

Describing rotations

To describe a rotation you give a centre, angle and direction. Very often you can find the centre using tracing paper and trial and error. Trace the object and rotate the tracing paper using different centres of rotation. Spotting the centres improves with practice.

> **Tip**
>
> Always give the direction and angle from the object to the image.

WORK IT OUT 38.5

Which of these transformations are reflections, which are rotations and which are translations? How did you make your decision?

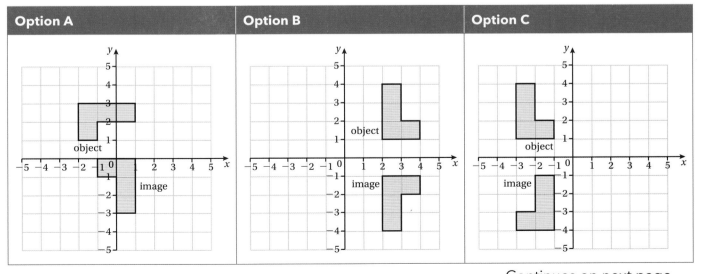

Continues on next page …

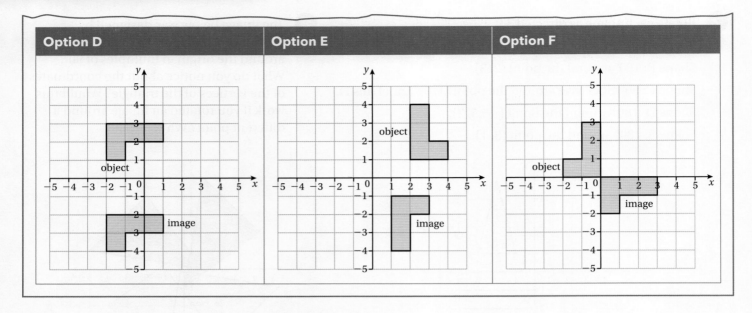

EXERCISE 38H

1. Describe each of the following rotations.

    a

    b

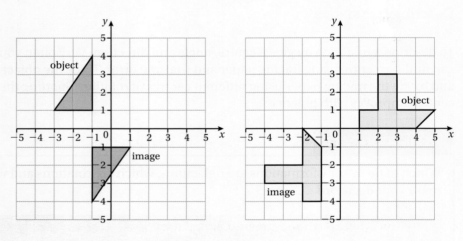

    c

    d

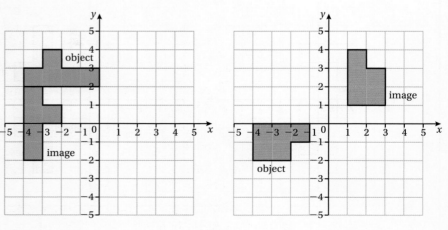

2 This section of wallpaper has been designed using rotations. A **coordinate** grid has been overlaid. Identify as many different rotations as you can.

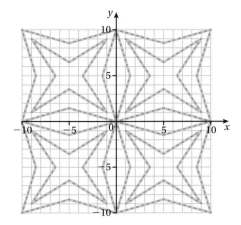

 Checklist of learning and understanding

Reflections
- Reflections change the orientation of a shape but the image remains congruent.
- To describe a reflection the equation of the mirror line needs to be given.
- The mirror line is the perpendicular bisector of any two corresponding points on the image and object.

Translations
- Translations leave the orientation of the shape unchanged but move it horizontally and/or vertically.
- Translations are described using vectors.

Rotations
- A rotation is a turn around a centre. Rotations are described by giving the coordinates of the centre of rotation, angle and direction of the rotation.

 Chapter review

 For additional questions on the topics in this chapter, visit GCSE Mathematics Online.

1 Which of the following statements are true? Explain your reasoning.
  a The images constructed by reflecting, rotating, or translating are congruent to the objects you started with.
  b The images constructed by reflecting, rotating, or translating are similar to the objects you started with.
  c The images constructed by reflecting, rotating, or translating are in the same orientation as the objects you started with.
  d The images constructed by reflecting, rotating, or translating have the same angles as the objects you started with.

Find answers at: cambridge.org/ukschools/gcsemaths-studentbookanswers

2 Describe fully the transformation from:

   a shape A to shape B   **b** shape B to shape C   **c** shape C to shape A.

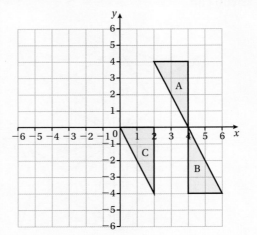

 3

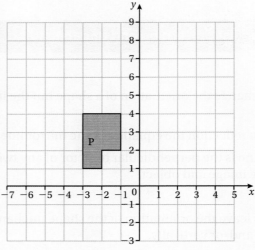

  a Translate shape P by the vector $\begin{pmatrix} 5 \\ -2 \end{pmatrix}$ *(2 marks)*

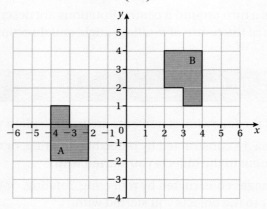

  b Describe fully the single transformation that maps shape A onto shape B.   *(3 marks)*

©*Pearson Education Ltd 2013*

38 Transformations

4 The triangle below is used to create a tessellating pattern. This is produced using multiple translations and one rotation. Explain how this could be done.

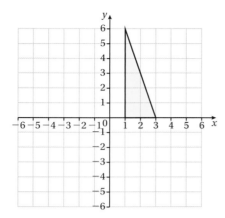

5 Rotate shape A 90° clockwise around the point (1, 2), label it B.

Reflect shape B in the line $y = x$, label it C.

Translate shape C through the vector $\begin{pmatrix} -1 \\ 1 \end{pmatrix}$, label it D.

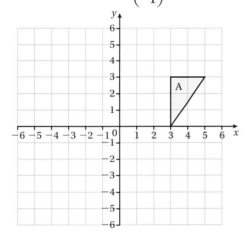

6 Look at the dancing figure below. Describe how the figure can be drawn using only transformations of shapes A, B, C, and D.

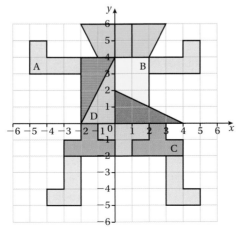

Glossary

A

Adjacent: next to each other; in shapes, sides that intersect each other.

Alternate angles: the angles on parallel lines on opposite sides of a transversal.

Angle of depression: when looking down, the angle between the line of sight and the horizontal.

Angle of elevation: when looking up, the angle between the line of sight and the horizontal.

Arithmetic sequence: a sequence where the difference between each term is constant.

B

Binomial: an expression consisting of two terms.

Binomial product: the product of two binomial expressions; for example, $(x + 2)(x + 3)$.

Bisect: to divide exactly into two halves.

Bivariate data: data that is collected in pairs.

C

Categorical data: data that has been arranged in categories.

Chord: a straight line that joins one point on the circumference of a circle to another point on its circumference. The diameter is a chord that goes through the centre of the circle.

Circumference: distance round the outside of a circle.

Coefficient: the number in front of a variable in a mathematical expression. In the term $5x^2$, 5 is the coefficient and x is the variable.

Co-interior angles: the angles within the parallel lines on the same side of the transversal. Co-interior angles are sometimes referred to as 'allied angles'.

Combined events: one event followed by another event producing two or more outcomes.

Common denominator: a number into which all the denominators of a set of fractions divide exactly.

Congruent: identical in shape and size.

Consecutive: following each other in order. For example 1, 2, 3 or 35, 36, 37.

Consecutive terms: terms that follow each other in a sequence.

Constant: in algebra, a constant is a value that does not change. It is usually a number. It could be a letter with a fixed value, like π.

Continuous data: data that can have any value.

Conversion factor: the number that you multiply or divide by to convert one measure into another smaller or larger unit.

Coordinates: an ordered pair (x, y) identifying a position on a grid.

Correlation: a relationship or connection between data items.

Corresponding angles: angles that are created at the same point of the intersection when a transversal crosses a pair of parallel lines.

D

Decay: the reduction in a quantity over time or some other measure such as distance. The opposite of growth.

Degree of accuracy: the number of places to which you round a number.

Dependent events: events in which the outcome is affected by what happened before.

Denominator: the number at the bottom of a fraction.

Dependent variable: data that is measured in an experiment.

Depreciation: the loss in value of an object over a period of time.

Direct proportion: two values that both increase in the same ratio.

Discrete data: data that can be counted and can only take certain values.

Displacement: a change in position.

E

Elevation view: a view of an object from the front, side or back.

Equally likely: having the same probability of happening.

Estimate: an approximate answer or rough calculation.

Evaluate: to calculate the numerical value of something.

Equivalent: having the same value, two ratios are equivalent if one is a multiple of the other.

Event: the thing to which you are trying to give a probability.

Exchange rate: the value of one currency used to convert that currency to an equivalent value in another currency.

Expanding: multiplying out an expression to get rid of the brackets.

Expression: a group of numbers and letters linked by operation signs.

Exterior angles: angles produced by extending the sides of a polygon.

Extrapolation: using values within a known range to estimate an unknown value that lies outside of that range.

F

First difference: the result of subtracting a term from the next term.

Formula: a general rule or equation showing the relationship between unknown quantities; the plural is formulae.

Function: a set of instructions for changing one number (the input) into another number (the output).

G

Geometric sequence: a sequence where the ratio between each term is constant.

Gradient: a measure of the steepness of a line.

Gradient $= \dfrac{\text{change in } y}{\text{change in } x}$.

H

Hypotenuse: the longest side of a right-angled triangle; the side opposite the 90° angle.

I

Identity: an equation that is true no matter what values are chosen for the variables.

Image: the new shape (once the object has been transformed).

Independent events: events that are not affected by what happened before.

Improper fraction: a fraction where the numerator is greater than (or equal to) the denominator.

Index (plural indices): a power or exponent indicating how many times a base number is multiplied by itself.

Index notation: writing a number as a base and index, for example 2^3.

Inequality: a mathematical sentence in which the left side is not equal to the right side.

Integers: whole numbers belonging to the set $\{\ldots -3, -2, -1, 0, 1, 2, 3, \ldots\}$, they are sometimes called directed numbers because they have a negative or positive sign.

Interior angles: angles inside a two-dimensional shape at the vertices or corners.

Interpolation: using two known values to estimate an unknown value.

Inverse proportion: a relation between two quantities such that one increases at a rate that is equal to the rate that the other decreases.

Irregular polygon: a polygon that does not have equal sides or equal angles.

Isometric grid: special drawing paper based on an arrangement of triangles.

L

Line of best fit: a line drawn through a group of points showing the general direction to represent the trend.

Line of symmetry: a line that divides a plane shape into two identical halves, each the reflection of the other.

Linear equation: an equation where highest power of the unknown is 1 for example $x + 3 = 7$ (x means x^1, but you don't need to write the '1').

Locus (plural loci): a set of points that satisfy the same rule.

Lower bound: the smallest value that a number (given to a specified accuracy) can be.

M

Midpoint: the centre of a line; the point that divides the line into two equal halves.

Mirror line: a line equidistant from all corresponding points on a shape and its reflection.

Mixed number: a number comprising an integer and a fraction.

Mutually exclusive: events that cannot happen at the same time.

N

Number line: a line marked with positions of numbers showing the valid values of a variable.

Numerator: the number at the top of a fraction.

O

Object: the original shape (before it has been transformed).

Orientation: the position of a shape relative to the grid.

Outcome: a single result of an experiment or situation.

Outlier: data value that is much larger or smaller than others in the same data set.

P

Parabola: the symmetrical curve produced by the graph of a quadratic function.

Parallel vectors: occur when one vector is a multiple of the other.

Perimeter: the distance around the boundaries (sides) of a shape.

Perpendicular bisector: a line perpendicular to another that also cuts it in half.

Plan view: the view of an object from directly above.

Plane shape: a flat, two-dimensional shape.

Polygon: a closed plane shape with three or more straight sides.

Polyhedron (plural polyhedra): a solid shape with flat faces that are polygons.

Polynomial: an expression made up of many terms with positive powers for the variables.

Population: the name given to a data set.

Position-to-term rule: operations applied to the position number of a term in a sequence in order to generate that term.

Prime factor: a factor that is also a prime number.

Product: the result of multiplying numbers and/or terms together.

Proportion: the number or amount of a group compared to the whole, often expressed as a fraction, percentage or ratio.

Pythagorean triple: three non-zero numbers (a, b, c) for which $a^2 + b^2 = c^2$.

Q

Quadratic: an expression with a variable to the power of 2 but no high power.

Quadratic equation: an equation that contains a variable squared term, like x^2, but no variable term with a power greater than 2. $x^2 = 4$ and $x^2 + 2x - 6 = 0$ are quadratic equations. $x^3 - 1 = 0$ and $6x + 7 = 35$ are not quadratic equations.

Quadratic expression: an expression in which the highest power of x is x^2.

R

Radius (plural radii): distance from the centre to the circumference of a circle. One radius is half of the diameter of the circle.

Random: not predetermined.

Ratio: a comparison of different parts or amounts in a particular order; the relationship between two or more groups or amounts, explaining how much bigger one is than another.

Reciprocal: the value obtained by inverting a fraction. Any number multiplied by its reciprocal is 1.

Reflection: an exact image of a shape about a line of symmetry.

Regular polygon: a polygon with equal straight sides and equal angles.

Representative sample: a smaller quantity of data that represents the characteristics of a larger population.

Right prism: a prism with sides perpendicular to the end faces (base).

Roots: the individual values of x in a quadratic equation.

Rotational symmetry: symmetry by turning a shape around a fixed point so that it looks the same from different positions.

Round to significant figures: round to a specified level of accuracy from the first significant figure.

Rounding: writing a number with zeroes in the place of some digits.

S

Sample: a set of data collected from a population.

Sample space: a list or diagram that shows all possible outcomes from two or more events.

Scalar: a numerical quantity (it has no direction).

Scale factor: a number that scales a quantity up or down.

Second difference: the difference between each term in the first difference.

Semi-circle: half of a circle.

Sequence: a number pattern or list of numbers following a particular order.

Set: a collection. The brackets { } is are shorthand for 'the set of'. For example, {2, 4, 6, 8} is the set of the numbers 2, 4, 6, 8, which represents the even numbers.

Significant figure (s.f.): a position in a number used to decide the level of accuracy. The first significant figure is the first non-zero digit when reading a number from the left.

Simultaneous equations: a pair of equations with two unknowns that can be solved at the same time.

Solution: both possible values of x in a quadratic equation.

Subject: the variable which is expressed in terms of other variables; it is the variable on its own on one side of the equals sign. In the formula $s = \dfrac{d}{t}$, s is the subject.

Substitute: to replace letters with numbers.

Surd: if $\sqrt[n]{a}$ is an irrational number, then $\sqrt[n]{a}$ is called a surd.

T

Term: a combination of letters and/or numbers. Each number in a sequence is called a term.

Term-to-term rule: operations applied to any number in a sequence to generate the next number in the sequence.

Theorem: a statement that can be demonstrated to be true by accepted mathematical operations.

Transversal: a straight line that crosses a pair of parallel lines.

Truncation: cutting off all digits after a certain point without rounding.

U

Unit fraction: a fraction with numerator 1 and denominator a positive integer.

Unknown: part of an equation which is represented by a letter.

V

Variable: a letter representing an unknown number.

Vector: a quantity that has both magnitude and direction.

Vertically opposite angles: angles that are opposite one another at an intersection of two straight lines. Vertical here means of the same vertex or point, not up and down.

X

x-intercept: the point where a line crosses the x-axis when $y = 0$.

Y

y-intercept: the point where a line crosses the y-axis when $x = 0$.

Index

2D shapes *see* two-dimensional shapes
'3 4 5 rule' 457
3D objects *see* three-dimensional objects
12-hour time system 134
24-hour time system 134

acceleration 250
accuracy
    levels of 119
    limits of 124-25
acute-angled triangle 23
addition
    algebraic expression 69-70
    decimals 98
    fractions 83
    inequality 508
    inverse operations 11
    vector 337-38
adjacent shapes 27
algebraic expressions
    addition 69-70
    brackets 72
    direct proportion 498-99
    division 70-71
    expansion 72
    factorising 74
    multiplication 70-71
    solving problems 74
    subtraction 69-70
algebraic notation 67
alternate angles 381
angle of depression 476
angle of elevation 476
angles
    alternate 381
    around point 379
    bisecting 327
    co-interior 382
    corresponding 381
    exterior 383, 386-87
    interior 385
    marking equal sides 19
    measuring and drawing 320
    parallel lines and 381-82
    sides and 18-19
    on straight line 379
    sum of polygon 385
    sum of triangle 383
    vertically opposite 380
angle side angle (ASA) 441
approximate values
    levels of accuracy 119
    rounded values 117

rounding decimals 118
rounding whole numbers 117
significant figure 120
truncation 121-22
area
    circle 177, 254
    composite shapes 179-80
    definition 131
    parallelogram 173-74, 254
    polygons 171-75
    problem solving 182
    rectangle 171, 254
    sector 178
    squares 171
    trapezium 174-75, 254
    triangle 171-72, 254
arithmetic sequence 232
ASA *see* angle side angle
average, selection of 538-39

bar charts
    composite 523-24
    display data 519-20
    multiple 523-24
bar graphs 521
bearings 145-46
binomial
    definition 188
    squaring 190
binomial product 188
bisectors
    angle 327
    line 324-25
    perpendicular 567
    shapes 27
bivariate data 548
brackets
    expanding expression 72
    group operations 8-9

calculations
    inverse operations 11
    negative and positive integers 6
    order of operations 8-9
    problem-solving framework 4
calculators
    index notation 106
    standard form and 414-15
categorical data 520
centre of enlargement 428-29
charts
    composite bar 523-24
    multiple bar 523-24
    pie 526-27

chord 321
circle
    area of 177, 254
    circumference of 158, 254
    parts of 321
    sectors of 161-62
circumference
    of circle 158, 254
    definition 158, 321
    problems involving 163
    problem-solving framework 159
class
    median 541
    modal 541
class intervals 540
coefficient 193, 349
co-interior angles 382
combined events
    definition 393
    tables and grids 393
    tables to list probabilities favourable outcomes 394
    theoretical probability 400-405
    tree diagram 398
    Venn diagrams 395-96
common denominator 80
common factor 55
common multiple 55
compare
    percentages fractions and decimals 276-77
    ratios 293-94
    two or more sets of data 543-44
complement of set 396
composite bar chart 523-24
composite bar graph 521
composite shapes
    area 179-80
    perimeter 156
composite solids 269
compound interest 484-85
compound units of measurement
    density 139
    pressure 140
    speed 137-38
cones 266-67
congruent shape 31
congruent shapes 571
congruent triangles
    definition 441
    problem-solving framework 445-46
    properties 442
consecutive number 56
consecutive terms 232

constant 193, 349
continuous data 530
conversion factor 130
converting degrees
    Celsius into degrees Fahrenheit 255
    Fahrenheit into degrees Celsius 255
coordinates 345
correlation in data 548
corresponding angles 381
cosine ratio 468, 474-75
cube number 55, 239
cubes
    properties 30-31, 38
    surface area of 254, 261-62
cuboids
    properties 30-31, 38
    surface area of 254, 261-62
    volume of 254
cylinders
    prisms and 265
    surface area of 265
    volume of 254

data
    analysing grouped 540-41
    bivariate 548
    categorical 520
    comparing two or more sets 543-44
    continuous 530
    correlation 548
    discrete 520
    displaying bar charts 519-20
    guidelines 521
    organise tables 519
    outlier 549, 551
    time-series 529-30
decay 488
decimals
    adding and subtracting 98
    comparing 94
    comparing with percentages and fractions 276-77
    converting fractions to 95
    converting into fractions 94
    estimating and reasoning 97
    multiplying and dividing 98
    rounding 118
denominator 80
density 139
dependent events 404
dependent variable 549
depreciation 487-88
diagrams
    drawing and labelling 18-19
    scatter see scatter diagrams
    tree see tree diagrams
    Venn see Venn diagrams
difference 3
difference of two squares identity 191
direct proportion
    algebraic and graphical representations 498-99

    definition 494
    unitary method 496
discrete data 520
displacement 250, 336
distance-time graphs
    description 558
    speed in 559
division
    algebraic expression 70-71
    decimals 98
    fractions 84
    inequality 508-9
    law of indices 108
    standard form 416-17
divisor see factor

elevation view 48
enlargement
    centre of 428-29
    definition 427
    describing 432
    fractional scale factor 430
equally likely outcomes 303
equations
    kinematics see kinematics equations
    linear see linear equations
    quadratic see quadratic equations
    simultaneous see simultaneous equations
    solving by graphs 222
equilateral triangle 23
equivalent fractions 80
equivalent ratio 290
estimate
    decimals 97
    definition 122
evaluate 249
even number 55
events
    definition 301
    dependent 404
    independent 403
    mutually exclusive 401-2
expanding expression 72
experimental probability
    empirical evidence 307
    organising outcomes 309-10
    predicting outcomes 306
exponent 369
expression
    algebraic 69-71
    definition 67
    expanding 72
    factorising 74
    one quantity as fraction of another 88
exterior angle 383, 386-87
extrapolation 550

factor 55
factorising
    difference of two squares 196
    expression 74

    quadratic expressions 192-95
Fibonacci sequences 239
first difference 232
formulae
    changing subject of 251-52
    definition 246
    problem-solving framework 247-48
    subject 246
    substituting values into 249-50
    working with 254-55
    writing to represent real-life contexts 246-47
fractional scale factor 430
fractions
    adding 83
    comparing with percentages and decimals 276-77
    converting decimals to 94
    converting into decimals 95
    dividing 84
    equivalent 80
    improper 81
    multiplying 82
    of quantities 87-88
    reciprocal 84
    subtracting 83
    unit 84
frequency distribution mean of 541
frequency tree 310
front elevation 48
functions
    definition 237
    generating sequence using 237-38
    in graphs 345
    reciprocal 371

geometrical constructions
    bisecting angle 327
    bisecting line 324-25
    constructing perpendiculars 325-26
    measuring and drawing angles 320
    pair of compasses 321
    problem-solving framework 330-31
geometric sequence 232, 239
golden ratio 296
gradient
    definition 346
    triangles to interpret 560
    and y-intercept 349
gradient-intercept 349
graphs
    advantages of 545
    bar 521
    composite bar 521
    distance-time 558
    horizontal lines 362-63
    line 521
    misleading 545-46
    pie 521
    plotting 345
    plotting sketching and recognising 373
    real-world contexts 557-58

sketching quadratic 367
solving equations 222
steepness of 559
vertical lines 362-63
group charge 557

HCF *see* higher common factor
higher common factor (HCF) 59-60
horizontal lines 362-63
hypotenuse 453

identity 67
image 566
improper fraction 81
independent events 403
index
   definition 104
   negative 106-7
   zero 106-7
index notation
   on calculator 106
   definition 104
inequality
   addition and subtraction 508
   definition 508
   multiplication and division 508-9
   number lines 509-10
   solving 510
   symbolic representation 508
   working with 512
integers
   definition 6
   negative 6
   positive 6
interior angle 385
interpolation 550
intersection of set 395
inverse proportion 501-2
isoceles triangle 23
isometric drawings 45
isometric grid 44-45

kinematics equations
   acceleration 250
   definition 249
   displacement 250
   velocity 249
kites properties 27

law of indices
   division 108
   multiplication 108
   powers of indices 109
LCM *see* lowest common multiple
linear equations
   definition 203
   forming and solving 207-8
   properties 203-4
   unknown on both sides 206
linear functions plotting 345
linear sequence 239

line graph
   in data 521
   problem-solving framework 530-31
   time-series data 529-30
   see also graphs
line of best fit 550
line of symmetry 20-21
locus 328-29
locus of points 328-29
lowest common multiple (LCM) 59

map scale 141
mean
   definition 538
   of frequency distribution 541
measurement
   compound units of 137-40
   standard units of 130-36
median 538
median class 541
mirror line 566
   see also symmetry
misleading graphs 545-46
mixed number 81
modal class 541
mode 538
money 135
multiple bar chart 523-24
multiple of number 55
multiplication
   algebraic expression 70-71
   decimals 98
   fractions 82
   inequality 508-9
   inverse operations 11
   law of indices 108
   by scalar 338
   standard form 416-17
mutually exclusive events 304, 401-2

negative index 106-7
nets 38-39
nth term 234-35
number lines 509-10
numbers
   consecutive 56
   cube 55, 239
   even 55
   linear 239
   mixed 81
   multiple of 55
   prime 55
   properties 55
   quadratic 239
   square 55
   square root 55
   triangular 239
numerator 80

object 566
obtuse-angled triangle 23
odd number 55

operations
   additive inverse 11
   multiplicative inverse 11
   order of 8-9
order of operations 8-9
order of rotational symmetry 21
orientation 571
outcomes
   definition 301
   organising 309-10
   predicting 306
outlier 549, 551

pair of compasses 321
parabola
   definition 364
   features of 365-66
parallel lines 18, 353
   angles and 381-82
parallelogram
   area of 173-74, 254
   properties 27
parallel vector 338
pentagonal prism 38
percentages
   calculations 279
   comparing with fractions and decimals 276-77
   finding original values 284
   increasing/decreasing amount 283
   one quantity as of another 281
   probability 302
perimeter
   composite shapes 156
   definition 152
   finding lengths 155
   formulae to finding 153
   problems involving 163
   of rectangle 254
perpendicular bisector 567
perpendicular lines 18
pie chart 526-27
pie graph 521
place value 57
plane shape 16
   see also specific shapes, e.g. squares, triangles, *etc.*
plan view 48
polygon
   angles and 385-87
   area of 171-75
   definition 16
   irregular 16
   naming 17
   regular 16
polyhedron 30
polynomials 369-70
population 517
position-to-term rule 234
powers and roots working with 109-11
powers of indices 109

pressure 140
prime factors 58–59
prime number 55
principal 483
prisms
    bases of other shapes 262–63
    and cylinders 265
    definition 31
    from parallel lines 44
    pentagonal 38
    problem-solving framework 264
    rearranging formulae 263
    right 261
    surface area 261
    triangular 38
    volume 261
probability
    calculating 303–4
    event not happening 304
    experimental 306–10
    mixed problems 311–12
    mutually exclusive events 304
    in numbers 301
    as percentage 302
    problem-solving framework 312
    theoretical 303
problem-solving framework
    calculations 4
    factorising quadratic expression 193–95
    formulae 247–48
    line graphs 530–31
    measurement 134–35
    prisms 264
    probability 312
    Pythagoras' theorem 460–61
    similar triangles 424
    simultaneous equations 217–18, 220–21
    standard form 417
    triangles 24–25
    trigonometric ratios 477–78
product 3, 67
proportion
    definition 290
    direct 494, 496, 498–99
    inverse 501–2
    vs. ratio 290
pyramids
    definition 31
    from parallel lines 44
    properties 31–32
    square-based 38
    surface area of 271
    triangular 38
    volume of 271
Pythagoras' theorem
    finding length any sides 454–55
    hypotenuse 453
    problem-solving framework 460–61
    proving right-angled triangle 457
    statement 453
Pythagorean triple 457

quadratic equations
    definition 210
    forming and solving 213
    solving by factorising 211–12
quadratic expressions
    definition 189, 364
    factorising 192–93
quadratic graphs
    characteristics 365–66
    sketching 367
quadratic numbers 239
quadrilateral
    angle sum of 28
    classification 27
    properties 27–28
quarter-circle 162
quotient 3

radius (plural radii) 321
random outcomes 303
random sampling 517–18
range 538
ratio
    comparing 293–94
    cosine 468, 474–75
    definition 290, 494
    description 289–90
    equivalent 290
    golden 296
    proportion vs. 290
    sharing in 292
    sine 468, 474–75
    tangent 468, 474–75
    trigonometric 468–69
reciprocal fractions 84
reciprocal functions 371
rectangle
    area of 171, 254
    perimeter of 254
    properties 28
reflections
    describing 569
    in transformation 566
reflection symmetry 20
representative sample 517
rhombus 27
RHS see right angle hypotenuse side
right-angled triangles 23
    finding unknown angles 472
    finding unknown sides 470
    naming sides of 467
    Pythagoras' theorem 457
    ratio of sides 468
    solving 470
    see also Pythagoras' theorem; triangles
right angle hypotenuse side (RHS) 441
right prism 261
roots 210
rotation
    definition 576–77
    describing 579

rotational symmetry 21
rounding
    decimals 118
    definition 117
    significant figure 120
    whole numbers 117

sample
    definition 517
    random 517–18
    representative 517
sample space 393
SAS see side angle side
scalar 338
scale drawing
    constructing 144
    definition 141
scale factor 141
scalene triangle 23
scatter diagrams
    advantages 548
    in data 521
    line of best fit 550
semi-circle 162, 321
sequences
    arithmetic 232
    definition 232
    Fibonacci 239
    geometric 232, 239
    linear 239
    special 239
sets
    notation 396–97
    union of 396
shapes
    congruent 571
    plane 16
    see also specific shapes, e.g. squares, triangles, etc.
side angle side (SAS) 441
side elevation 48
side side side (SSS) 441
significant figure 120
significant figures 415
similar shapes 434
similar triangles see triangles
    description 423
    finding unknown lengths 423
    problem-solving framework 424
    ratio of sides 468
simple arithmetic progression 239
simple interest 483
simultaneous equations
    definition 215
    forming and solving 220–21
    problem-solving framework 217–18, 220–21
    solving by elimination 218–19
    solving by substitution 216–17

sine ratio 468, 474–75
slant height 267
solids 30
   composite 269
solution 210
special sequence 239
speed
   definition 137
   distance-time graphs 559
   properties 137–38
spheres 267
square-based pyramid 38
square numbers 55, 239
square root number 55
squares
   area of 171
   difference of two squares identity 191
   properties 28
SSS *see* side side side
standard form
   calculators and 414–15
   converting to ordinary numbers 413
   expressing numbers 411–13
   multiplying and dividing numbers 416–17
   problem-solving framework 417
   writing numbers 411
standard units of measurement
   area 131
   conversion factor 130
   money 135
   problem-solving framework 134–35
   time 133–34
   volume 131
straight-line graphs
   characteristics of 346–51
   gradient 346, 349
   interpretation 355
   parallel lines 353
   plotting linear functions 345
   x-intercept 346, 348, 350
   y-intercept 346, 348–50
   see also graphs
subject 246
substitute 249
substitution 69
subtraction
   algebraic expression 69–70
   decimals 98
   fractions 83
   inequality 508
   vector 337–38
sum 3
surface area
   of cuboid 254
   cylinders 265
   prisms 261
   pyramids 271

symmetry
   line of 20–21
   reflection 20
   rotational 21

tables organise data 519
table tree 309
tangent ratio 468, 474–75
term 67, 232
term-to-term rule 232
theoretical probability
   combining rules 403
   definition 303
   dependent events 404
   events that are not mutually exclusive 402
   formula 400
   independent events 403
   mutually exclusive events 401–2
   problem-solving framework 405
three-dimensional objects
   cylinders 43–44
   isometric grid 44–45
   nets and 38–39
   prisms 43–44
   pyramids 44
   solids 30
time
   12-hour and 24-hour systems 134
   units 133–34
time-series data 529–30
transformation
   definition 566
   reflections 566, 569
   rotation 576–77, 579
   translation 571, 575
translation
   definition 571
   describing 575
transversal 381
trapezium
   area of 174–75, 254
   properties 27
travel graphs *see* distance-time graphs
tree diagram 398
triangles
   angles and 383
   area of 171–72, 254
   problem-solving framework 24–25
   properties 24
   types 23
   using gradient to interpret 560
   see also right-angled triangles; similar triangles
triangular numbers 239
triangular prism 38
triangular pyramid 38
trigonometric ratios
   angle of depression 476
   angle of elevation 476

   description 468–69
   exact values of 474–75
   problem-solving framework 477–78
truncation 121–22
two-dimensional shapes 16

union of set 396
unique factorisation theorem 58
unitary method 496
unit fractions 84
units
   area 131
   compound 137–40
   standard 130–35
   time 133–34
   volume 131
unknown 203
   on both sides of linear equations 206
   right-angled triangles 470, 472
   similar triangles 423

variables 67
   dependent 549
vector
   addition 337–38
   column 336
   definition 336
   mixed practice 340
   parallel 338
   subtraction 337–38
velocity 249
Venn diagrams
   complement of set 396
   description 395
   intersection of set 395
   union of set 396
vertical lines 362–63
vertically opposite angles 380
volume
   cone 255
   cuboid 254
   cylinder 254
   definition 131
   prisms 261
   pyramids 271

whole numbers rounding 117

x-intercept
   definition 346
   roots of quadratic equation 366
   and y-intercept 348, 350

y-intercept
   definition 346
   gradient and 349
   and x-intercept 348, 350

zero index 106–7

Acknowledgements

Questions from Edexcel past question papers ©Pearson Education Ltd.

These questions are indicated by .

Questions from Cambridge IGCSE® Mathematics reproduced with permission of Karen Morrison and Nick Hamshaw.

The authors would like to thank Fran Wilson for her work on GCSE Mathematics Online.

Cover © 2013 Fabian Oefner www.fabianoefner.com; p1 (top) Henry Gan/Photodisc/Thinkstock; p1 Denis Kuvaev/Shutterstock; p14 (top) Monarx3d/iStock/Thinkstock; p26 kilukilu/Shutterstock; p27 (top) HUANG Zheng/Shutterstock; p27 (bottom) Marafona/Shutterstock; p 32 © Owen Franken/Corbis; p33 Ieva Geneviciene/Shutterstock; p36 (top) alfimimnill/iStock/Thinkstock; p53 (top) Yuriy S/iStock/Thinkstock; p53 Tyler Olson/Shutterstock; p65 (top) agsandrew/iStock/Thinkstock; p65 wavebreakmedia/Shutterstock; p78 (top) SDivin09/iStock/Thinkstock; p78 Chameleons Eye/Shutterstock; p92 (top) Andrey Popov/iStock/Thinkstock; p92 Leah-Anne Thompson/Shutterstock; p94 William West/Staff/Getty Images; p96 © Jumana el Heloueh/Reuters/Corbis; p103 (top) Jeffrey Collingwood/Hemera/Thinkstock; p103 Candy Box Images/Shutterstock; p115 (top) Joe McDaniel/iStock/Thinkstock; p115 Dmitry Kalinovsky/Shutterstock; p122 Pixsooz/Shutterstock; p128 (top) Mike Watson Images/moodboard/Thinkstock; p128 Dorling Kindersley/Getty Images; p134 Taiga/Shutterstock; p150 (top) David Chapman/Design pics/Valueline/Thinkstock; p150 itman_47/Shutterstock; p151 dkART/Shutterstock; p160 iceink/Shutterstock; p169 (top) imagean/iStock/Thinkstock; p169 Federico Rostagno/Shutterstock; p175 Stefano Spezi/Shutterstock; p187 (top) Leigh Prather/iStock/Thinkstock; p187 Dariush M/Shutterstock; p201 (top) Alison Bradford Photography/iStock/Thinkstock; p201 Andrey Popov/Shutterstock; p230 (top) © PhotoAlto/Alamy; p230 © Monty Rakusen/Cultura/Corbis; p233 © Roger Bamber/Alamy; p240 (top) murengstockphoto/Thinkstock; p240 (bottom) Shaiith/Thinkstock; p244 (top) deyangeorgiev/iStock/Thinkstock; p244 Leah-Anne Thompson/Shutterstock; p253 © PCN Photography/Alamy; p256 Anton Gvozdikov/Thinkstock; p259 (top) Phil Ashley/Photodisc/Thinkstock; p259 govicinity/Shutterstock; p260 (top) Arvind Balaraman/Shutterstock; p260 (bottom) lsantilli/Shutterstock; p265 Dmitry Kalinovsky/Shutterstock; p266 Sergio Bertino/Shutterstock; p267 © Daniel Mogan/Alamy; p269 (left) Courtesy of Newcastle International Airport; p269 (right) © 2ebill/Alamy; p272 (top) Chameleons Eye/Shutterstock; p272 (bottom) kravka/Shutterstock; p273 cpphotoimages/Shutterstock; p277 (top) shutter_m/iStock/Thinkstock; p288 (top) Eugen Weide/Hemera/Thinkstock; p288 T. Fabian/Shutterstock; p290 Anteromite/Shutterstock; p292 zentilia/Shutterstock; p293 Adam Gilchrist/Shutterstock; p295 Valentyn Volkov/Shutterstock; p299 (top) Chemik11/iStock/Thinkstock; p301 lev radin/Shutterstock; p318 (top) Huskyomega/iStock/Thinkstock; p318 Franz Pfluegl/Shutterstock; pp318/320/321 (top) Yulia Glam/Shutterstock; p321 (bottom) Milos Luzanin/Shutterstock; p329 (left) Vector House/Shutterstock; p329 (right) UMB-O/Shutterstock; p333 Gemenacom/Shutterstock; p334 (top) sutichak/iStock/Thinkstock; p334 Michael Winston Rosa/Shutterstock; p340 Herbert Kratky/Shutterstock; p343 (top) KeremYucel/iStock/Thinkstock; p343 Dan Breckwoldt/Shutterstock; p360 (top) shaunnessey/iStock/Thinkstock; p360 © Jenny E. Ross/Corbis; p364 herreid/Thinkstock; p377 (top) Mark_Stillwagon/iStock/Thinkstock; p377 Mikio Oba/Shutterstock; p381 Dmitry Kalinovsky/Shutterstock; p384 Ryan Lewandowski/Shutterstock; p391 (top) maury75/iStock/Thinkstock; p391 angellodeco/Shutterstock; p409 (top) agsandrew/iStock/Thinkstock; p409 Sergey Kamshylin/Shutterstock; p421 (top) Comstock/Stockbyte/Thinkstock; p421 Kletr/Shutterstock; p432 Lucy Clark/Shutterstock; p439 (top) Pietro_Ballardini/iStock Editorial/Thinkstock; p439 Ross Strachan/Shutterstock; p450 (top) © Emma Smales/VIEW/Corbis; p450 Oleksandr Kalinichenko/Thinkstock; p457 racorn/Shutterstock; p464 © Rob Wilkinson/Alamy; p465 (top) Digital Vision./Photodisc/Thinkstock; p465 GECO UK/Science Photo Library; p469 © Andrew Ammendolia/Alamy; p482 (top) shutter_m/iStock/Thinkstock; p482 Pressmaster/Shutterstock; p485 Anatoliy Lukich/Shutterstock; p492 (top) Karol Kozlowski/iStock/Thinkstock; p492 michaeljung/Shutterstock; p493 (top) larik_malasha/Shutterstock; p493 (bottom) terekhov igor/Shutterstock; p494 ermess/Shutterstock; p 95 Maksim Kabakou/Shutterstock; p496 Rtimages/Shutterstock; p497 Christian Delbert/Shutterstock; p506 (top) © Tabor Gus/Corbis; p506 © Olaf Doering/Alamy; p515 (top) shutter_m/iStock/Thinkstock; p515 Chad McDermott/Shutterstock; p517 © Detail Nottingham/Alamy; p536 (top) marekuliasz/iStock/Thinkstock; p536 Brian A. Jackson/Shutterstock; p555 (top) zentilia/iStock/Thinkstock; p555 Zryzner/Shutterstock; p564 (top) ninjacpb/iStock/Thinkstock; p566 Pal Teravagimov/Shutterstock.